T5-ARV-583

Development, Maturation, and Senescence of Neuroendocrine Systems

Development, Maturation, and Senescence of Neuroendocrine Systems

A Comparative Approach

Edited by

Martin P. Schreibman

Department of Biology
Brooklyn College of the City University of New York
Brooklyn, New York

Colin G. Scanes

Department of Animal Science
Cook College
Rutgers University
New Brunswick, New Jersey

ACADEMIC PRESS, INC.
Harcourt Brace Jovanovich, Publishers

San Diego New York Berkeley Boston London Sydney Tokyo Toronto

Academic Press Rapid Manuscript Reproduction

Academic Press, Inc.
San Diego, California 92101

United Kingdom Edition published by
Academic Press Limited
24–28 Oval Road, London NW1 7DX

Library of Congress Cataloging in Publication Data

Development, maturation, and senescence of neuroendocrine systems : a comparative approach / edited by Martin P. Schreibman and Colin G. Scanes.
p. cm.
Includes index.
ISBN 0-12-629060-1 (alk. paper)
1. Neuroendocrinology. 2. Endocrine glands--Aging.
3. Physiology, Comparative. 4. Endocrine glands--Aging--Animal models. I. Schreibman, Martin P. II. Scanes, C. G.
QP356.4.D48 1989
599'.0188--dc19 88-34429
CIP

Printed in the United States of America
89 90 91 92 9 8 7 6 5 4 3 2 1

Contents

PART III
Neuroendocrinology and the Environment

Contributors

Numbers in parentheses indicate the pages on which the authors' contributions begin.

Kurt D. Ackerman (381), Department of Neurobiology and Anatomy, University of Rochester, School of Medicine, Rochester, New York 14642

Denise L. Bellinger (381), Department of Neurobiology and Anatomy, University of Rochester, School of Medicine, Rochester, New York 14642

Howard A. Bern (289), Zoology Department and Cancer Research Laboratory, University of California, Berkeley, California 94720

Christopher L. Brown (289), Zoology Department and Cancer Research Laboratory, University of California, Berkeley, California 94720

Rocco V. Carsia (353), Department of Animal Sciences, Rutgers—The State University, New Brunswick, New Jersey 08903

James Norman Dent (63), Department of Biology, University of Virginia, Charlottesville, Virginia 22901

Walter W. Dickhoff (253), School of Fisheries, University of Washington, Marine Fisheries Service, Seattle, Washington 98112

R. S. Donhan (173), Physiology and Anatomy Program, School of Life and Health Services, University of Delaware 19716

David L. Felten (381), Department of Neurobiology and Anatomy, University of Rochester, School of Medicine, Rochester, New York 14642

Suzanne Y. Felten (381), Department of Neurobiology and Anatomy, University of Rochester, School of Medicine, Rochester, New York 14642

Caleb E. Finch (397), Andrus Gerontology Center, and the Department of Biological Sciences, University of Southern California, Los Angeles, California 90089-0191

Milton Fingerman (23), Department of Biology, Tulane University, New Orleans, Louisiana 70118

Penny M. Hopkins (23), Department of Zoology, University of Oklahoma, Norman, Oklahoma 73701

T. H. Horton (173), Physiology and Anatomy Program, School of Life and Health Services, University of Delaware, Newark, Delaware 19716
C. Janse (43), Department of Biology, Vrije University, 1007 MC, Amsterdam, The Netherlands
J. Joosse (43), Department of Biology, Vrije University, 1007 MC, Amsterdam, The Netherlands
Kelley S. Madden (381), Department of Neurobiology and Anatomy, University of Rochester, School of Medicine, Rochester, New York 14642
Sasha Malamed (353), Department of Anatomy, University of Medicine and Dentistry of New Jersey, Robert Wood Johnson Medical School, Piscataway, New Jersey 08854
Henrietta Margolis Nunno (97) , Department of Biology, Brooklyn College of the City University of New York, Brooklyn, New York 11210
F. M. Anne McNabb (333), Department of Biology, Virginia Polytechnic Institute and State University, Blacksburg, Virginia 24061
Charles V. Mobbs (223), Rockefeller University, New York, New York 10021
Peter W. Nathanielsz (155), Laboratory for Pregnancy, and Newborn Research, Department of Physiology, College of Veterinary Medicine, Cornell University, Ithaca, New York 14853
David O. Norris (63), Department of Environmental, Population and Organismic Biology, University of Colorado, Boulder, Colorado 80309
Sergio R. Ojeda (195), Division of Neuroscience, Oregon Regional Primate Research Center, Beaverton, Oregon 97006
Mary Ann Ottinger (135), Department of Poultry Science, University of Maryland, College Park, Maryland 20742
Lynn M. Riddiford (9), Department of Zoology, University of Washington, Seattle, Washington 98185
Colin G. Scanes (3,93,269,307), Department of Animal Sciences, Rutgers—The State University of New Jersey, New Brunswick, New Jersey 08903
Martin P. Schreibman (3,93,97), Department of Biology, Brooklyn College of the City University of New York, Brooklyn, New York 11210
M. H. Stetson (173), Physiology and Anatomy Program, School of Life and Health Services, University of Delaware, Newark, Delaware 19716
Paola S. Timiras (275), Department of Physiology–Anatomy, University of California, Berkeley, California 94720
James W. Truman (9), Department of Zoology, University of Washington, Seattle, Washington 98185
Henryk F. Urbanski (195), Division of Neuroscience, Oregon Regional Primate Research Center, Beaverton, Oregon 97006

Preface

Aging is a continuum that spans birth (or even earlier, as, for example, during gestation) and death and has as its major components development, maturation, and senescence. Although studies on the senescence of neuroendocrine systems are in the early stages, research has accelerated at an amazing pace. This acceleration is especially true for studies on neuroendocrine development and maturation. It is therefore appropriate at this time to summarize and evaluate the progress that has been made and to evaluate directions for future study and research.

Aging is reflected in physiological processes that change with time under the influences of interaction between genome and environment. Model systems for the study of development, maturation, senescence, and longevity should therefore utilize animals that are genetically defined and capable of being maintained under controlled environmental conditions. These models should also be structurally and physiologically simple to permit greater access to an analysis of basic neuroendocrine phenomena. The chapters in this volume reflect this philosophy and present new approaches, new models, and new interpretations of "aging" phenomena. The book is organized into three parts: (1) a comparison of neuroendocrine systems: (2) development and aging of reproductive systems and functions: and (3) neuroendocrinology and the environment. Each has an introduction written by the editors and several chapters related to the theme being considered.

The various chapters have been written by established senior investigators or by younger scientists whose work has been exciting and innovative and who have already made significant contributions to their respective fields of study. In most instances, the authors have provided background for and then summarized and evaluated recent literature pertaining to neuroendocrine systems of specific invertebrate and vertebrate groups, basic neuroendocrine-regulated phenomena, or structures and functions of organs making up the neuroendocrine system. Additionally, they have, where appropriate, suggested new models for study and proposed new directions for future research. We are indeed fortunate to have been able to assemble such qualified people whose expertise cuts across phylogenetic lines and areas of interest.

The special and unique aspects of this treatise include:

- All phases of the aging continuum—maturation, development, and senescence are addressed.
- A fundamental , physiological component of animals (the reproductive system), which many believe is an essential contributor to senescence is discussed.
- It offers a comparative approach since it deals with invertebrate and vertebrate forms as diverse as molluscs and humans.
- Alternative animal models for the study of aging (e.g. , molluscs, fish, insects) are presented.
- It examines systems that have not been previously considered, or have been dealt with in only a cursory fashion (e.g., the relationship between genetics and aging, nerve cells as models for aging studies, neurohumoral hysteresis, aging of secondary immune organs).

Some of the material in this book was presented at a symposium sponsored by the Division of Comparative Endocrinology during the annual meeting of the American Society of Zoologists in New Orleans, December 27–28, 1987. The present volume contains expanded presentations and additional chapters.

It is our intention that this book serve students at various levels of study—the advanced undergraduate, the graduate, the postgraduate, and the senior scientist. We hope all will read and extract different messages, information, and insights.

Martin P. Schreibman
Colin G. Scanes

I
Comparison of Neuroendrocrine Systems

1

AN INTRODUCTION TO THE NEUROENDOCRINE SYSTEM

Colin G. Scanes and Martin P. Schreibman

Department of Animal Sciences
Rutgers - The State University of New Jersey
New Brunswick, N.J. 08903

Department of Biology
Brooklyn College of the
City University of New York
Brooklyn, New York 11210

The concept of the neurosecretory neuron was introduced by Ernst Scharrer in 1928 (reviewed Scharrer and Scharrer, 1963). A neurosecretory cell may be described as a modified neuron physiological and morphological characteristics typical of both neurons and endocrine cells. It is capable of releasing biologically active factor(s) into the circulatory system and, thereby, to target tissues. These "factors", therefore, can justifiably considered to be neurohormones. In neurosecretory cells, the sites of synthesis and release of neurohormones are anatomically separated; synthesis occurs in the cell body and release at the axon ending. The neurohormones are predominantly peptides but non-peptide hormones do occur (e.g., dopamine). Neurosecretory neurons are found in virtually all multicellular phyla in the animal kingdom. Indeed, Berta Scharrer (1975) noted:

"The remarkable parallelisms of neuroendocrine phenomena throughout the animal kingdom bespeak their funda-

Development, Maturation, and Senescence of
Neuroendocrine Systems: A Comparative Approach

mental significance which...needs to be viewed in an evolutionary perspective".

We attempt to adhere to this philosophy in the present volume. There are chapters on the development and/or senescence of neuroendocrine systems in molluscs, crustaceans, insects, as well as in the various classes of vertebrates.

In this introduction, we will discuss briefly: (1) problems with definitions and concepts in neuroendocrinology; (2) the APUD concept; and (3) homeostatic controls in neuroendocrine-regulated physiological systems (as for example the brain-pituitary axis). In the chapters that follow, we have given more specific considerations to the development and senescence of various neuroendocrine systems in a variety of animals and physiological processes.

1. Problems with Definitions and Concepts in Neuroendocrinology

While there is little doubt that peptides or amines can be produced by modified neurons, there are questions as to whether a particular neuronal product is, in fact, a hormone (Barrington, 1979). Traditionally, for the peptide or amine to be considered a hormone it must be released into the blood stream and be transported to the target tissue where it exerts a specific effect. It would appear, based on this accepted definition, that the humoral agents produced by neurosecretory cells are, indeed, neurohormones. The absence of definite evidence of these characteristics for neurohormones has in the past led to the use of such terms as "principles" or "factors". In the present volume (as it is in common practice), putative neurohormones will be referred to simply as neurohormones or hormones.

The different types of chemical communication systems (hormone, neurohumors and local effectors such as paracrine and autocrine factors) display many similarities. Neurohormone may act as neurotransmitters and/or neuromodulators. Furthermore, the same peptide or amine may have roles as a neurotransmitter, a neuromodulator, neurohormone, a hormone, a paracrine factor and an autocrine factor. Additionally, there are numerous examples of multiplicity of functions, and sites of localization for specific neurohormones. For example, somatostatin (SRIH), whose major function is as a hypophysiotropic (hypothalamic releasing) factor; (reviewed Jackson, 1979), also serves as a neuromodulator and neurotransmitter in the nervous system (reviewed Jackson, 1979). Pancreatic SRIF appears to exert a paracrine (local) effect on the endocrine cells of the pancreatic islets. In addition, pancreatic SRIF is a major source of circulating

SRIF and exerts a marked effect on lipolysis in avian species (Strosser et al., 1983). Thus, SRIF of gut origin may be presumed to have paracrine (diffuses from origin to stimulate cells in proximity), endocrine and possible autocrine (stimulates cell of origin) activities.

Similarly in mammals, thyrotropin releasing hormone (TRH) a modified tripeptide (pyro glu-His-Pro-NH_2) is found in the hypothalamus, posterior pituitary gland, extra hypothalamic brain (cerebrum, diencephalon, olfactory lobe, cerebral cortex, brain stem) and the motor neurons of the spinal cord. In amphibians, the highest concentrations of TRH are found in the hypothalamus and in the skin, moderately high concentrations are present in the extra-hypothalamic brain, spinal cord and retina, and low concentrations were detected in other organs, including the gastro-intestinal tract, lung and gonads (reviewed Jackson, 1979).

Vasoactive intestinal peptide (VIP) distribution is another example of a neuropeptide with a widespread distribution (hypothalamus, extra-hypothalamic brain, gut, pancreas, gonads). Thus neurohormones may show multiplicity of location which may also reflect on their multiplicity of functions. This may simply be the result of nature's economical use of the same peptide for different functions at disparate locations. As Barrington (1979) indicated, "there is a marked element of opportunism in evolution". Thus, it is not (necessarily) obligatory for the cells producing the same peptides to be anatomically, embryologically or physiologically related. An alternative view is that there is a diffuse (neuro) endocrine system (Pearse and Takor, 1979), the APUD system.

2. The APUD System

Pearse (1968) advanced the notion that a series of amine- and peptide-producing endocrine cells shared a number of features. These cells are capable of amines (or amine precursors) uptake and decarboxylation. This led to the naming of the series the APUD system. The APUD system includes the endocrine cells of gut and pancreas, the chromaffin cells of the adrenal medulla and the C-cells of the thyroid in mammals and the ultimobranchial bodies in birds and lower vertebrates. These cells share cytochemical and morphological characteristics with the neuroendocrine cells of the hypothalamus and posterior pituitary gland. In addition, Pearse (1969; 1975) proposed a common embryonic origin, the neural crest and neuroectoderm for the APUD system and also the hypothalamus/posterior pituitary gland. Components of the APUD system certainly have an origin in the neural crest/neuroectoderm. These include the C-cells and the adrenal medulla. However, there is strong evidence that this is not the case for the endocrine cells of the gut and pancreas.

It should be added that there is no universal acceptance for the APUD hypothesis. Some investigators suggest that the possession of APUD characteristics and specific enzymes reflect common metabolic needs and not similar embryonic origins. The concept of paraneurons (Fujita et al., 1980) was proposed to recognize that paracrine secreting cells have properties in common with nerve cells although they are not neurons. They also have APUD characteristics although their embryonic origin is unknown. It has been postulated that paracrine hromones, neurohormones and traditional hormones have a common ancestry. This suggestion is supported by the observations that regulatory peptides are present in both vertebrates and invertebrates and that neurotransmitters may serve as neurohormones in the same organism (see description of SRIH above).

3. First and Higher Order Neuroendocrine Systems

The neuroendocrine tissues can be classified by the control system involved into first order and higher order systems. This is illustrated in the figure below. Obviously, finer control can be influenced with the higher order system.

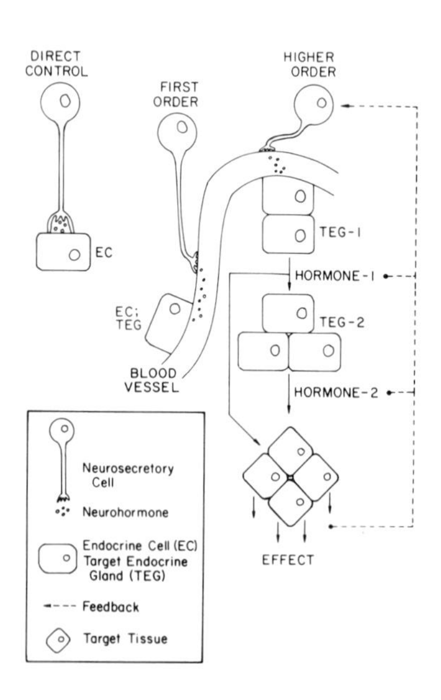

Figure 1 First and Higher Order Neuroendocrine Systems

References

Barrington, E.J.W. (1979) In: Hormones and Evolution (pp VII-XXI). Academic Press, London.
Jackson, I.M.D. (1979) In: Hormones and Evolution (pp. 723-790). (Ed. E.J.W. Barrington, Academic Press, New York).
Pearse, A.G.E. (1968) Proc. R. Soc. B., 170, 71-80.
Pearse, A.G.E. (19689) J. Histochem. Cytochem., 17, 303-313.
Pearse, A.G.E. (1975) Folia. Anat. Jugoslavica 4, 5-20.
Pearse, A.G.E. and Takor, T.T. (1979) Fed. Proc. 38, 2288-2294.
Rawdon, B.B. (1984) J. Exp. Zool. 232, 659-670.
Sawyer, W.H. and Pang, P.K.T. (1979) Hormones and Evolution (pp 493-523) (Ed. E. J. W. Barrington), Academic Press, London.
Scharrer, B. (1975) Amer. Zool., 15(Suppl. 1), 7-11.
Scharrer, E. and Scharrer, B. (1963) Neuroendocrinology, Columbia University Press, New York.
Strosser, M.T., DiScala-Guernot, D., Koch, B., and Mialhe, P. (1983) Biochim. Biophys. Acta. 763, 191-196.

2

DEVELOPMENT OF THE INSECT NEUROENDOCRINE SYSTEM

James W. Truman
Lynn M. Riddiford

Department of Zoology
University of Washington
Seattle, Washington

I. INTRODUCTION

The study of the development of the insect neuroendocrine system poses a number of difficulties. Unlike endocrine tissues which form discrete, homogeneous glands, insect neurosecretory cells (NSCs) are scattered in stereotyped locations throughout the nervous system. They are a heterogeneous group of cells that produce over a dozen different peptide hormones (Orchard and Loughton, 1985). The insect neurosecretory system is also faced with the challenge of metamorphosis. The larval and adult forms are often highly specialized for growth in one case and reproduction in the other. In addition, the 2 stages may live in radically different environments resulting in disparate metabolic and homeostatic challenges. How then do the neuroendocrine systems of the two stages compare? Are the larval NSCs carried over into the adult stage? If so, are their properties altered?

Although the embryonic development of the central NSCs has not been directly examined, the knowledge gained over the past 10 years on the development of the insect CNS in general has direct applicability to the neuroendocrine system. Techniques of intracellular dye injection have provided a detailed description of neuronal form including that of NSCs. The determination of which hormones are produced in particular cells and the development of immunological and molecular probes for these products allow a level of analysis that was not possible a few years ago. This review will examine some of the issues that pertain to the development of the insect neuroendocrine system. Due to space limitations, we will focus primarily on Lepidoptera and in particular on the tobacco hornworm, *Manduca sexta.*

II. THE INSECT NEUROSECRETORY SYSTEM.

The insect neuroendocrine system has both central and peripheral components and has been recently reviewed by Orchard and Loughton (1985). The NSCs of the central nervous system (CNS) are found both in discrete groups in the brain and distributed throughout the segmental ganglia (Figure 1). The brain NSCs extend axons to the *corpora cardiaca-corpora allata* (CC-CA) complex which is located along the aorta caudal to the brain. Three groups of NSCs are evident in the brain: the group I cells are situated laterally and project to the ipsilateral CC-CA complex; the group II cells and group III cells have contralateral projections and their cell bodies are located medially and laterally respectively. Besides serving as a neurohemal release site, the CC also contains an intrinsic set of NSCs. The CA is an endocrine gland that secretes juvenile hormone, but, at least in Lepidoptera, it also serves as a release site for some of the brain NSCs (Agui *et al.*, 1980).

The segmental NSCs are associated with segmental neurohemal organs, the parasympathetic or perivisceral organs (PVOs; Raabe, 1965). The number and position of the cells projecting to the PVOs varies but in moths there are 2 groups of NSCs that supply the abdominal PVOs: 1) a group of 4 lateral neurons which send axons that terminate in the ipsilateral PVO, and 2) a set of 8 midline NSCs that project to both sides of the PVO (Figure 1).

A number of peptide hormones have been assigned to specific NSCs in *Manduca* (Figure 1). Dissection of single NSC somata and the assay of their contents in an *in vitro* bioassay showed that at least one of the 2 group III cells contains prothoracicotropic hormone (PTTH) activity (Agui *et al.*, 1979). The identification of the eclosion hormone (EH) cells in the group Ia NSCs was based on the bioassay of the contents of dissected lateral cells, positive immunoreactivity to antisera raised against EH, and the ability of these cells to release EH activity when stimulated by intracellular microelectrodes (Copenhaver and Truman, 1986a). Immunocytochemistry showed that the intrinsic cells of the CC in *Manduca* are a homogeneous set of cells that contain a member of the adipokenetic hormone (AKH) family (J. Witten, unpublished). The sequence of this AKH-like peptide from adult *Manduca* has recently been elucidated (Ziegler *et al.*, 1985). In the ventral ganglia the identity of the cells that contain the tanning hormone, bursicon, was established by assaying the contents of dissected cell bodies in a sensitive bioassay (Taghert and Truman, 1982). Neurons containing the cardioacceleratory peptides (CAPs) were first demonstrated through evoking the release of CAP bioactivity by intracellular stimulation (Tublitz and Truman, 1985b). Later a monoclonal antibody that precipitates CAP bioactivity confirmed that these neurons contain the CAPs (Tublitz *et al.*, 1986). This knowledge of the identity of the neurons that make particular hormones has facilitated analysis of the development of these neuroendocrine systems.

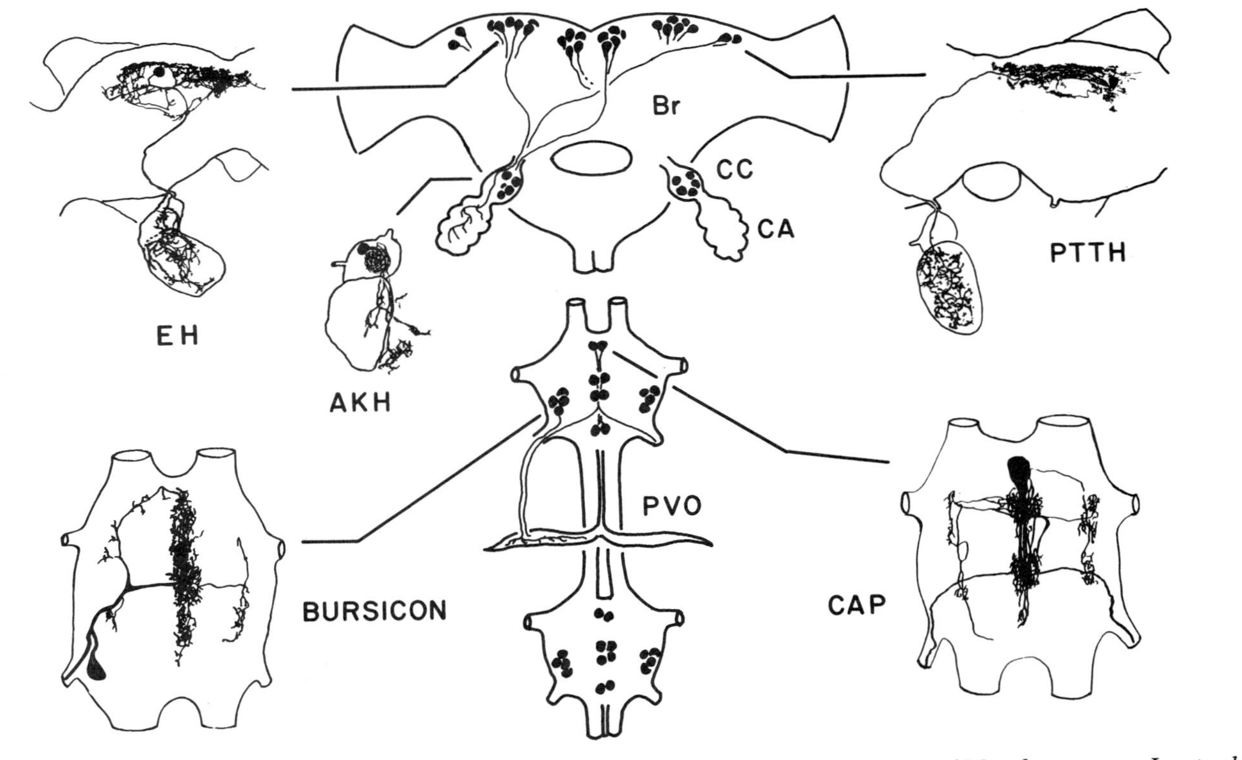

Fig. 1. Positions of the neurosecretory cells in the cephalic and segmental ganglia of Manduca sexta. Insets show the anatomy of particular NSCs as revealed by intracellular dye injections. Br, brain; CA, corpus allatum; CC, corpus cardiacum; PVO, perivisceral organ. Hormone abbreviations as in text. (Modified from Copenhaver and Truman, 1986b; Taghert and Truman, 1982; and Tublitz and Truman, 1985b).

III. EMBRYONIC DEVELOPMENT OF THE NEUROENDOCRINE SYSTEM

1. Fluctuations in neurohormone titers during embryogenesis.

Studies of hormone titers during embryogenesis have focussed primarily on their possible roles in regulating embryonic development. The best characterized titers are for the ecdysteroids which show peaks of free hormone that are correlated with times of cuticle production (for a review, see Hoffmann and Lagueux, 1985). The steroids that are found early in embryogenesis result from hormone-conjugates that were stored in the egg by the female during oogenesis. Those found towards the end of embyrogenesis may arise from the embryo's own prothoracic glands.

The only two neurosecretory hormones that have been studied in any detail are PTTH and EH, both of which are associated with molting in postembryonic stages. In the embryos of the giant silkmoth, *Hyalophora cecropia*, EH bioactivity was first detected in dissected embryos about midway (50%) through embryogenesis (Truman *et al.*, 1981). Activity dropped around the time that the first stage larva shed its embryonic cuticle and then remained low until hatching. In the commercial silkmoth, *Bombyx mori*, EH titers were determined from egg extracts that were prepared at various times during embryogenesis (Fugo *et al.*, 1985). Activity first appeared at about 70% of development, increased through the remainder of embryogenesis, and then dropped abruptly at hatching. Since in postembryonic stages each secretion of EH results in a marked depletion in stored hormone (Truman *et al.*, 1981), the sudden drops seen in the embryo suggest that EH is released during embryogenesis. The reasons for the differences in the time of hormone appearance and the timing of the drop in the two species is not known. Further data are clearly needed to clarify the role of EH in embryonic development.

In *Bombyx* there are two forms of PTTH (PTTH-S and PTTH-B) and the titers of both have been examined during embryogenesis (Chen *et al.*, 1987; Fugo *et al.*, 1987). In newly laid eggs PTTH-S activity is found at low levels whereas PTTH-B is not detectable. Titers of both forms remain unchanged until about 70% of embryonic development when both begin to increase and continue to do so until hatching. The titers of the PTTHs in *Bombyx* show no gross fluctuations that are correlated to the pattern of ecdysteroid secretion in the embryo. Studies on *Manduca* also show the presence of PTTH activity in the brain of the developing embryo (Dorn, *et al.*, 1987). However, it is not clear in this species whether the activity is due to only one or to both forms of *Manduca* PTTH.

Based on his histological observations of embryos of the grasshopper *Locustana pardalina*, Jones (1953) postulated that the embryonic prothoracic glands were controlled by brain NSCs. The only experimental evidence supporting such an interaction comes from studies on the stick insect, *Clitumnus extradentatus* (Cavallin and Fournier, 1981). Lesions of the protocerebral lobes, thereby destroying the brain NSCs, resulted in a derangement of the ecdysteroid titers late in embryogenesis.

Overall, the timing of appearance of neurohormones in insect embryos is in line with that seen for neurotransmitters and neuromodulators. For example, in the grasshopper embryo, serotonin can first be detected by immunoreactivity and radioenzyme assays at about 55% of embryogenesis (Taghert and Goodman, 1984), octopamine expression is first seen at 65% (Goodman, *et al.*, 1980), and the neuropeptide proctolin appears at about 60-70% (Keshishian and O'Shea, 1985).

2. **Development of neurosecretory cells.**

The development of the insect CNS has been under extensive study for the past 10 years and has focussed on the neurons of the segmental nervous system. Each segmental ganglion arises from a fixed number of stem cells that segregate from the neural ectoderm of the early embryo (Bate, 1976; Bate and Grunewald, 1981). In the locust embryo these stem cells include 7 midline precursor cells and 61 neuroblasts per segment. Each midline precursor cell undergoes a single symmetrical division to form 2 daughter neurons. The neuroblast, by contrast, undergoes repeated unequal divisions to form a chain of ganglion mother cells, each of which divides once to form 2 daughter neurons. Thus, a given neuroblast may give rise to well over 100 neurons. Although the lineage patterns of neurons in the embryonic brain have not been worked out in detail, it is clear that most brain neurons also arise from neuroblast lineages.

Experimental studies on grasshopper embryos have revealed some of the factors involved in determining neuronal fate. For example, most of the segmental ganglia of the grasshopper have 2 pairs of neurons that contain serotonin. In each hemiganglion, the 2 serotonin cells (S1 and S2) arise from the division of the first ganglion mother cell produced by neuroblast 7-3 (Figure 2). After laser ablation of neuroblast 7-3 on one side of a forming ganglion, S1 and S2 failed to appear on the lesioned side but developed normally on the control side (Taghert and Goodman, 1984). Importantly, the progeny of surrounding neuroblasts did not regulate to replace the progeny of the absent neuroblast. These experiments show that the development of specific neuronal characteristics in this region of the CNS depends primarily on the lineage of the cell (i.e., on its neuroblast of origin). The same relationship would be expected to hold for the NSCs in the brain.

Although the neuroblasts that produce the various NSC groups in the embryo have not yet been identified, a number of obvious questions exist. Do the cells in each cluster arise from a single neuroblast? If so, how then is hormone type related to birth order within the lineage? Do the neuroblasts that generate neurosecretory neurons also make conventional neurons?

3. Development of the peripheral neurosecretory cells.

Peripheral NSCs are found in the ganglia of the stomatogastric system and also scattered in stereotyped locations along peripheral nerves (Orchard and Loughton, 1985). In the anterior ganglia of the stomatogastric system in *Manduca*, the pattern of neurogenesis is similar to that seen for the CNS (Copenhaver and Taghert, 1988). Neuroblasts emerge from the dorsal region of the developing foregut and generate columns of cells which will become the neurons of the frontal and hypocerebral ganglia. For these cells, mitotic ancestry seems to be the dominant factor in determining cellular fate.

A different mechanism for regulating differentiation appears at work for the most posterior cells of the stomatogastric system, those that make up the enteric plexus, a loose network of peptide-containing cells situated over the anterior midgut (Copenhaver and Taghert, 1988). The enteric neurons arise from columnar epithelium that evaginates from the posterior foregut. These cells, which are mainly post-mitotic, first migrate circumferentially to encircle the foregut and then stream posteriorly to cover the muscle bands along the anterior midgut. The characteristic neuropeptides of these cells are

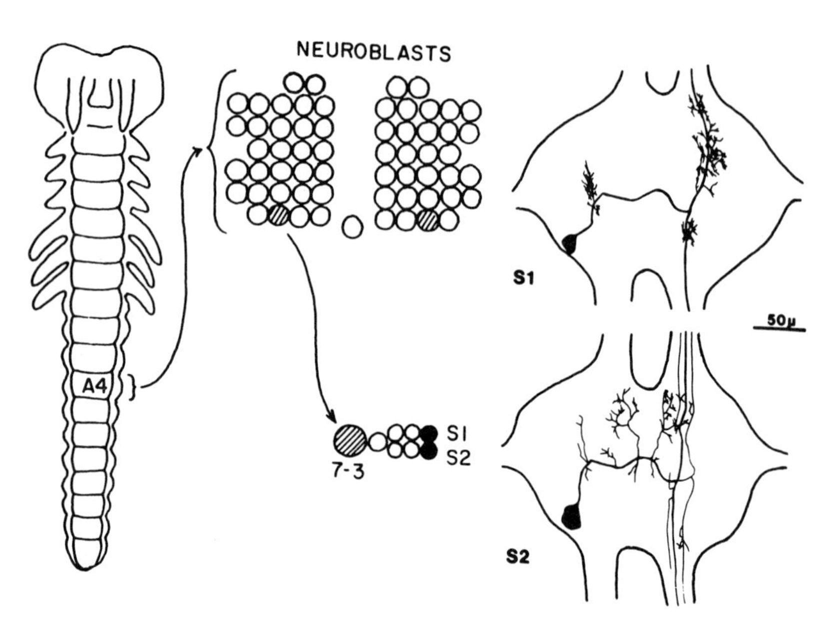

Fig. 2. Origin of the serotonin-containing neurons in the ventral CNS of the grasshopper. Left, a grasshopper embryo showing the set of neuroblasts that will produce the fourth abdominal ganglion. Within the lineage of cells produced by neuroblast 7-3 (cross hatched) are the two serotonin cells (S1 and S2; black) which arise from the division of the first ganglion mother cell produced by 7-3. Right, morphology of S1 and S2 in the late embryo. (Modified from Taghert and Goodman, 1984).

only expressed after the migration phase is complete. This type of development has similarities with that seen during the differentiation of the vertebrate neural crest (Le Douarin, 1982) and suggests that factors encountered during or after migration may be involved in determining neuronal characteristics.

IV. POSTEMBRYONIC DEVELOPMENT OF THE NEUROENDOCRINE SYSTEM.

The neuroendocrine system formed during embryogenesis may be further modified during postembryonic life so that the system eventually used by the adult is quite different from that found in the larva. This is especially true in the case of insects with complete metamorphosis in which the larval stage is radically different from that of the adult. There is evidence for at least three ways in which neurosecretory systems are modified during postembryonic life. A given neurosecretory cell may alter the types or ratios of hormones that it secretes, new neurosecretory cells may be added, or cells that are born during embryogenesis may mature during metamorphosis to become a functional endocrine axis in the adult.

1. Alterations in the hormones secreted by a NSC.

The ability of a cell to change the neurohormones that it secretes as the animal matures is an intriguing possibility but at present there exists little direct evidence for such changes. In the case of *Manduca*, a large and a small form of PTTH are thought to be released at different times in the insect's life history (Bollenbacher *et al.*, 1984). However, it is not clear whether the group III neurons (Figure 1) release both. An alternate possibility is that these neurons release only one form of PTTH and the cells that release the other form have yet to be identified. The latter possibility seems reasonable because monoclonal antibodies raised against the small form of PTTH in *Bombyx* bind to some of the medial NSCs rather than to cells in Group III (Ishizaki *et al.*, 1987).

The only direct evidence for a NSC altering the character of its hormonal product during postembryonic life involves the intrinsic cells of the CC of locusts, cells which secrete AKH. Locusts produce 2 forms of AKH, AKH I and AKH II, which are colocalized to the same neurons and even to the same secretory granule (Diederen *et al.*, 1987). In the newly hatched hopper, the ratio of AKH I to II is about 1:1. The ratio changes to about 4:1 in the adult (Hekimi and O'Shea, 1986). The functional significance of this shift in the ratios of the two peptides is not known.

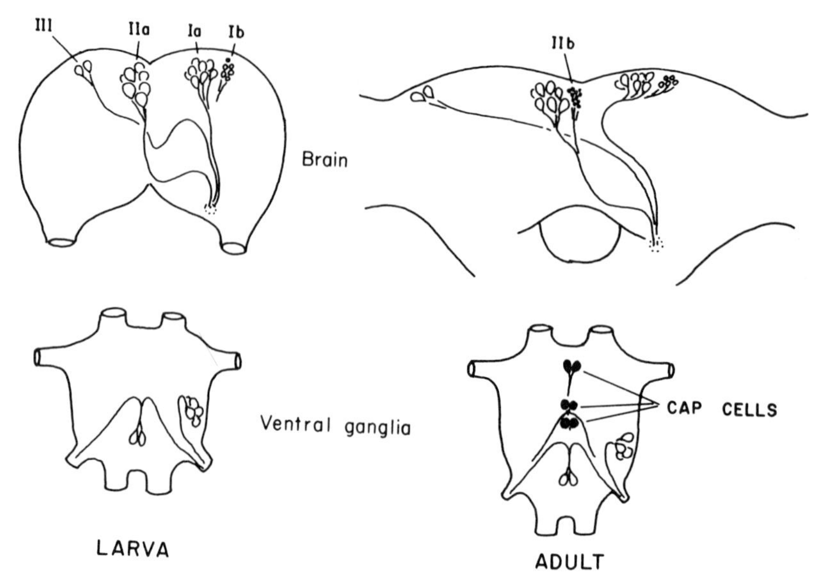

Fig. 3. Addition of new neurosecretory cells to the brain and ventral ganglia of Manduca sexta during metamorphosis. See text for further details.

2. **Addition of new neurosecretory cells.**

The addition of new NSCs during postembryonic life is best documented in *Manduca* and is evident from a comparison of the cephalic and segmental neuroendocrine systems of larvae and adults (Figure 3). New cells include the group IIb cells in the brain and the CAP cells in the segmental ganglia.

Backfills of the axons projecting to the CC-CA complex of the adult reveal a compact cluster of small NSCs, the group IIb cells (Copenhaver and Truman, 1986b). These cells show the Tyndall coloration typical of neurosecretory cells. Also, they stain avidly with Eosin Y, a feature shared by the other cephalic and segmental NSCs in adult *Manduca*. Recently, they have been shown to react positively to antisera raised against gastrin/CCK (Homberg, *et al.*, 1987). Some cells of the group project to the CC where their axons terminate; others have arbors that stay confined within the brain (Copenhaver and Truman, 1986b).

In the larval brain, no cells that correspond to the group IIb cells are found when the nerves to the CC-CA are backfilled (Copenhaver and Truman, 1986b). Rather their position is occupied by mitotic neuroblasts and associated clusters of immature neurons (J. Truman, unpublished). Similar neuroblasts have been studied in detail in the ventral CNS of larval

Manduca (Booker and Truman, 1987a). They become mitotically active early in larval life but the cells that they produce do not immediately develop into functional neurons. Instead they arrest early in differentiation so that, as the larva grows, each neuroblast becomes associated with a growing cluster of postmitotic, immature neurons. These immature cells only finish their differentiation into functional neurons when confronted with the endocrine cues that bring about metamorphosis (Booker and Truman, 1987b). This addition of new neurons to the ventral CNS of the larva can be prevented by injecting larvae with the DNA synthesis inhibitor, hydroxyurea (Truman and Booker, 1986). This drug blocks the birth of new neurons while not affecting post-mitotic cells. After treated larvae complete metamorphosis, their brains show the normal set of large NSCs (which are embryonic in origin) but the group IIb cells are absent (J. Truman, unpublished). This result supports the hypothesis that the group IIb cells arise by neuroblast activity during the course of larval growth.

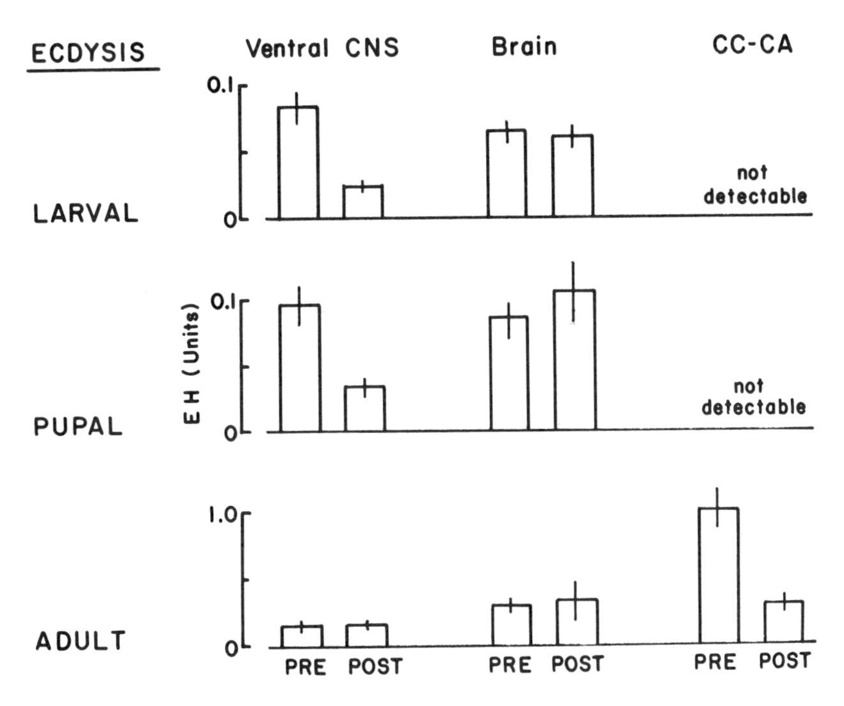

Fig. 4. Eclosion hormone titers present in various regions of the nervous system during the times of ecdysis to the fifth larval stage, the pupa, and the adult. Tissues were extracted 4-5 hrs before ecdysis (PRE) or immediately after ecdysis (POST). Note the change in scale for the adult ecdysis.

Three pairs of adult-specific NSCs are also found situated along the midline of each abdominal ganglion. As indicated above, these cells release the cardioacceleratory peptides (CAPs; Tublitz and Truman, 1985b). Birth dating experiments involving treatment of larvae with ^{3}H-thymidine at various times showed that these cells are born midway through larval life (R. Booker, unpublished). These cells are most likely the progeny of midline precursor cells, similar to those found in the embryo (Bate and Grunewald, 1981). In the last (5th) stage larva, the presumptive CAP cells are quite small and lack peripheral axons, but, with the onset of metamorphosis, they start to enlarge. By the time of pupation, each has extended an axon which branches and projects to the periphery through the paired ventral nerves (Taylor and Truman, 1974). At this time, however, the amount of cytoplasm around the nucleus remains meager. With the onset of adult development, these neurons show an extensive elaboration of cytoplasm accompanied by the appearance of CAP activity in the ventral CNS (Tublitz and Truman, 1985c). These cells are first used shortly after the emergence of the adult to cause the circulatory adjustments which aid in the expansion of the new adult wings (Tublitz and Truman, 1985a).

3. **Maturation of new neuroendocrine centers.**

In *Manduca* there appears to be some NSCs that are embryonic in origin but are not involved in hormone release until the adult stage. Such a situation may occur for the EH cells in the brain. Eclosion hormone is released at the end of each molt to trigger ecdysis, but depending on the stage, secretion occurs from one of 2 discrete sites (Truman *et al.*, 1981). One is somewhere in the ventral ganglia and is used for both larval and pupal ecdyses. The other, which is used for adult eclosion, is cephalic and comprised of 5 pairs of group Ia NSCs (Figure 1; Copenhaver and Truman, 1986a). During larval and pupal ecdyses, the ventral ganglia show dramatic depletions of stored EH coincident with the appearance of circulating peptide before ecdysis (Figure 4). By contrast, the brain center shows no such depletions (except for possibly in the first stage larva; Truman *et al.*, 1981). After metamorphosis, by contrast, the ventral CNS site appears to become dormant and the brain center takes over the secretion of EH (Figure 4). This shift is thought to be related to the regulation of the adult release by a brain-centered, circadian clock.

The switch in the centers that release EH during metamorphosis poses a number of interesting questions. What is the fate of the ventral EH neurons in the adult? What is the state of the brain EH cells in the larva and how are they activated in the adult? What regulates the shift from one endocrine center to the other?

Only indirect inferences can be made concerning the fate of the ventral EH cells of the larva because they have not yet been identified. Eclosion hormone bioactivity accumulates in the ventral CNS through metamorphosis and into adult life (Truman *et al.*, 1981). Consequently, it appears likely that the ventral EH neurons persist through metamorphosis and continue to synthesize EH. Their possible function in the adult is unknown.

Information on the brain EH cells in immature stages comes from scattered bits of information. In the adult the Ia group consists of 9 large cells, 5 of which are immunopositive for EH (Copenhaver and Truman, 1986a,b). These are the only immunopositive cells in the brain and, thus, are presumably responsible for all of the EH activity that can be extracted from the brain. Backfills of the nerves leading from the brain to the CC-CA complex in larvae and pupae show that all 9 Ia cells are present in these immature stages and that all send axons to the CC-CA (Copenhaver and Truman, 1986b). Intracellular dye fills of these cells in the pupal stage show that their axons branch in the CA or the CC but that the extent of branching is relatively sparse (Carrow, *et al.*, 1984). The brains of larvae and pupae contain substantial EH (Truman *et al.*, 1981). We assume that in these stages, as in the adult, EH is confined to the subset of group Ia cells, but this assumption has not been tested using immunocytochemistry. The EH neurons in the brain of larvae and pupae differ from those in the adult in one interesting aspect. Just prior to adult emergence, the CC-CA complex (which contains the terminals of the brain EH neurons) has 3 fold more EH than does the brain (Figure 4). By contrast, no EH activity was detected in the CC-CA complex of larvae or pupae (Copenhaver and Truman, 1982; Truman *et al.*, 1981). Thus, the brain EH cells of immature stages have axons that extend to the CC-CA complex and they apparently synthesize EH, but this hormone fails to appear at the normal storage and release sites. Whether this failure results from a lack of transport or from degradation of the peptide during transit is not known. Eclosion hormone activity first appears in the CC-CA within a few days after the start of adult development (Truman *et al.*, 1981).

The factors that regulate the shift from one EH release center to the other are presumably the hormones that cause metamorphosis. Whether their effects are directly on the neuroendocrine cells or indirect through actions on other tissues remains to be determined.

V. SENESCENCE OF THE INSECT NEUROENDOCRINE SYSTEM

Senescent changes that one might expect in the neurosecretory system include an impairment in the production or secretion of neurohormones or the actual death of NSCs. A suggestion of the impairment of secretory capacity with age comes from an ultrastructural study of the brain of ageing adult *Drosophila* (Herman, *et al.*, 1971). Neurosecretory cell bodies from old flies showed a marked loss of free ribosomes and rough endoplasmic reticulum and an increase in the frequency of autophagic vacuoles. It will be interesting in the future to see this question addressed using immunological or molecular probes against particular neurohormones to quantify the loss of function in specific NSCs.

In terms of cell death, this is definitely seen in the endocrine system with the degeneration of the prothoracic glands during metamorphosis (*e.g.*, Herman and Gilbert, 1966). However, we have not found reports of similar deaths among the NSCs either during or after metamorphosis. In *Manduca* a period of endocrine regulated neuronal death follows the emergence of the adult moth (Truman and Schwartz, 1984). Approximately half of the abdominal neurons die; neurons whose functions are finished at emergence and are not needed in the adult. Among those that survive through this period of cell death are the bursicon neurons, a fact that is surprising since their only known function is to cause cuticular tanning at adult emergence. Either these cells assume some new function in the adult, which is still to be discovered, or they and the other NSCs in *Manduca* share some property which makes them immune to the signals that kill conventional neurons.

VI. CONCLUSIONS

A key to understanding the development of the neuroendocrine system of insects is the unambiguous identification of the cells that produce particular hormones. This then allows the formulation of developmental questions in the context of the origin, function and fate of specific cells. This task is becoming easier due to the influx of techniques from other fields including neurobiology, immunology, and molecular biology. Examples given in this essay show that these techniques have already made a significant impact and they will become even more prominent as the field expands.

The growing body of data relating to the development of the neuroendocrine system in insects is bringing a number of questions into focus. In the embryo, a major task relates to understanding the developmental mechanisms that generate the diversity of neurosecretory cell types. Also, the role of the neuroendocrine system in regulating embryonic development remains to be defined. In postembryonic stages, the relationship of the larval to the adult NSCs presents a number of intriguing questions. It is clear that many of the same NSCs are found in both stages, but we know little about how the physiology of these cells may differ under the two conditions. Also, cells may arise that are unique to one stage or the other. Consequently, the neuroendocrine system of the adult is a mosaic of cells that have diverse developmental histories. Understanding how these histories are regulated and their impact on the various stages represents one of the challenges for the future.

ACKNOWLEDGMENTS

We thank Drs. P.F.Copenhaver and P.H.Taghert for making available a preprint of their data on the embryonic development of the peripheral nervous system.

REFERENCES

Agui, N., Granger, N.A., Gilbert, L.I., and Bollenbacher, W.E. (1979). Proc. Natl. Acad. Sci. USA 76, 5694.
Agui, N., Bollenbacher, W.E., Granger, N.A., and Gilbert, L.I. (1980). Nature 285, 669.
Bate, C.M. (1976). J. Embryol. Exp. Morph. 35, 107.
Bate, C.M., and Grunewald, E.B. (1981). J. Embryo. Exp. Morphol. 61, 317.
Bollenbacher, W.E., Katahira, E., O'Brian, M., Gilbert, L.I., Thomas, M.K., Agui, N., and Baumhover, A.H. (1984). Science 224, 1243.
Booker, R., and Truman, J.W. (1987a). J. Comp. Neurol. 255, 548.
Booker, R., and Truman, J.W. (1987b). J. Neurosci. 7, 4107.
Carrow, G.M., Calabrese, R.L., and Williams, C.M. (1984). J. Neurosci. 4, 1034.
Cavallin, M., and Fournier, B. (1981). J. Insect Physiol. 27, 527.
Chen, J.H., Fugo, H., Nakajima, M., Nagasawa, H., and Suzuki, A. (1987). J. Insect Physiol. 33, 407.
Copenhaver, P.F., and Taghert, P.H. (1988). In "Endocrinological Frontiers in Physiological Insect Ecology". (F. Sehnal and A. Zabza, eds.) University of Wroclaw Press, Wroclaw, Poland. In press.
Copenhaver, P.F., and Truman, J.W. (1982). J. Insect Physiol. 28, 695.
Copenhaver, P.F., and Truman, J.W. (1986a). J. Neurosci. 6, 1738.
Copenhaver, P.F., and Truman, J.W. (1986b). J. Comp. Neurol. 249, 186.
Diederen, J.H.B., Maas, H.P., Pel, H.J., Schooneveld, H., Jansen, W.F., and Vullings, H.G.B. (1987). Cell Tiss. Res. 249, 379.
Dorn, A., Gilbert, L.I., and Bollenbacher, W.E. (1987). J. Comp. Physiol. B 157, 279.
Fugo, H., Saito, H., Nagasawa, H., and Suzuki, A. (1985). J. Insect Physiol. 31, 293.
Fugo, H., Chen, J.H., Nakajima, M., Nagasawa, H., and Suzuki, A. (1987). J. Insect Physiol. 33, 243.
Goodman, C.S., O'Shea, M., McCaman, R.E., and Spitzer, N.C. (1980). Science 204, 1219.
Hekimi, S. and O'Shea, M. (1986) Soc. Neurosci. Abstr. 12, 948.
Herman, M.M., Miquel, J., and Johnson, M. (1971). Acta Neuropath. 19, 167-183.
Herman, W.S., and Gilbert, L.I. (1966). Gen. Comp. Endocr. 7, 275.

Hoffmann, J.A., and Lagueux, M. (1985). In "Comprehensive Insect Physiology, Biochemistry, and Pharmacology" (G.A.Kerkut and L.I.Gilbert, eds), Vol. 1, p. 435. Pergamon, Oxford.
Homberg, U., Kingan, T.G., and Hildebrand, J.G. (1987). Soc. Neurosci. Abstr. 13, 235.
Ishizaki, H., Mizoguchi, A., Hatta, M., Suzuki, A., Nagasawa, H., Kataoka, H., Isogai, A., Tamura, S., Fujino, M., and Kitada, C. (1987). In "Molecular Entomology" (J.H.Law, ed.), p. 119. Alan R. Liss, New York.
Jones, B.M. (1953). Nature (London) 172, 551.
Keshishian, H., and O'Shea, M. (1985). J. Neurosci. 5, 1005.
Le Douarin, N.M. (1982). "The Neural Crest". Cambridge University Press, Cambridge, England.
Orchard, I., and Loughton, B.G. (1985). In "Comprehensive Insect Physiology, Biochemistry, and Physiology" (G.A.Kerkut and L.I.Gilbert, eds.), Vol. 7, p.61. Pergamon, Oxford.
Raabe, M. (1965). C. R. Acad. Sci. Paris. 260, 6710.
Taghert, P.H., and Goodman, C.S. (1984). J. Neurosci. 4, 989.
Taghert, P.H., and Truman, J.W. (1982) J. Exp. Biol. 98, 385.
Taylor, H.M., and Truman, J.W. (1974). J. Comp. Physiol. 90, 367.
Truman, J.W., and Booker, R. (1986). J. Neurobiol. 17, 613.
Truman, J.W., and Schwartz, L.M. (1984). J. Neurosci. 4, 381.
Truman, J.W., Taghert, P.H., Copenhaver, P.F., Tublitz, N.J., and Schwartz, L.M. (1981). Nature (London) 291, 70.
Tublitz, N.J., and Truman, J.W. (1985a). J. Exp. Biol. 114, 381.
Tublitz, N.J., and Truman, J.W. (1985b) Science 228, 1013.
Tublitz, N.J., and Truman, J.W. (1985c). J. Exp. Biol. 116, 395.
Tublitz, N.J., Copenhaver, P.F., Taghert, P.H., and Truman, J.W. (1986). Trends NeuroSci. 9, 359.
Ziegler, R., Eckart, K., Schwartz, H., and Keller, R. (1985). Biochem. Biophys. Res. Comm. 133, 337.

3

DEVELOPMENT, MATURATION AND AGING IN THE CRUSTACEAN NEUROENDOCRINE SYSTEM

Penny M. Hopkins
Department of Zoology
University of Oklahoma
Norman, Oklahoma

Milton Fingerman
Department of Biology
Tulane University
New Orleans, Louisiana

I. INTRODUCTION

The neuroendocrine mechanisms that control physiological processes in adult crustaceans have received considerable attention. The controls of very early and very late life stages, however, are not as well understood. Interest in the controls of larval development has increased in past years and the literature on larval crustacean endocrine controls is growing. The number of taxa in which most studies have been carried out is very limited. A large portion of the research on adult or larval forms has been done on the order Decapods (see Bowman and Abele, 1982, for details of classification scheme). This summary of crustacean neuroendocrine development and aging will be limited, therefore, to the crustacean order Decapoda and the two suborders Natantia and Reptantia.

II. DESCRIPTION OF CRUSTACEAN NEUROSECRETORY SYSTEMS.

A. Anatomy

1. Neurosecretory Centers.

Neurosecretory cells are distributed widely throughout the crustacean central nervous system. Nerve cells with neurosecretory cytologic characteristics have been found in virtually all ganglia examined. Most of our understanding about the anatomy, development and functions of the neuroendocrine systems in crustaceans comes from studies of higher malacostracans. Other groups of crustaceans have homologous yet less studied neurosecretory systems.

At least 11 different types of neurosecretory cell bodies have been identified in the eyestalk optic ganglia, brains (or supraesophageal ganglia), circumesophageal connectives, and ventral thoracic and abdominal ganglia of decapod crustaceans (Enami, 1951b; Bliss and Welch, 1952; Knowles, 1953; Bliss et al., 1954; Durand, 1956; Knowles and Carlisle, 1956; Matsumoto, 1958, 1962; Fingerman and Aoto, 1959; Goldstone and Cooke, 1971; Evans et al., 1976; Kravitz et al., 1976; Livingstone et al.,1981). Figures 1,2 and 3 show the distribution of these cell bodies in the eyestalk (ES), brain and thoracic ganglia (TG) of natantians, macrurans and brachyurans.

One of the richest accumulations of neurosecretory cell bodies is found in the eyestalk. The nervous elements in the ES of decapod crustaceans are extensions of the protocerebrum of the brain. Each ES contains discrete ganglia; the medulla terminalis (MT), medulla interna (MI) and medulla externa (ME) plus the lamina ganglionaris (LG) (Fig.1). Neurosecretory-appearing cell bodies have been identified in all of these ganglia. The first group of neurosecretory cells identified in the crustacean ES was named the "X-organ" by Hanström (1934). Some confusion arose when another group of neurosecretory cells was later given the same name by Welsh (1941). This confusion was resolved by calling Hanström's original group of cells the pars distalis X-organ (PDXO) or the sensory pore X-organ (SPXO) and Welsh's group of cells the medulla terminalis X-organ (MTXO) or the pars ganglionaris X-organ (PGXO)(Carlisle and Passano, 1953; Carlisle and Knowles, 1959). The term "X-organ" refers generically to any well-defined, compact group of neurosecretory cells in the eyestalk. Thus, the major group of neurosecretory cells in the ME is called the MEXO (Bellon-Humbert et al., 1981) and that of the MI, the MIXO (Fig.1).

2. Neurohemal Structures.

Crustacean neurosecretory cells release their secretions from specific "neurohemal

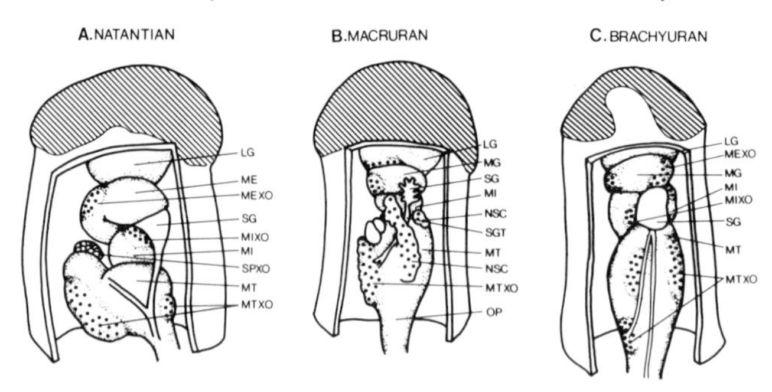

Fig. 1. Distribution of neurosecretory cell bodies in eyestalk ganglia of: A. Natantians, B. Macrurans and C. Brachyurans. LG = lamina ganglionaris; ME = medulla externa; MI = medulla interna; MT = medulla terminalis; MEXO = medulla externa X-organ; MTXO = medulla terminalis X-organ; SG = sinus gland; SGT = nerve tract leading to SG; SPXO = sensory pore X-organ; MIXO = medulla interna X-organ; OP = optic peduncle; NSC = neurosecretory cell bodies; black dots represent cell bodies. (This Figure and Figures 2 and 3 redrawn from Bliss and Welsh, 1952; Bliss et al., 1954; Durand, 1956; Matsumoto, 1958; Fingerman and Aoto, 1959; Carlisle and Knowles, 1959; Hubschman, 1963; Fingerman and Oguruo, 1967; Smith and Naylor, 1972)

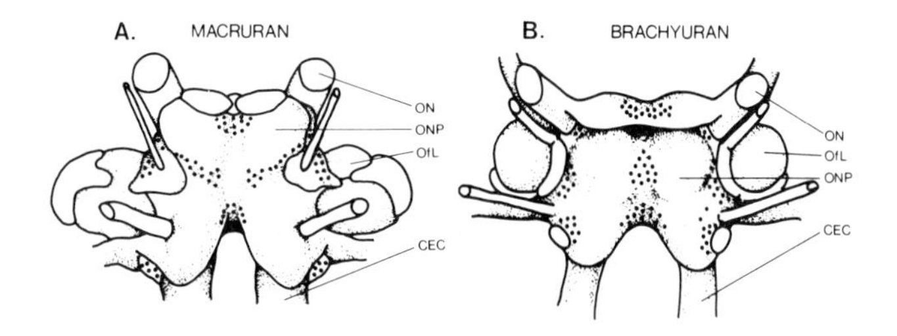

Fig. 2. Distribution of neurosecretory cell bodies in brains of: A. Macrurans and B. Brachyurans. OfL = olfactory lobe; ONP = optic neuropile; ON = optic nerve to eyestalks; CEC = circumesophageal connectives.

organs. Neurohemal organs are collections of axon terminals from neurosecretory cell bodies located at some other site. In crustaceans, the major neurohemal structures are the sinus gland, the pericardial organs and the post-commissural organs.

The best studied neurohemal structure in crustaceans is the sinus gland (SG). In crustaceans with stalked eyes, the SG is located on the surface of the MI or MT on the dorso-lateral side of the ES (Fig.1). The epineurium which covers the entire ES ganglia is invaginated and thickened at the SG. The shape of the SG varies with the group. The shapes range from a mere thickening of the epineurium to a single cup-shaped indentation to a series of cup-shaped sinuses. Axons from various parts of the ES and nervous system come together at the SG to release neurohormones directly into the blood space (Fig. 4A). At first the SG was thought to be a gland in its own right. In 1951, Bliss and Passano independently demonstrated that the SG is a neurohemal organ. They showed that severance of the large SG tract leading from the MT disrupted the transport of neurohormones from the MTXO to the SG. There appears to be around six neurosecretory cell types that terminate in the SG. Most of these come from the MTXO, but there is some evidence that other cell groups from the ES and from more distal portions of the nervous system also

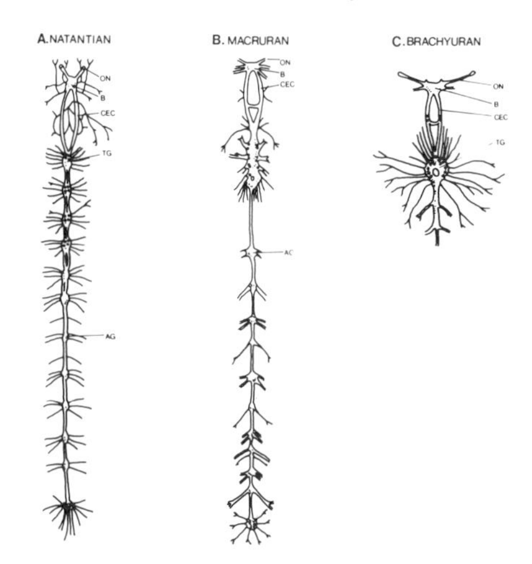

Fig. 3. Distribution of neurosecretory cell bodies in thoracic and abdominal ganglia. ON = optic nerve; B = brain; AG = abdominal ganglia; TG = thoracic ganglia; CEC = circumesophageal connectives; N = nerve.

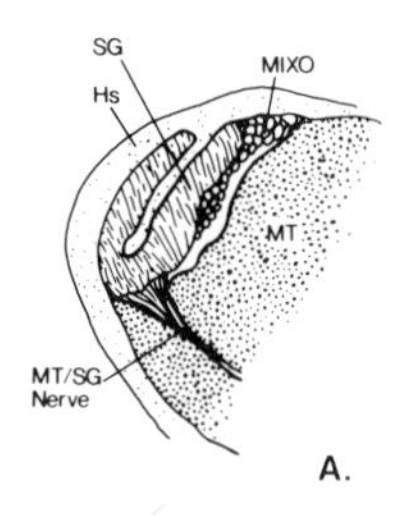

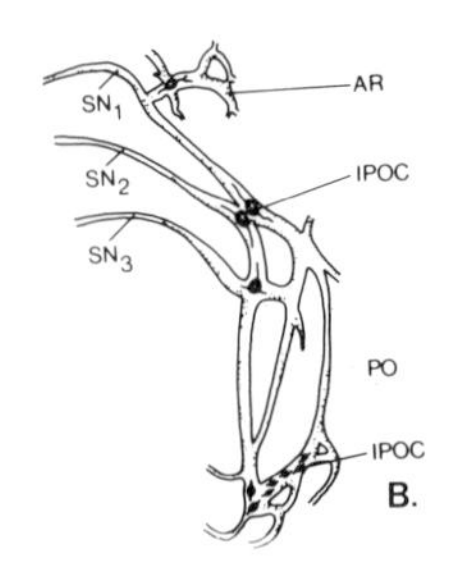

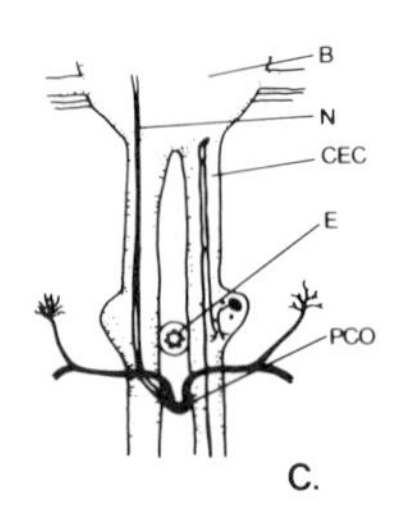

Fig. 4. Neurohemal structures in crustaceans. A. Sinus gland, B. Pericardial organ and C. Post-commissural organ. HS = hemal space; SG = sinus gland; MT = medulla terminalis; MIXO = neurosecretory cells from medulla interna; MT/SG = nerve tract from MT to SG; E = esophagus; AR = anterior ramification; SN 1, 2, 3 = segmental nerves from the anterior TG; IPOC = intrinsic PO cells; B = brain; CEC = circumesophageal connectives; (B redrawn from Maynard and Maynard, 1962; and C redrawn from Carlisle and Knowles, 1959; Fingerman, 1966; and Cooke and Sullivan, 1982).

send their axons to the SG (Bliss and Welsh, 1952; Matsumoto, 1958; Smith and Naylor, 1972; Bellon-Humbert et al., 1981). In addition to the axonal endings, the SG of the crab *Carcinus maenas* also contains intrinsic secretory cells that are similar to the intrinsic cells seen in the corpus cardiacum of insects (May and Golding, 1983). In some crustaceans lacking stalked eyes, e.g. barnacles, a SG is not present. When a SG is present it usually lies in close association with the optic lobes of the brain (Chiang and Steel, 1985). The hormones released from the SG control most of the major physiological processes in crustaceans.

The pericardial organs (PO) are located in the pericardial blood sinus surrounding the crustacean heart (Alexandrowicz, 1953). They are the largest neurohemal structures in the Crustacea (Fig. 4B). The PO's produce and release biogenic amines and at least two peptide neurohormones (Alexandrowicz and Carlisle, 1953). Both of the neuropeptides released in the PO's have cardioexcitatory activity (see below). Although they release catecholamines into the circulation, the analogy of the PO's to the vertebrate adrenal medulla is limited (Cooke and Sullivan, 1982). The PO's receive their neurosecretory input through the segmental or second root nerves. The anatomy of the PO's varies greatly from group to group. In some, the PO's appear to be plexuses while in other groups they consist of two trunks connected by bars (Alexandrowicz, 1953; Maynard, 1960,1961b; Sullivan et al., 1977). There are at least 7 different types of neurosecretory terminals identified in the PO's (Cooke and Sullivan, 1982). The neurosecretory cell bodies contributing axons to these tracts are found mainly in the TG (Maynard, 1961a,b). In addition, some intrinsic cells are located within the PO's (Alexandrowicz, 1953; Matsumoto, 1958; Maynard, 1961a).

Post-commissural organs (PCO) are neurohemal structures located behind the esophagus and above the circumesophageal connectives of the nerve cord (Knowles, 1953; Maynard, 1961a,b; Fingerman, 1966). The neurosecretory cell bodies whose axons terminate in the PCO's are thought to be located in the brain (Carlisle and Knowles, 1959). The hormones of the PCO's have not been thoroughly identified. Extracts of the PCO's contain chromatophorotropic factors (Fig. 4C).

B. Neuroendocrine hormones of crustaceans.

Considering the vast number of neurosecretory-appearing cells located in the crustacean ganglia, it should come as no surprise that many neuroendocrine hormones have been implicated in the control of various physiological events. Gross removal or destruction of neurosecretory cells has been the usual way in which the functions of arthropod neurosecretory centers have been identified. Removal of the stalked eyes of the decapod crustaceans has been in fact the most widely used technique with these organisms.

1. Hormones from the ES only.

Eyestalk removal is a standard method of ablating the MTXO-SG complex. ES removal in adult crustaceans results in a number of physiological responses many of which (but not all) can be reversed with implantation of ES tissue or injection of ES extracts.

a. Crustacean hyperglycemic hormone (CHH). Hormones with a diabetogenic (blood glucose elevating) effect are found in the ES of most crustaceans. These hormones are group-specific: CHH's from crayfish, crabs and shrimps are not interactive (Kleinholz and Keller, 1973; Keller, 1977). CHH is produced in the MT-X-organ (MTXO) of the crab, *C. maenas* (Jaros et al., 1979) and the crayfish *Astacus leptodactylus* (Gorgels-Kallen and van Herp, 1981) and released from the SG. The molecular weight of the CHH's range from 6-7 kd (Table I).

b. Gonad-inhibiting hormone (GIH). This neuropeptide has been found in the ES/SG complex of several crustaceans (see Adiyodi, 1982, for review). GIH has been shown to control both the androgenic gland in males and the ovary in females.

TABLE I. Neurohormones from Crustaceans

Source	Name	MW (kd)	First seen in
Eyestalks Only	Hyperglycemic Hormone (CHH)	6-7	n.d.
	Gonad-inhibiting (GIH)	2-5	Puberty-both sexes
	Molt-inhibiting (MIH)	8.7,7.2	Late larval instars
	Neurodepressing (NDH)	1.2	n.d.
	Retinal Pigment Adapting (RPLAH, RPDAH)	n.d.	n.d
Eyestalks and brain	Red Pigment Concentrating Hormone (RPCH)	1.0	Embryo
	Ionoregulators:		
	Larval Pattern	n.d.	Early larval instars
	Adult Pattern	n.d	Early juvenile
	Water regulators:		
Eyestalk	Larval Pattern	n.d.	Late larval instars
TG and brain	Adult Pattern	1.0	n.d.
TG and brain	Gonad-stimulating (GSH)	n.d.	At puberty in females
Brain	Gonad-Maintenance (GMH)	n.d.	At puberty in males
Brain(?)	Cardioexcitatory	0.6	n.d.

ES removal during the reproductive season in adult decapods results in hypertrophy of the male androgenic gland and in ovarian development in females (Bomirski et al., 1981).

c. Molt-inhibiting hormone (MIH). ES/SG complex in the source of a hormone which exerts a chronic inhibition upon ecdysteroid production by the Y-organs. ES removal results in an increase in circulating levels of the steroid ecdysteroids. Two putative MIH's have been recently isolated. Both MIH's are peptides. One MIH has an inhibitory effect on ecdysteroid production by in vitro Y-organs taken from the crab *C. maenas* in terminal anecdysis (Webster and Keller, 1986) while the other MIH has the effect of lowering circulating ecdysteroids in vivo in eyestalkless juvenile lobsters *Homarus americanus* (Chang et al., 1987). The molecular weights of these MIH's are given in Table I.

d. Neurodepressing hormone (NDH). Spontaneous and responsive electrical activity of portions of the CNS of *C. maenas, Procambarus clarkii* and *P. bouvieri,* and *Panulirus interruptus* can be blocked by extracts of whole ES or of the SG alone (Arechiga et al., 1977; Huberman et al., 1979; Mancillas et al., 1980). NDH is a peptide of about 1.2 kd (Huberman et al., 1979) and is proposed as a modulator of CNS circadian activity.

e. Retinal pigment light adapting and dark adapting hormones (RPLAH and RPDAH). The ability of decapod crustacean retinal pigments (distal and reflecting) to adapt to environmental lighting changes is controlled by at least two ES hormones (see Fingerman, 1987, for review).

2. Hormones from multiple or unclear sources.

a. Red pigment concentrating hormone (RPCH). RPCH was the first crustacean neuropeptide to be isolated and characterized (Fernlund and Josefsson, 1972). This octapeptide is similar to some vertebrate neurohormones in having a terminal pyroGlu. The molecular weight is around 1 kd. It has not been established whether vertebrate-like neurophysins are associated with this or other crustacean neuropeptides. RPCH (or RPCH-like activity) has been isolated from SG, various ES ganglionic X-organs, brain, circumesophageal connectives and TG (see Rao, 1985 and Fingerman, 1987 for reviews). The PCO's are also rich sources of color controlling hormones.

b. Salt and water balance hormones. The brain and ES are the source of a hormone which stimulates sodium uptake in hyposmotic media by the crab *U. pugilator* (Heit and Fingerman, 1975; Davis and Hagadorn, 1982). Hormones controlling water homeostasis appear to be produced in the brain and TG of the crab *C. maenas* (Berlind and Kamemoto, 1977).

c. Gonad stimulating hormone (GSH). This neurohormone is produced by the brain and TG (Ōtsu, 1963; Gomez, 1965; Eastman-Reks and Fingerman, 1984). In female fiddler crabs, *U. pugilator,* GSH is present in highest titers during the reproductive season (Eastman-Reks and Fingerman, 1984). GSH stimulates both male and female gonads.

d. Gonadal maintenance hormone (GMH). Following activation of the androgenic gland in males, a substance is produced in the brain of the shrimps *P. serratus* and *Crangon crangon* and the crayfish *A. leptodactylus* that is necessary to maintain the adult male reproductive system (Touir, 1977; Amato and Payen, 1978).

e. Cardioexcitatory hormones (Proctolin). Two neuropeptides that cause an increase in amplitude of heart beat in crustaceans are released from the PO's. One is a trypsin-sensitive compound and the other is similar to the insect hormone proctolin (Sullivan, 1979; Strangier et al., 1986). The trypsin sensitive compound is produced in the neurosecretory cells of the TG (Terwilliger, 1967).

III. DEVELOPMENT OF NEUROENDOCRINE SYSTEMS

A. Summary of patterns

Crustaceans typically go through a series of distinct developmental stages (Fig. 5). Eggs are laid by adult females and are immediately released to the environment or retained on specialized brooding structures, usually on the abdominal appendages. Postembryonic development of crustaceans can be "direct" or "indirect" (see Williamson, 1982; Gore, 1985, for detailed discussions). Most decapod crustaceans (and indeed most crustaceans in general) go through indirect development. This means that the embryos hatch as larvae and go through a series of changes in form. Larval crustaceans make up a large proportion of the zooplankton of the oceans. Before settling down to a benthic existence, most crustacean larvae go through a true metamorphosis.

Larval forms of crustaceans are called nauplii, zoeae and megalopae. The megalopae instar is a distinct form that is considered a transitional form. The megalopae utilize abdominal appendages (or pleopods) for locomotion. Since true pleopods are found only in the class Malacostraca this larval form is found only in that class. In the decapod crustaceans, early naupliar development occurs in the

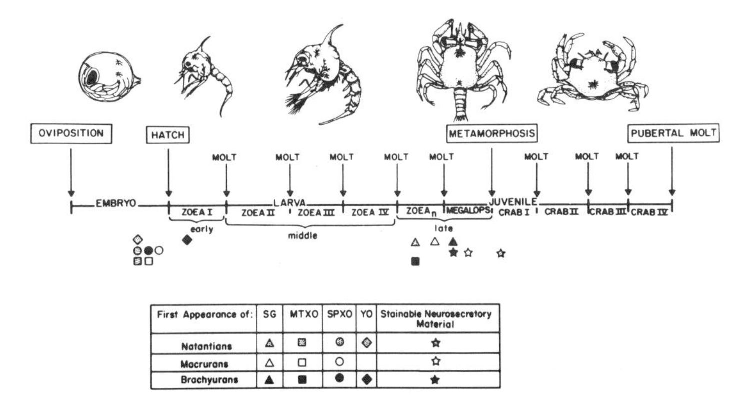

Fig. 5. Generalized developmental sequence for decapod crustaceans. (Drawings at top redrawn from Costlow, 1968; Cole and Lang, 1980; and Bliss, 1982).

egg. In freshwater crayfish and crabs, the zoeal and megalopal instars also occur within the egg (Matsumoto, 1958; Williamson, 1985). The majority of decapod eggs hatch into zoeae which then develop into megalopae. The number of larval instars is quite variable between groups.

Larval decapods molt into juvenile forms that resemble the adult. A series of juvenile instars lead to the pubertal molt. At the pubertal molt juvenile morphology changes to adult morphology. Usually, the pubertal molt results in the postmolt appearance of external reproductive structures.

Natantians have variable numbers of larval molts. In the process of these molts, thoracic appendages and abdominal segments are added (Costlow, 1968). The reptantians also have variable numbers of molts during larval development. In the macrurans (lobsters and crayfish) the changes are gradual - the molt from larva to juvenile involves only minor morphological changes. In the brachyurans (true crabs) the larval forms are very different from the juvenile. The larval instars do not resemble adult forms much and the molt from megalopae to juvenile is usually an abrupt change.

B. Developmental changes in decapod crustaceans(Fig. 5).

1. Molting.

a. Natantians. The development of the natantian neurosecretory centers is fairly straightforward. The SPXO and a single "giant cell" (associated with the MI) appear during embryonic development in *Palaemon,* but the sinus gland does not appear until the larval V instar (Hubschman, 1963). The Y-organ is seen midway through the embryonic development of *P. serratus* (Spindler et al., 1987), but the MTXO and other ganglionic X-organs do not appear until metamorphosis (Bellon-Humbert et al., 1978) and stainable neurosecretory material does not appear until after metamorphosis in *C. crangon* (Dahl, 1957). Eyestalk removal has no effect on the frequency of molts in larval shrimps *Palaemonetes* and *Palaemon macrodactylus* (Hubschman, 1963 and Little, 1969). Eyestalk removal hastens metamorphosis in *Palaemonetes* but delays it in *Palaemon.*

b. Macrurans. The lobster, *H. americanus,* has functional motor neurons early in embryonic development (Cole and Lang, 1980) and the SPXO and MTXO develop late in embryonic instars (Pyle, 1943). The SG, however, is not formed until the larval III instar (Pyle, 1943; Charmantier et al., 1984). The first appearance of the SG is a thickening of the epineurium of the eyestalk ganglia. A nerve tract extends from the developing SG to the distal side of the medulla terminalis. On the basis of staining characteristics, Pyle concluded that the SG is not functional in the larval lobster and that the SG/MTXO complex does not become functional until the IV (megalopae) instar (Pyle, 1943; Charmantier et al., 1984).

Despite the apparent lack of functional ES neurosecretory centers, ES removal in II instar larvae (as well as IV instar larvae) accelerates molting in larval lobsters (Snyder and Chang, 1986a; Rao et al., 1973). ES removal is most effective in shortening the intermolt duration in larval lobsters if performed in the latter half of larval development (Charmantier et al., 1985). Ecdysteroids are detectable in larval lobsters from I larval instar (Chang and Bruce, 1981). In II larval instar lobsters, ES removal did not have much effect on circulating levels of ecdysteroid until the last (megalopal or IV) larval instar. Thus in macrurans it appears that the ES-MIH system becomes effective in the middle of the larval instars. Ecdysteroids are found in the first post-hatching larval instars and throughout all the larval instars of the lobster. Ecdysteroids of larval lobsters show precise cycling; low titers at the beginning and end of each instar with a peak of ecdysteroids in the middle (Snyder and Chang, 1986b). Extracts of juvenile lobster ES cause an inhibition of proecdysis and a decrease in ecdysteroids in larval lobsters (Snyder and Chang, 1986b). It appears that the larval macruran Y-organ is active and possibly responsive to some kind of control(s) prior to the functional organization of the ES neurosecretory centers that control it in the adult.

c. Brachyurans. The development of the brachyuran neurosecretory system is different than that of the natantians and macrurans. In those crabs that have an SPXO (such as *Pinnotheres maculatus*), it develops very early (Pyle, 1943), while the SG and stainable neurosecretory materials are not seen until late larval instar in the crab, *P. dehaani* (Matsumoto, 1958).

The Y-organs develop and begin cyclic activity very early in the crabs, *Hyas araneus* and *Cancer anthonyi* (Anger and Spindler, 1987 and McConaugha, 1980). Indeed, ecdysteroids are produced in embryonic as well as larval crab instars (McCarthy and Skinner, 1979; Lachaise and Hoffman, 1982; Chaix and DeReggi, 1982; Spindler and Anger, 1986).

Eyestalk removal early in larval development (II instar) has no effect on subsequent larval instar durations, but does result in abbreviated megalopal instars. ES removal in II instar *Rhithropanopeus harrisii* larvae slightly shortens the C_4-D_0 period of IV instar larvae and is more effective in shortening the subsequent megalopal instar (Freeman and Costlow, 1980). In a later study on the same animal, ES removal in IV instar larvae had no effect on the duration of the megalopal instar but did affect subsequent juvenile intermolt durations (McConaugha and Costlow, 1987). Earlier studies by Costlow (1963;1966) had led him to conclude that early larval molts, prior to the megalopal molt, were not accelerated by ES removal and that the XO-SG complex is not activated in the crabs *Callinectes sapidus, Sesarma reticulatum* or *R. harrisii* until the megalopal or I crab instars. In the crab *C. sapidus* anecdysial duration from megalopae to I crab is accelerated, but only if ES are removed within 12 hours after the final zoeal molt (Costlow, 1963). Freeman and Costlow (1983) concluded that in the crab *R. harrisii,* MIH is present but in very low amounts during late larval instars then rises in the megalopal instar. In the brachyurans, stainable neurosecretory material is seen prior to metamorphosis.

During embryonic and larval instars, brachyurans have the ability to produce ecdysteroids and the cyclic appearance of the Y-organs suggests that some overriding control is exerted as early as I larval instar. But the ES ablation data suggest that the adult-like MIH control mechanism is not developed until late larval instars. Toward the end of the larval period, the "diecdysial" cycle of molting changes to an "anecdysial" cycle. These two terms describe the time spent between preparations for succeeding molts. If there is a long period between the postecdysial and subsequent proecdysial periods it is called an anecdysial period. If, however, postecdysis slips "imperceptibly" into the next proecdysis it is called a diecdysis (Carlisle and

Knowles, 1959). Diecdysis is characteristic of rapidly molting forms such as the early larval instars and anecdysis is more typical of adult forms (Freeman and Costlow, 1980). I instar *R. harrisii* take about 40 hours from hatching to proecdysis (D_0). By IV instar those same crabs take 50-70 hours. This is the point at which diecdysis switches to anecdysis. The time from ecdysis to D_0 remains fairly constant in this crab during the megalopal through II crab instars (Freeman and Costlow, 1983). Molting in all stages appears to be directed by the ecdysteroids but the neurosecretory control of ecdysteroid production appears to change from anecdysis to diecdysis. The controls (if any) of diecdysis in embryonic and early larval instars is unclear. The switch from the diecdysial pattern of molting to the anecdysial pattern coincides with the first signs of functional development of the ES neurosecretory centers and the SG.

2. Metamorphosis.

Metamorphosis is a change in morphology that occurs at the end of the larval instars. Metamorphosis appears to be controlled by different mechanisms than molting. Eyestalkless late larval instar crabs *R. harrisii* molt slightly sooner than do controls, but metamorphic changes do not keep pace with the accelerated molt cycle (Costlow, 1966). The result is that following what should be the metamorphic molt, the crabs retain some larval characteristics giving the appearance of supernumerary larval stages. Extracts of whole III instar crabs *R. harrisii* contain a metamorphosis inhibiting factor that delays a key event in the metamorphosis of the crab (Freeman and Costlow, 1983). These authors suggest that this factor is produced in the brain or TG, since ES removal accelerates molt but not metamorphosis.

3. Reproduction.

The development of the sexual organs begins early in decapod development. Proliferation of genital rudiments begins in early post-embryonic stages of the crayfish. In the brachyurans, genital organs begin developing during zoeal and megalopal instars, but remain undifferentiated until III juvenile crab stage. Sexual dimorphism of the external genitalia occurs later than dimorphism of internal organs. Control of development in genetic males is directed by the mesodermal androgenic gland and its hormone. The androgenic gland hormone (AGH) induces the differentiation of the male genital tract (Payen, 1982). Neuroendocrine control of the onset of sexual differentiation is located in the ES and brain. An inhibitor substance from the ES (GIH) restrains the activity of the androgenic gland (AG) in genetic males until the onset of puberty. ES removal in male crabs during zoeal, megalopal or juvenile instars results in hypertrophied AG and subsequent precocious spermatogenesis (Payen et al., 1971). The cessation of AG inhibition and the subsequent release of androgenic gland hormone (AGH) at puberty, induces a second neuroendocrine substance from the male brain (the gonadal maintenance hormone, MGH) which is necessary to maintain the structural integrity of the male genital system. Cauterization of the protocerebrum of male shrimps, *P. serratus* or *C. crangon*, results in degeneration of the testis and AG. Reimplantation of male brain tissue (but not female brain tissue) reverses the effects of cauterization (Touir, 1977). Thus the onset of puberty in males is triggered by the withholding of an androgenic inhibitor hormone (GIH) from the ES. Once the inhibition is lifted the AG begins to develop and to release AGH which directs the development of secondary male characters and induces the production of GMH in the brain which is necessary for the maintenance of the male reproductive tract.

Control of development in genetic females is a permissive effect due to the lack of AGH which allows the ovary to develop and produce its own hormone(s). In genetic females (larvae or juveniles) ES removal has no effect on the rate of development of the ovaries (Payen, 1969). Transplantation of AG's into young females results in no inhibition of the development of the AG. Thus, prepubertal females lack both the ES inhibitor of AG activity (GIH) and the brain hormone (GMH) necessary for maintenance of the male reproductive tract.

In conclusion, the neuroendocrine complex of male and female decapods are functionally differentiated early in larval development. The genetic male produces two neurohormones neither of which are found in females. The ES inhibitor is produced and released very early in male development. It prohibits early sexual development. The second male neurosecretory hormone is produced in the brain but appears after puberty and relies on AGH induction. In genetic females the production of GIH by the ES is delayed until after puberty.

4. Color change.

The pigments within the crustacean chromatophore cells migrate in response to environmental light and background color. The pigmentary responses are mediated by a family of neuroendocrine hormones called the chromatophorotropins (Fingerman, 1987). The development of the neuroendocrine system that controls pigment cells is different from the development of neuroendocrine control of molt. Chromatophores are present in all developmental instars of decapods from egg through adult and are responsive to background changes and variations in incident light. The neurohormones which control the dispersion of pigments within the chromatophores are present before hatching. Chromatophorotropin hormones have been isolated from eggs of the crabs *S. reticulatum* (Costlow and Sandeen, 1961) and *Ocypode macrocera* (Rao, 1968) and from the ES of larval *S. reticulata* (Costlow, 1961). Extracts of larval tissues have effects on adult crab chromatophores (Costlow, 1961; Costlow and Sandeen, 1961; Rao, 1968) and extracts of adult tissues have effects on larval chromatophores (Broch, 1960; Rao, 1967). The production and release of the neuropeptide chromatophorotropins in the adult is not limited to the ES, but is distributed throughout the CNS (see above). Since the CNS of decapods begins to develop sooner that do the stalked eyes (Matsumoto, 1958), perhaps control of chromatophorotropins is located in the brain and thoracic neurosecretory centers during early development and switches to the ES neurosecretory centers later in development.

Chromatophorotropins are first seen in eggs of the crab *S. reticulata* on the 11th day of embryonic development (Costlow and Sandeen, 1961). From that time on the production of chromatophorotropins increases through the larval and juvenile instars. The levels and number of different hormones increases as the crab *O. macrocera* matures. The least number of hormones were found in the eggs and zoeae whereas the greatest amounts were found in the megalopae (Rao, 1968).

5. Osmoregulation.

The neuroendocrine mechanism that controls osmoregulation in larval crabs *R. harrisii* becomes differentiated well in advance of metamorphosis (Costlow, 1968). Larval osmoregulation is different from the pattern of osmoregulation in adult decapods and patterns of larval osmoregulation vary according to instar. The adult pattern of osmoregulation begins during early juvenile instars. Since larvae are planktonic, they may require different osmoregulatory controls than benthic adults. Perhaps, the hyperosmoregulation of larvae increases their density so they are not swept out of estuaries (Kalber, 1970). Regardless, there is a switch in osmoregulatory control in at least two crab species (*R. harrisii* and *C. sapidus*). The change involves both the sites of neurosecretory hormone production and the function of the hormone(s). The switch occurs at the same time that the SG of the ES appears to become functional and neurosecretory material within the neurosecretory cells of the brain and ES of the crab *P. dehaani* becomes stainable (Matsumoto, 1958).

As larval lobsters *H. americanus* go from larval III instar to IV instar their sodium regulation changes from an isoionic regulatory pattern to a slightly hypertonic. ES removal in IV or V instar lobsters causes them to revert back to the isoionic pattern (Charmantier et al., 1984). Reimplantation of IV or V ES tissue corrects the ionic control in IV or V eyestalkless lobsters but not in III instar animals. These authors suggest that there is a change in osmoregulatory control that occurs between III (the last larval) and IV (megalopal) instars. Ionic control in larval instars is located in neurosecretory sites in the brain whereas post larval control is also located in ES centers and begins to appear only as those centers begin to develop (Pyle, 1943; Charmantier et al., 1984).

Larval *R. harrisii* (II instar) hyperosmoregulate in 10-30% sea water. Since ES removal reduces this ability, it is assumed that there are neurosecretory centers in the ES which control osmoregulation in larvae (Kalber and Costlow, 1966). As the larvae enter late larval instars (III,IV and megalopal) ES removal no longer affects osmoregulation. In adult decapods, the control of salt and water balance seems to be distributed between the ES and the brain/TG (Table I).

The ability to regulate ionic concentrations of blood is tied closely to the ability to control water uptake. Larval decapods seem to be able to control water movement and the control center appears to be in the ES (Table I). ES removal in III instar larvae of *R. harrisii* results in larger IV instar larvae (Costlow, 1966). ES removal in larvae of shrimp *Palaemonetes vulgaris* and *P. macrodactylus* produce animals that are significantly larger than controls at metamorphosis (Hubschman, 1963; Little, 1969). ES removal in juvenile lobsters *H. americanus* and crabs *R. harrisii* results in post molt animals that are larger than controls (Mauviot and Castell, 1976; Freeman and Costlow, 1983). These increases in size are due in part to greater than normal uptake of water at ecdysis. The neurohormones involved in this physiological phenomenon are probably different from those controlling ordinary homeostatic ion movements because all larval instars seem able to ionoregulate, whereas the increases in size are restricted to late larval instars. Also, the ability to osmoregulate is undisturbed during proecdysis in eyestalkless *R. harrisii* larvae - the period in which water uptake begins. In juvenile *R. harrisii* the increase in size can also be attributed to increase in tissue growth (Freeman et al., 1983).

6. Conclusions.

Larvae of decapod crustaceans have the ability to respond to environmental conditions in ways that are similar to adult responses. The structural components of the ES neurosecretory system (the ganglionic X-organs and the sinus gland) do not develop until late larval and megalopal instars. Because no stainable neurosecretory material is demonstrable in the neurosecretory centers during early development, the consensus of the investigators in this field is that the neurosecretory/neurohemal centers of the ES are not functional until late larval/early juvenile development. Yet, the abilities to osmoregulate and to adapt to light and background changes occurs very early in development. ES removal in some (but not all) larvae results in shortening of the intermolt duration - just as in adults. Some of these early physiological responses (such as osmoregulation and color responses) are mediated by hormones other than ES hormones. The remaining puzzle is that of the control of ecdysteroid levels in embryos and larvae. Ecdysteroids and the Y-organs exhibit the same cyclic activity as seen in adults. Either there is some control located outside of the ES in these early developmental instars or the ES neurosecretory products of early instars have different staining properties than those of adults.

IV. MATURATION AND SENESCENCE OF NEUROENDOCRINE SYSTEMS.

A. Neuroendocrine control of molt and growth.

At the decapod pubertal molt, female and male adult reproductive and growth patterns become established. Molting and reproduction appear to be mutually exclusive in many decapods (Adiyodi, 1982). This is thought to be due to the extreme demands that both molting and female reproduction place on organic reserves. The timing of molting and reproduction appears to be controlled by the neuroendocrine system of the ES and brain. In some decapods, reproductive activity occurs during the anecdysial portion of the molt cycles throughout adult life. Some decapods simply suspend molting activity after the pubertal molt to avoid the conflict altogether.

1. Adult molt patterns.

Adult growth and molting patterns are established at the pubertal molt. These patterns can be either "indeterminate" or "determinate" (Hartnoll, 1985). The indeterminate pattern consists of continuous post-pubertal molting that is terminated by death (Table II). The best examples of this type of growth pattern are the very large lobsters infrequently collected off the North American coast. The assumption is that left alone these animals will molt and grow indefinitely (Table II,1). While there is no age-related termination of molting and growth in these animals, there is an indication that in some animals the intermolt periods become longer with age (Passano, 1960). Whether these prolongations of the molt cycle can be attributed to aging of the neuroendocrine controls of molting or to other factors is unknown. As an animal becomes larger it may take longer to reach the critical level of nutrient reserves necessary to trigger a new molt. In the crab *U. pugilator,* the mean length of the molt cycle under constant environmental conditions does not change significantly as the animal ages (Hopkins, 1982). ES removal in animals which have an indeterminate growth pattern usually results in a shortened anecdysial cycle. The longer anecdysial cycles seen in older lobsters are due to increasingly longer or constant periods of MIH production. Aging or senescence of the neuroendocrine cells does not seem to be a factor in the indeterminate molting pattern. The neuroendocrine complexes in these animals remain functional and constant until death.

The determinate pattern of molting and growth includes a very definite end to adult molting cycles (Table II, 4,5). This cessation of adult molting has been called "terminal anecdysis" (Carlisle and Dohrn, 1953) and appears to be mediated by distinct changes in the adult endocrine system (Hartnoll, 1972). At first glance this appears to be a result of neuroendocrine senescence. Terminal anecdysis, however, can be reversed in some crabs by ES removal and in the other crabs it appears that Y-oran degeneration is the cause of terminal anecdysis. Thus, there appears to be three patterns for control of adult molting cycles: The first one (indeterminate pattern) is a result of continual, cyclic release of MIH until death. The second pattern is one in which there appears to be a continuous, non-cyclic release of MIH (determinate pattern). The third pattern is one in which the role of MIH ends with the onset of the terminal ecdysis. The determinate patterns of post-pubertal molting, like the indeterminate pattern, give no indication of neuroendocrine senescence. In those majid crabs that cease molting immediately following the pubertal molt (Table II, 5) it appears that the Y-organs degenerate with age. Whether the neuroendocrine centers that control the Y-organ also degenerate is not known.

TABLE II. Molt and Reproductive Patterns in Crustaceans

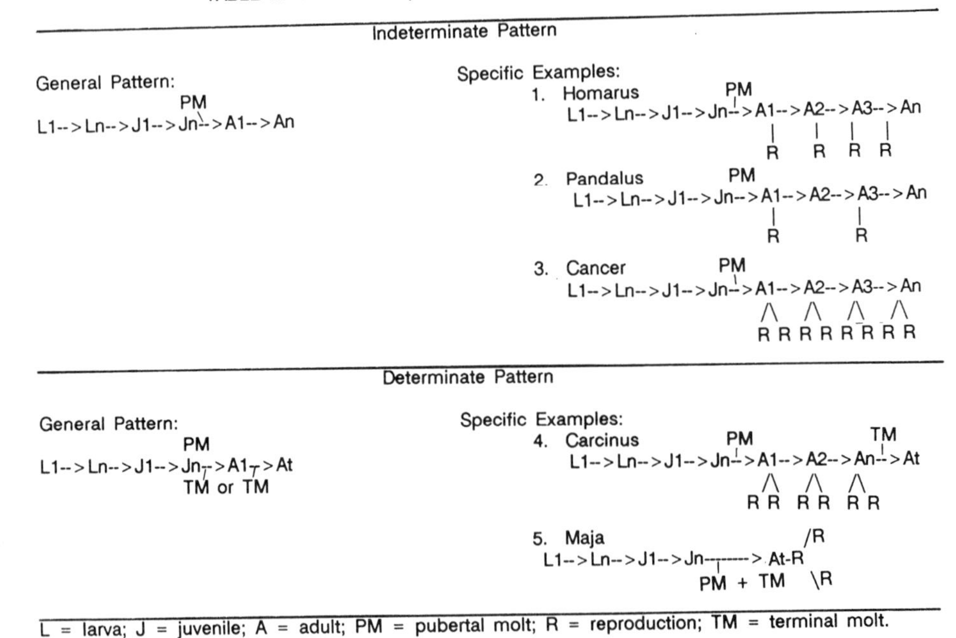
Indeterminate Pattern

General Pattern:
PM
L1-->Ln-->J1-->Jn-->A1-->An

Specific Examples:
1. Homarus
PM
L1-->Ln-->J1-->Jn-->A1-->A2-->A3-->An
R R R R

2. Pandalus
PM
L1-->Ln-->J1-->Jn-->A1-->A2-->A3-->An
R R

3. Cancer
PM
L1-->Ln-->J1-->Jn-->A1-->A2-->A3-->An
R R R R R R R R

Determinate Pattern

General Pattern:
PM
L1-->Ln-->J1-->Jn-->A1-->At
TM or TM

Specific Examples:
4. Carcinus
PM TM
L1-->Ln-->J1-->Jn-->A1-->A2-->An-->At
R R R R R R

5. Maja
L1-->Ln-->J1-->Jn-------->At-R
PM + TM
/R
\R

L = larva; J = juvenile; A = adult; PM = pubertal molt; R = reproduction; TM = terminal molt.

2. Reproduction.

Superimposed upon these molting patterns are variable female reproductive patterns. Mating in many decapod crustaceans is limited to the post-molt period when the female is soft shelled. In most crustaceans this means that only a single clutch of eggs can be brooded during a given cycle and molting episodes separate brooding episodes. In fact, some natantians and macrurans may skip a cycle between clutches (Table II). The brachyurans differ from other decapods in that the females can store sperm. This allows the true crabs to reproduce independently of the molt cycle. Thus, most female crabs produce more than one clutch/molt cycle (Table II, 3,4,5) or since the sperm are not shed during molt, to spawn during a non-mating cycle.

The function of MIH in late larval forms is to prolong the period in which feeding can occur to allow optimal growth at molt. At puberty, MIH begins to subserve a secondary function - that of prolonging the intermolt period so that a complete reproductive cycle can intervene between molts. At this point in development, molting control begins to be coordinated with reproductive control. In the indeterminate molting pattern, the anecdysial period is protracted (by MIH). This ability to prolong intermolt allows for more than one reproductive cycle per molt cycle. In terminal anecdysis constant release of MIH (or degeneration of the Y-organs) allows for continuous reproductive activity. Indeterminate molting is characterized by alternating periods of anecdysis and proecdysis. Anecdysis is dominated by MIH whereas proecdysis is dominated by the steroidal ecdysteroids.

Eyestalk removal in the adult decapod can precipitate proecdysis by removing the neurosecretory hormone MIH or it can cause gonadal development. Adult ES neurosecretory centers contain more than one inhibitory principle. They also produce a hormone that chronically inhibits the development of the gonads. This hormone (Gonadal Inhibiting Hormone - GIH) may be produced prior to the pubertal molt in males only. In adult males and females that are seasonal breeders and molters, MIH and GIH are produced at different times of the year so that during certain seasons ES removal will induce molting and at other times it will induce ovarian development.

This pattern of alternating molting and reproductive activity is predominantly found in the indeterminate molting crustaceans. In the animals which have the determinate molt pattern, the picture is less clear. ES removal in post pubertal and pre-terminal anecdysial female *C. maenas* in any season results only in ovarian growth and no molting (Demeusy, 1965). ES removal in mature specimens of the majid crab *Libinia emarginata* enhances reproductive activity but does not induce molting (Hinsch, 1972). Thus, following the pubertal molt in female *C. maenas* and *L. emarginata,* there is no MIH in the ES only GIH. Yet, ES removal in juvenile instars and terminal anecdysis, result in molting in female *C. maenas* but no response in *L. emarginata.*

3. Osmoregulation.

The differences between larval and adult control of iono- and osmoregulation are give in Table I. There is no indication in the literature that after the pubertal molt there is any change in the neurosecretory control of osmoregulation as a function of aging.

4. Color change.

Aging does seem to have an effect on the neurosecretory controls of color change in crustaceans. Older crabs, *Carcinus, Sesarma, Uca* and *Ocypode,* lose their capacity to respond to changes in background color (Stephenson, 1932; Enami, 1951a; Rao, 1968; Rao and Fingerman, 1968). In *Carcinus*, at least this appears to be due to a "scarcity or absence" of the chromatophore hormones (Stephenson, 1932). Also, it has been demonstrated in the crayfish *Procambarus clarki* that there is more neurohormone in the CNS of juveniles that in the CNS of adults (Fingerman et al., 1964). In the crab *Sesarma hematocleir* it has been shown that the number of neurosecretory centers is reduced in adults when compared to juveniles (Enami, 1951). Thus, aging in the crustacean neurosecretory chromatophorotropin system occurs through a reduction in the number of neurosecretory centers producing chromatophorotropins and a reduction in the amount produced in each center.

Acknowledgements. The authors wish to express their gratitude to Ms. Coral McCallister, Dr. Frank Sonleitner, Dan Hough and Dr. Gary Schnell for their expert and kindly help and advice.

REFERENCES

Adiyodi, RG (1982) Reproduction and its control. In: "The Biology of Crustacea" (DE Bliss, ed) Vol. 9 pp. 147-215. Acad. Press, NY.

Alexandrowicz, JS (1953) Nervous organs in the pericardial cavity of the decapod Crustacea. J. Mar. Biol. Assoc. U.K. 31: 563-580.

Alexandrowicz, JS and DB Carlisle (1953) Some experiments on the function of the pericardial organs in Crustacea. J. Mar. Biol. Assoc. U.K. 32: 175-192.

Amato, GD and GG Payen (1978) Mise en évidence du côntrole endocrine des différentes étapes de la spermatogenèse chez l'Ecrevisse *Pontastacus leptodactylus leptodactylus* (Eschschultz 1823) Crustacé, Decapode, Reptantia. Gen. Comp. Endocrinol. 36: 487-496.

Anger, K and K-D Spindler (1987) Energetics, moult cycle and ecdysteroid titers in spider crab (*Hyas araneus*) larvae starved after the Do threshold. Mar. Biol. 94: 367-375.

Arechiga, H, A Huberman and A Martinez-Palomo (1977) Release of a neurodepressing hormone from the crustacean sinus gland. Brain Res. 128: 93-108.

Bellon-Humbert, C, MJP Thijssen and F van Herp (1978) Development, location and relocation of sensory and neurosecretory sites in the eyestalks during larval and postlarval life of *Palaemon serratus* (Pennant) J. Mar. Biol. Assoc. U.K. 58: 851-868.

Bellon-Humbert, C, F van Herp, GECM Strolenberg and JM Denuce (1981) Histological and physiological aspects of the medulla externa X-organ, a neurosecretory cell group in the eyestalk of *Palaemon serratus* Pennant (Crustacea, Decapoda, Natantia). Biol. Bull. 160: 11-30.

Berlind, A and FI Kamemoto (1977) Rapid water permeability changes in eyestalkless euryhaline crabs in isolated, perfused gills. Comp. Biochem. Physiol. 58A: 383-385.

Bliss, DE (1951) Metabolic effects of sinus gland or eyestalk removal in the land crab, *Gecarcinus lateralis*. Anat. Rec. 111: 502-503.

Bliss, DE and JH Welsh (1952) The neurosecretory system of brachyuran Crustacea. Biol. Bull. 103: 157-169.

Bliss, DE, JB Durand and JH Welsh (1954) Neurosecretory systems in decapod Crustacea. Z. Zellforsch. Mikrosk. Anat. 39: 520-536.

Bliss, DE (1982) "Shrimps Lobsters and Crabs. Their fascinating life story." New Century Publ. Co. N.J.

Bomirski, M, M Arendarezyk, M Kawinska and LH Kleinholz (1981) Partial characterization of crustacean gonad-inhibiting hormone. Intl. J. Invert. Reproduct. 3: 213-219.

Bowman, TE and LG Abele (1982) Classification of the recent Crustacea. In: "The Biology of Crustacea" (DE Bliss, ed). Vol. 1. Acad. Press, N.Y.

Broch, ES (1960) Endocrine control of the chromatophores of the zoeae of the prawn *Palaemonetes vulgaris*. Biol. Bull. 119: 305-306.

Carlisle, DB (1953) Studies on *Lysmata seticaudata* Risso (Crustacea Decapoda). VI. Notes on the structure of the neurosecretory system of the eyestalk. Pubbl. Staz. Zool. Napoli 24: 435-447.

Carlisle, DB and PFR Dohrn (1953) Studies on *Lysmata seticaudata* Risso (Crustacea Decapoda). II. Experimental evidence for a
growth-accelerating factor obtainable from eyestalks. Pubbl. Staz. Zool. Napoli 24: 69-83.
Carlisle, DB and LM Passano (1953) The x-organ of Crustacea. Nature 171: 1070-1071.
Carlisle, DB and F Knowles (1959) "Endocrine control in crustaceans." Cambridge Univ. Press. N.Y.
Chaix, J-C and M DeReggi (1982) Ecdysteroid levels during ovarian development and embryogenesis in the spider crab *Acanthonyx lunulatus*. Gen. Comp. Endocrinol. 47: 7-14.
Chang, ES and MJ Bruce (1981) Ecdysteroid titers of larval lobsters. Comp. Biochem. Physiol. 70A: 239-241.
Chang, ES, MJ Bruce and RW Newcomb (1987) Purification and amino acid composition of a peptide with molt-inhibiting activity from the lobster, *Homarus americanus*. Gen. Comp. Endocrinol. 65: 56-64.
Charmantier, G, M Charmantier-Daures and DE Aiken (1984) Neuroendocrine côntrol of hydromineral regulation in the American lobster, *Homarus americanus* H. Milne-Edwards, 1837. Gen. Comp. Endocrinol. 54: 20-34.
Charmantier, G, M Charmantier-Daures and DE Aiken (1985) Intervention des pédoncules oculaires dans le controle de la métamorphose chez *Homarus americanus* H. Milne-Edwards 1837 (Crustacea Decapoda). C.R. Acad. Sci. Paris 300: 271-276.
Chiang, RG and CGH Steele (1985) Structural organization of neurosecretory cells terminating in the sinus gland of the terrestrial isopod, *Oniscus ascellus*, revealed by paraldehyde fuchsin and cobalt backfilling. Can. J. Zool. 63: 543-549.
Cole, JJ and F Lang (1980) Spontaneous and evoked postsynaptic potentials in an embryonic neuromuscular system of the lobster, *Homarus americanus*. J. Neurobiol. 11: 459-470.
Cooke, I and MW Goldstone (1970) Fluorescence localization of monoamines in crab neurosecretory structures. J. Exp. Biol. 53: 651-658.
Cooke, IM and RE Sullivan (1982) Hormones and neurosecretion. In: "The Biology of Crustacea" (DE Bliss, ed). Vol. 3. pp 205-290. Acad. Press. N.Y.
Costlow, JD (1961) Fluctuations in hormone activity in Brachyura larvae. Nature 192: 183-184.
Costlow, JD (1963) Regeneration and metamorphosis in larvae of the blue crab, *Callinectes sapidus* Rathbun. J. Exp. Zool. 152: 219-223.
Costlow, JD (1966) The effect of eyestalk extirpation on larval development of the mud crab, *Rhithropanopeus harrisii* (Gould). Gen. Comp. Endocrinol. 7: 255-274.
Costlow, JD (1968) Metamorphosis in crustaceans. In: "Metamorphosis - a problem in developmental biology " (W Etkin and LI Gilbert, eds). pp. 3-41 Appleton-Century Crofts. N.Y.
Costlow, JD and MI Sandeen (1961) The appearance of chromatophorotropic activity in the developing crab, *Sesarma reticulatum*. Amer. Zool. 1: 443.
Dahl, E (1957) Embryology of the X-organs in *Crangon allmanni*. Nature 179: 482.
Davis, CW and IR Hagadorn (1982) Neuroendocrine control of Na balance in the fiddler crab *Uca pugilator*. Amer. J. Physiol. 242: R505-R513.
Demeusy, N (1965) Croissance somatique et fonction de reproduction chez la femelle du décapode brachyoure *Carcinus maenas* Linné. Arch. Zool. Exp. Gen. 106: 625-664.
Durand, JB (1956) Neurosecretory cell types and their secretory activity in the crayfish. Biol. Bull. 111: 62-76.
Eastman-Reks, S and M Fingerman (1984) Effects of neuroendocrine tissue and cylic AMP on ovarian growth *in vivo* and *in vitro* in the fiddler crab, *Uca pugilator*. Comp. Biochem. Physiol. 79A: 675-684.
Enami, M (1951a) The sources and activities of two chromatophorotropic hormones in crabs of the genus *Sesarma*. I. Experimental analysis. Biol. Bull. 100: 28-43.
Enami, M (1951b) The sources and activities of two chromatophorotropic hormones in crabs of the genus *Sesarma*. II. Histology of incretory elements. Biol. Bull. 101: 241-258.
Evans, PD, EA Kravitz, BR Talamo and BG Wallace (1976) The association of octopamine with specific neurons along lobster nerve trucks. J. Physiol. 262: 51-70.
Fernlund, P and L Josefsson (1972) Crustacean color-change hormone: Amino acid sequence and chemical synthesis. Science 177: 173-175.
Fingerman, M (1966) Neurosecretory control of pigmentary effectors in crustaceans. Amer. Zool. 6: 169-179.
Fingerman, M (1987) The endocrine mechanisms of crustaceans. J. Crust. Biol. 7: 1-24.
Fingerman, M and T Aoto (1959) Effects of eyestalk ablation upon neurosecretion in the supraesophageal ganglia of the dwarf crayfish *Cambarellus shufeldti*. Trans. Amer. Microsc. Soc. 79: 68-74.
Fingerman, M and C Oguro (1967) The neuroendocrine system in the head of the crayfish *Faxonella clypeata*. Trans. Amer. Microsc. Soc. 86: 178-183.
Fingerman, M, Y Yamamoto and CW Jacob (1964) Differences between the chromatophore systems of juvenile and adult specimens of the crayfish *Procambarus clarki*. Amer. Zool. 4: 413.

Freeman, JA and JD Costlow (1980) The molt cycle and its hormonal control in *Rhithropanopeus* larvae. Devel. Biol. 74: 479-485.

Freeman, JA and JD Costlow (1983) Endocrine control of spine epidermis resorption during metamorphosis in crab larvae. Roux's Arch. Dev. Biol. 192: 362-365.

Freeman, JA, TL West and JD Costlow (1983) Postlarval growth in juvenile *Rhithropanopeus harrisii*. Biol. Bull. 165: 409-415.

Goldstone, M and I Cooke (1871) Histochemical localization of monoamines in the crab central nervous system. Z. Zellforsch. Mikrosk. Anat. 116: 7-19.

Gore, RH (1985) Molting and growth in decapod larvae. In: "Larval Growth" (AM Wenner, ed). pp. 1-65. Balkema Press, Boston, MA.

Gorgels-Kallen, JL and F van Herp (1981) Localization of crustacean hyperglycemic hormone (CHH) in the X-organ sinus gland complex in the eyestalk of the crayfish *Astacus leptodactylus* (Nordmann 1842). J. Morph. 170: 347-355.

Hanström, B (1934) Über das Organ-X eine inkretorische Gehirndrüse der Crustacean. Psychiat. Neurol. Bl. Anat. 3: 1-14.

Hartnoll, RG (1972) The biology of the burrowing crab, *Carystes cassivelaunas*. Bijdr. Dierk. 42: 139-155.

Hartnoll, RG (1985) Growth, sexual maturity and reproductive output. In: "Factors in Adult Growth" (AM Wenner, ed) pp. 101-128. Balkema Press. Boston, MA.

Heit, M and M Fingerman (1975) The role of an eyestalk hormone in the regulation of the sodium concentration of the blood of the fiddler crab, *Uca pugilator*. Comp. Biochem. Physiol. 50A: 277-280.

Hinsch, GW (1972) Some factors controlling reproduction in the spider crab, *Libinia emarginata*. Biol. Bull. 143: 358-366.

Hopkins, PM (1982) Growth and regeneration patterns in the fiddler crab, *Uca pugilator*. Biol. Bull. 163: 301-319.

Huberman, A, H Arechiga, A Cimet, J de la Rosa and C Aramburo (1979) Isolation and purification of neurodepressing hormone from eyestalks of *Procambarus bouvieri*. European J. Biochem. 99: 203-208.

Hubschman, JH (1963) Development and function of neurosecretory sites in the eyestalks of larval *Palaemonetes* (Decapoda: Natantia). Biol. Bull. 125: 96-113.

Jaros, PP and R Keller (1979) Immunocytochemical identification of hyperglycemic hormone-producing cells in the eyestalk of *Carcinus maenas*. Cell Tissue Res. 204: 379-385.

Kalber, FA (1970) Osmoregulation in decapod larvae as a consideration in culture techniques. Helgol. Wiss. Meeresunters. 20: 697-706.

Kalber, FA and JD Costlow (1966) The ontogeny of osmoregulation and its neurosecretory control in the decapod crustacean, *Rhithropanopeus harrisii*. Amer. Zool. 6: 221-229.

Keller, R (1977) Comparative electrophoretic studies of crustacean neurosecretory hyperglycemic and melanophore-stimulating hormones from isolated sinus glands. J. Comp. Physiol. 122: 359-373.

Kleinholz, LH and R Keller (1973) Comparative studies in crustacean neurosecretory hyperglycemic hormones. I. The initial survey. Gen. Comp. Endocrinol. 21: 554-564.

Knowles, FGW (1953) Neurosecretory pathways in the prawn *Leander serratus*. Nature 171: 131.

Knowles, FGW and DB Carlisle (1956) Endocrine control in the Crustacea. Biol. Rev. 31: 396-473.

Kravitz, EA, PD Evans, B Talamo, B Wallace and G Battelle (1976) Octopamine neurons in lobsters: Location, morphology, release of octopamine and possible physiological roles. Cold Spring Harbor Symp. Quant. Biol. 40: 127-133.

Lachaise, F and JA Hoffmann (1982) Ecdysteroids and embryonic development in the shore crab, *Carcinus maenas*. Hoppe-Seyler's Z. Physiol. Chem. 363: 1059-1067.

Little, G (1969) The larval development of the shrimp, *Palaemon macrodactylus* Rathbun, reared in the laboratory, and the effect of eyestalk extirpation on development. Crust. 17: 69-87.

Livingstone, MS, SF Schaeffer and EA Kravitz (1981) Biochemistry and ultrastructure of serotonergic nerve endings in the lobster: Serotonin and octopamine are contained in different nerve endings. J. Neurobiol. 12: 27-54.

Mancillas, JR, JF McGinty, AI Selverston, H Karten, FE Bloom (1981) Immunocytochemical localization of enkephalin and substance P in retina and eyestalk neurones of lobster. Nature 293: 576-578.

Matsumoto, K (1958) Morphological studies on the neurosecretion in crabs. Biol. J. Okayama Univ. 4: 103-176.

Matsumoto, K (1862) Experimental studies of the neurosecretory activities of the thoracic ganglion of a crab, *Hemigrapsus*. Gen. Comp. Endocrinol. 2: 4-11.

Mauviot, JC and JD Castell (1976) Molt and growth-enhancing effects of bilateral eyestalk ablation on juvenile and adult american lobsters (*Homarus americanus*). J. Fish. Res. Bd. Can. 33: 1922-1929.

May, BA and DW Golding (1983) Aspects of secretory phenomena within the sinus gland of *Carcinus maenas* (L). Cell Tissue Res. 228: 245-254.

Maynard, DM (1960) Circulation and heart function. In: "The Physiology of Crustacea" (TH Waterman, ed) pp. 161-226. Acad. Press, N.Y.
Maynard, DM (1961a) Thoracic neurosecretory structures in Brachyura I. Gross anatomy. Biol. Bull. 121: 316-329.
Maynard, DM (1961b) Thoracic neurosecretory structures in Brachyura II. Secretory neurons. Gen. Comp. Endocrinol. 1: 237-263.
Maynard, DM and E Maynard (1962) Thoracic neurosecretory structures in Brachyura III. Microanatomy of peripheral structures. Gen. Comp. Endocrinol. 2: 12-28.
McCarthy, JG and DM Skinner (1979) Changes in ecdysteroids during embryogenesis of the blue crab, *Callinectes sapidus* Rathbun. Devel. Biol. 69: 627-633.
McConaugha, JR (1980) Identification of the Y-organ in the larval stages of the crab, *Cancer anthonyi* Rathbun. J. Morphol. 164: 83-88.
McConaugha, JR and JD Costlow (1987) Role of ecdysone and eyestalk factors in regulating regeneration in larval crustaceans. Gen. Comp. Endocrinol. 66: 387-393.
Ōtsu, T (1963) Biohormonal control of sexual cycle in the freshwater crab, *Potamon dehaani*. Embry. 8: 1-20.
Passano, LM (1951) The X-organ-sinus gland neurosecretory system in crabs. Anat. Rec. 111: 502.
Passano, LM (1960) Molting and its control. In: "The Physiology of Crustacea" (TH Waterman, ed). Vol.1. pp. 473-536. Acad. Press. N.Y.
Payen, G (1969) Expériences de greffes de glandes androgènes sur le Crabe *Carcinus maenas* L. a Premiers résultats. C.R. Acad. Sci. Ser. D. 266: 1056-1058.
Payen, G (1974) Morphogenèse sexuelle de quelques Brachyoures au cours du développement embryonnaire, lavaire et post larvaire. Bull. Mus. Hist. Nat. Zool. 209: 201-262.
Payen, GG (1980) Experimental studies of reporduction in Malacostraca crustaceans. Endocrine control of spermatogenic activity. In: "Advances in Invertebrate Reproduction" (MW Clark and TS Adams, eds). pp. 187-196. Elsevier/North Holland, N.Y.
Payen, GG, JD Costlow and H Charniaux-Cotton (1971) Etude comparative de l'ultrastructure des glandes androgènes de Crabes normaux et pédonculectomisés pendant la vie larvaire ou après la puberté chez les especes: *Rhithropanopeus harrisii* (Gould) et *Callinectes sapidus* Rathbun. Gen. Comp. Endocrinol. 17: 526-542.
Pyle, RC (1943) The histogenesis and cyclic phenomena of the sinus gland and X-organ in Crustacea. Biol. Bull. 85: 87-102.
Rao, KR (1967) Responses of crustacean chromatophores to light and endocrines. Exp. 231: 1-4.
Rao, KR (1968) Variations in the chromatophorotropins and adaptive color change during the life history of the crab, Ocypode macrocera. Zool. Jb. Physiol. 74: 274-291.
Rao, KR (1985) Pigmentary effectors. In: "The Biology of Crustacea" (DE Bliss,ed). Vol. 9. pp 395-462. Acad. Press. N.Y.
Rao, KR and M Fingerman (1968) Dimorphic variants of the fiddler crab *Uca pugilator* and their chromatophore responses. Proc. Louisiana Acad. Sci. 31: 27-39.
Rao, KR, SW Fingerman and M Fingerman (1973) Effects of exogenous ecdysones on the molt cycles of fourth and fifth stage american lobsters, *Homarus americanus*. Comp. Biochem. Physiol. 44A: 1105-1120.
Smith, G and E Naylor (1972) The neurosecretory system of the eyestalk of *Carcinus maenas* (Crustacea:Decapoda). J. Zool. 166: 313-321.
Snyder, MJ and ES Chang (1986a) Effects of eyestalk ablation on larval molting rates and morphological development of the american lobster, *Homarus americanus*. Biol. Bull. 170: 232-243.
Snyder, MJ and ES Chang (1986b) Effects of sinus gland extract on larval molting and ecdysteroid titers of the american lobster, *Homarus americanus*. Biol. Bull. 170: 244-254.
Spindler, K-D and K Anger (1986) Ecdysteroid levels during the larval development of the spider crab, *Hyas araneus*. Gen. Comp. Endocrinol. 64: 122-128.
Spindler, K-D, A VanWormhoudt, D Sellos and M Spindler-Barth (1987) Ecdysteroid levels during embryogenesis in the shrimp, *Palaemon serratus* (Crustacea Decapoda): Quantitative and qualitative changes. Gen. Comp. Endocrinol. 66: 116-122.
Stephenson, EM (1932) Color changes in Crustacea. Nature 130: 931.
Strangier, J, H Dircksen and R Keller (1986) Identification and immunocytochemical localization of proctolin in pericardial organs of the shore crab, *Carcinus maenas*. Peptides 7: 67-72.
Sullivan, RE (1979) A proctolin-like peptide in crab pericardial organs. J. Exp. Zool. 210: 543-552.
Sullivan, RE, B Friend and DL Barker (1977) Structure and function of spiny lobster ligamental nerve plexuses: Evidence for synthesis, storage and secretion of biogenic amines. J. Neurobiol. 8: 581-605.
Terwilliger, RC (1967) The nature and localization of a cardioactive neurohormone found in the pericardial organs of *Cancer borealis*. PhD Thesis, Boston Univ. Boston, MA.
Touir, A (1977) Données nouvelles concernant l'endocrinologie sexuelle des Crustaces Décapodes

Natantia hermaphrodites et gonochoriques III Mise en évidence d'un controle neurohormonal du maintien de l'appareil genital mâle et des glandes androgènes exerce par le protocérébron median. C.R. Hebd. Acad. Sci. 285D: 539-542.
Webster, SG and R Keller (1986) Purification and amino acid compostition of the putative moult-inhibiting hormone (MIH) of *Carcinus maenas* (Crustacea: Decapoda). J. Comp. Physiol. 156: 617-624.
Welsh, JH (1941) The sinus gland and 24-hour cycles of retinal pigment migration in the crayfish. J. Exp. Zool. 86: 35-49.
Williamson, DI (1982) Larval morphology and diversity. In: "The Biology of Crustacea" (DE Bliss, ed). Vol. 1. pp 43-110. Acad. Press, N.Y.

4

AGEING IN MOLLUSCAN NERVOUS AND NEUROENDOCRINE SYSTEMS

C. Janse
J. Joosse

Department of Biology, Vrije Universiteit,
1007 MC, Amsterdam, The Netherlands.

I. MOLLUSCS AS EXPERIMENTAL ANIMALS

A. Systematics and General Biology

The total number of animal species is estimated to be as high as 1.2 million. The three largest groups are the arthropods, molluscs and vertebrates with 78 %, 7 % and 3 % of the species, respectively. Molluscs are thus one of the most successful animal groups. Gastropods, bivalves and cephalopods are the main molluscan groups. They have adapted to the three major habitats on earth, i. e. marine, terrestrial and fresh water. Representatives can be found world wide in such diverse habitats as the tropical seas, near the eternal snow line in Greenland, and in the deserts of Africa (Newell, 1964; Russell Hunter, 1964). This great adaptability may be related to the well developed behavioural and organ systems of molluscs which are controled by adequate nervous and neuroendocrine systems. Among molluscs gonochorism as well as hermaphroditism occurs. Which sexual lifestyle is used by molluscs depends on individual mobility and density of natural populations (Calow, 1983). Among gastropods and bivalves gonochorism as well as hermaphroditism occurs whereas cephalopods are exclusively gonochoristic.

B. Life Cycles and Breeding

Most molluscan species have a larval stage which strongly differs from the adult stage. Consequently during development they show metamorphosis. Cephalopods have, as an exception, a direct development. The larvae of most marine and some freshwater species of gastropods and bivalves are free living. In metamorphosis often special conditions (i. e., food) are

Development, Maturation, and Senescence of
Neuroendocrine Systems: A Comparative Approach

required. This makes the particular species difficult to breed in the laboratory. Nevertheless, breeding methods for such species have been developed for the hermaphroditic marine gastropods *Aplysia californica* (Kriegstein *et al.*, 1974; Capo *et al.*, 1979) and *Tritonia diomedea* (Kempf and Willows, 1977).

The larvae of pulmonate gastropods show metamorphosis within the egg. Young snails hatch from the egg capsules in their adult appearance. Pulmonates are, therefore, relatively easy to breed. In our laboratory the hermaphroditic pulmonate *Lymnaea stagnalis* is cultured in a semi-automatic system with continous water refreshment, which produces several ten-thousands of snails of different age a year (van der Steen *et al.*, 1969; Janse *et al.*, 1988).

In principle animals show a von Bertalanffy type of growth; after a period of rapid growth, growth rate decreases. This also applies to molluscs (Kooijman, 1988). The maximum obtainable size of molluscs differs considerably. Among cephalopods giants with a length of up to 16 m occur (Villee *et al.*, 1978). *Aplysia californica* can reach a weight of about 7 kg and a length of about 1 m (Kandel, 1979). The bivalve *Mytilus edulis* (Comfort, 1979) and the pulmonate snail *L.stagnalis* reach shell lengths of about 6 cm. Molluscs have a pre-reproductive period. In addition in a number of pulmonate species a post-reproductive period occurs in which female reproductive activity ceases. The post-reproductive period starts after growth has ceased (Comfort, 1957; 1979).

C. Maximum Ages

The maximum age among molluscs differs considerably. Comfort (1957) gives a great number of data on molluscan longevity, most of them concern animals under natural conditions. The shortest living gastropods occur among *Lymnaea* species; maximum life spans of 139 days to 3 years have been reported. The marine gastropod *Aplysia californica* has a maximum life span of about 10 months (Hirsch and Peretz, 1984). Land pulmonates belong to the longest living gastropods. Reports are known of *Helix* specimens living 15 years in breeding conditions (Comfort, 1957). In our laboratory the pond snail *L.stagnalis* has a maximum life span of 22 months (Janse *et al.* , 1988a).

Bivalves belong to the longest living invertebrates. The washboard mussel *Margeritana margeritifera* has been reported to attain an age of about 100 years (Comfort, 1957). As inferred from growth ridges the common mussel *Mytilus edulis* can reach an age of 8-10 years (Comfort, 1957). It goes without saying that species with a relatively short life span and relatively small body size are most suitable for experimental ageing studies.

II. THE NERVOUS AND NEUROENDOCRINE SYSTEM

A. Gross Anatomy

In molluscs an intimate relationship exists between the nervous and the neuroendocrine system, since the majority of the neuroendocrine cells are incorporated in the CNS. In addition, like in other animals, neuroendocrine and "common" neurons resemble each other in many aspects. They have similar integrative and electrical properties and in priciple similar mechanisms to release their products (cf., Joosse *et al.*, 1982; Joosse, 1986).

The nervous system of molluscs consists of a number of ganglia connected by connectives (between different ganglia) and commissures (between paired ganglia). In gastropods and bivalves the ganglia are clearly separated, in cephalopods they are fused to form a large brain mass. In this animal group the brain has reached the highest level of organization amongst the invertebrates (Young, 1971). In addition to the CNS molluscs have a peripheral nervous system which is able to mediate reflex behaviour (Janse, 1974; Mpitsos and Lukowiak, 1985).

The cell bodies of neurons in molluscan ganglia are situated underneath the perineurium. Molluscan neurons are essentially monopolar, in this aspect they resemble the vertebrate sensory neuron. They send their axons to the central region of the ganglion, the neuropil, where they make synaptic contacts with axons originating from neurons in the same ganglion or in other ganglia, or with axons from peripheral neurons. Chemical and electrical synapses occur. Clusters of synapses can be found on the axon hillock and/or on the branching axon. Axo-somatic synaptic contacts occur as well (Roubos, 1984). Molluscan and vertebrate neurons share a common plan, and their biophysical, biochemical and integrative properties are basically the same (Kandel, 1976).

B. Giant Neurons

The CNS of many molluscs contains giant neurons which are relatively easily accessible for different techniques thus permitting the combination of physiological, biochemical, morphological and molecular studies. This makes the molluscan CNS ideal model systems for the study of fundamental mechanisms of the nervous sysem. The use of such systems has been started by studies of the giant-fiber and synapse systems in the squid. In these preparations fundamental processes of action potential generation and synaptic transmission have been elucidated. At present, especially gastropod nervous systems are popular for studies of fundamental mechanisms underlying behaviour. In these animals the diameter of the nerve cell bodies may even exceed 1 mm. Individual neurons can be identified by their position, shape, and electrical characteristics. Moreover, in gastropods such as *Aplysia* and *Lymnaea* the neurons are differently coloured, which makes them easy to identify visually. This makes it possible in these animals to relate activities of individual nerve cells to specific behaviours. Both in *Lymnaea* and *Aplysia* the behavioural function of a great number of identified nerve cells is now known (Kandel, 1976; 1979; Benjamin, 1983; Winlow and Haydon, 1985; Geraerts *et al.*, 1988; van der Wilt *et al.*, 1987; 1988).

C. Motor and Sensory Systems

In studies of molluscan nervous systems much emphasis has been laid on motor systems. This is probably related to the fact that motor neurons are relatively easy to identify and to study. The best-known preparation which has evolved from these studies is the gill-preparation of *Aplysia.* Its reputation has been established by the studies of Kandel and coworkers who used it as a model system for studies on learning (Kandel, 1976; 1979).

Sensory systems in molluscs are more difficult to approach, especially with neurophysiological techniques. This is partly due to the fact that in molluscs the peripheral nervous system plays an important role in sensory processing (e.g. Janse, 1974). Sensory neurons which appeared readily accessible for intracellular recording techniques are the tactile neurons in *Aplysia* (Byrne *et al.*, 1974; Hawkins, 1987), statocyst sensory cells in different gastropods (cf., Janse *et al.* , 1988 b, c) and visual sensory cells in *Hermissenda crassicornis* (Alkon, 1980). In *Hermissenda* and *Aplysia* changes at the sensory level appeared to be of importance for simple forms of learning (Mpitsos and Lukowiak, 1985; Alkon *et al.* , 1986; Hawkins, 1987).

D. Neuroendocrine Systems

Neurohormones in molluscs are produced by neurosecretory cells which occur in groups or dispersed in the CNS. These neurons have release areas located underneath the perineurium of ganglia, connectives, commissures and peripheral nerves (Wendelaar Bonga, 1970; Joosse, 1979). Contrary to vertebrates, molluscs only possess a few (1-3) endocrine organs. Consequently nearly all neurohormones in molluscs act directly on non-endocrine targets. In the pond snail *Lymnaea stagnalis* only one endocrine organ is known, the paired Dorsal Bodies (DB). The DB are located on the cerebral ganglia, and control oocyte growth and maturation and secretory activity of the female accessory sex organs (Geraerts and Joosse, 1975). Like in vertebrates neurohormones in molluscs control physiological processes such as growth, reproduction, ionic and energy metabolism, blood volume and heart rate. In addition group specific functions such as shell growth are under hormonal control (for review see Joosse and Geraerts, 1983; Joosse, 1988).

E. Neurons with Classical Transmitters

Especially in gastropods, pharmacological studies have produced evidence of many substances being putative transmitters. ACh is at present the only transmitter substance which can fulfill all the requirements for being a neurotransmitter. In addition to ACh, noradrenaline, dopamine, 5-hydroxytryptamine, histamine, GABA and several amino acids are likely neurotransmitters in molluscs. These substances are present in the nervous system and responses of central neurons to these substances have been studied extensively (for a review see Rózsa, 1984). Indications of coexistence of different classical transmitters in one neuron have been obtained in *Helix* by Cottrell (1977).

F. Peptidergic Neurons

In the last decade in studies on the nervous system much attention has been paid to peptides. Neuropeptides occur in molluscs in a great molecular diversity (up to 100 in one species (cf., Joosse, 1986). Immunocytochemical studies have shown that a great number of these peptides are vertebrate-like. On the other hand peptides related to those first described in molluscs have been found in the vertebrate brain (Boer and van Minnen, 1988). Evidence is accumulating rapidly that neuropeptides function both at synaptic sites as well as in more diffuse ways of transmission in molluscs and vertebrates (Joosse, 1986). In molluscs neuropeptides play a role in neuroendocrine control systems (Joosse, 1986), in modulation of neuronal networks underlying behaviour (Murphy *et al.*, 1985; van der Wilt *et al.*, 1988), in the induction of complex behaviours such as egg laying in *Lymnaea* and *Aplysia* (Geraerts *et al.*, 1988), in synaptic control of muscles (Cottrell *et al.*, 1983, Schot *et al.*, 1983) or neurons (Boer *et al.*, 1984) and in neurite outgrowth and synapse formation (Bulloch, 1987). Coexistence of several peptides in one neuron has been shown in *Lymnaea* among many other neurons in the Caudo Dorsal Cells (CDC) and in *Aplysia* in the Bag Cells (Geraerts *et al.*, 1983; 1988; Scheller *et al.*, 1983). These multiple peptides function probably in the control of different aspects of egg laying behaviour of the respective snails (Geraerts *et al.*, 1988). Peptidergic neurons involved in modulation or induction of behaviour often seem to be multifunctional.

III. MOLLUSCS IN AGEING STUDIES OF THE CNS

A. Ageing and Survival

Ageing in a population is characterized by an increase of the age-specific death-rate with age (Comfort, 1979). In many natural populations death-rate is independent of age. In such populations the number of survivors decreases exponentially with age (Fig. 1A). Animal populations maintained in laboratory conditions often show an age-related increase in the age-specific death-rate. As a consequence the survival curves become more rectangular (Fig. 1B).

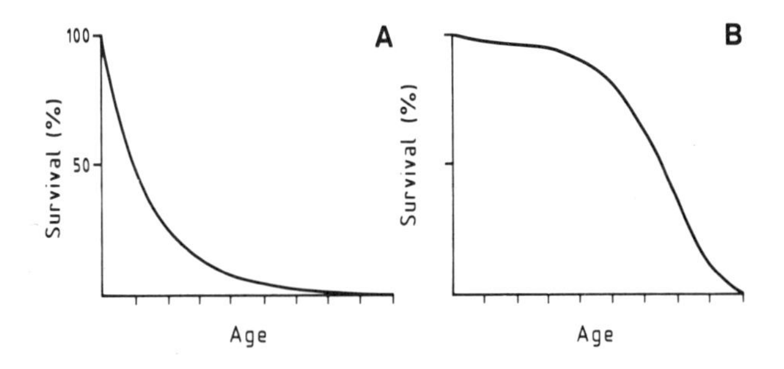

Fig.1. Schematic survival curves of a population in natural conditions (A) and in laboratory conditions (B) (After Comfort, 1979).

In *Lymnaea*, kept under laboratory conditions, the age-specific death-rate increases with age (Slob and Janse ,1988). In this species(Janse *et al.*, 1988a; Slob and Janse, 1988) and in other pulmonates (Comfort, 1979) rectangular survival curves occur in breeding conditions (Fig. 2). In *Lymnaea* survival curves could be fitted with the Gompertz and Weibull function (Slob and Janse, 1988). Fig. 2 shows that mortality in *Lymnaea* increases at an age of about 8 months, the median life span is reached at about 12 months and the maximum life span at about 22 months. Captured *Aplysia* kept under laboratory conditions also have a rectangular survival curve and an increase of the age-specific death-rate with age (Hirsch and Peretz, 1984). According to the data of Hirsch and Peretz death-rate in *Aplysia* increases at an age of about 3 months, the median age is about 5 months and the maximum age 10 months. From these data it is concluded that, in breeding conditions, *Lymnaea* and *Aplysia* age. For bivalves and cephalopods survival data from laboratory cultures are not available.

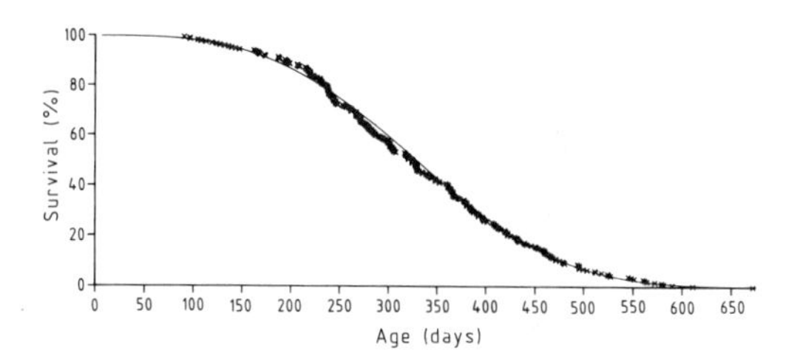

Fig. 2. Weibull function (continuous line) fitted through survival curve of Lymnaea stagnalis maintained in laboratory conditions (After Slob and Janse, 1988).

In *Lymnaea* death-rate can suddenly increase, probably due to a water-transmittable disease (Janse *et al.*, 1988a). This makes it necessary in ageing studies to discriminate between healthy and infected populations. Such a discrimination can be made on the basis of the shape and parameters of survival curves. With maximum-likelihood-methods predictions of survival curves and their parameters can now be made on the basis of the Weibull function, even with only 25% of the death-rate data (75% survival) (Slob and Janse, 1988). In this way populations which are used for ageing studies are now screened for abnormal death-rates.

With the mathematical methods described above effects of experimental conditions on death-rate can also be studied. An example is given in Fig. 3 which concerns an ongoing experiment on the effect of feeding conditions on ageing in *Lymnaea* . Survival data are shown of two populations hatched at the same time and kept under different feeding conditions, but in the same breeding tank in separate compartments (see Janse *et al.*, 1988a). On the basis of these data the total survival curves of the two populations could be estimated. Fig. 3 shows that animals fed ad libitum are expected to have a shorter life span (estimated median age 310 ± 8 (s.e.) days) than animals fed at a restricted diet (estimated median age 347 ± 9 (s.e.) age). However, confirmation of these estimations awaits the decease of the last animals of these groups. In rodents food restriction increases life span (Weindruch and

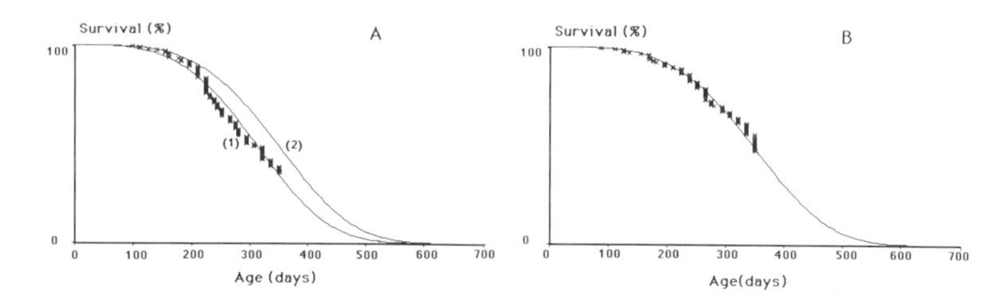

Fig. 3. Survival of Lymnaea stagnalis fed ad libitum (A) and fed on a restricted diet (B). Continuous lines are calculated curves. In A the calculated curves for snails fed ad libitum (1) and fed on a restriced diet (2) are given.

Walford, 1982) and diminishes the age-related decrease of nervous functions (Levin *et al.*, 1981). In *Lymnaea* further studies on these subjects will be done in the near future.

B. Reproductive Activity and Ageing

According to Kirkwood (1985) animal species keep a balance between energy investments in maintenance, growth and repair on the one hand and reproductive activity on the other hand. In his view reproductive strategy is related to longevity and therefore to ageing. According to this view it can be expected that the frequency of lethal damage to an organism due to ageing processes will increase at about the age at which reproduction decreases. It might thus be of importance to know the age of animals at which their post-reproductive period starts. In molluscs only in pulmonate snails a distinct post-reproductive period has been described (Comfort, 1979). In the cephalopod *Octopus* eating behaviour is inhibited after the production of eggs and the animals die after reproduction and subsequent brood care are finished (Wodinsky, 1977). *Aplysia* dies shortly after cessation of reproductive activity (Van Heukelem, 1979).

In cultured *Lymnaea*, female reproductive activity increases up to an age of about 8 months. Thereafter, female reproductive activity decreases with age (Fig. 4). As described above at about this age death-rate of *Lymnaea* increases notably in our cultures. Closer inspection of individual animals revealed that during reproductive decline the percentage of animals that ceases egg-laying increases with increasing age.

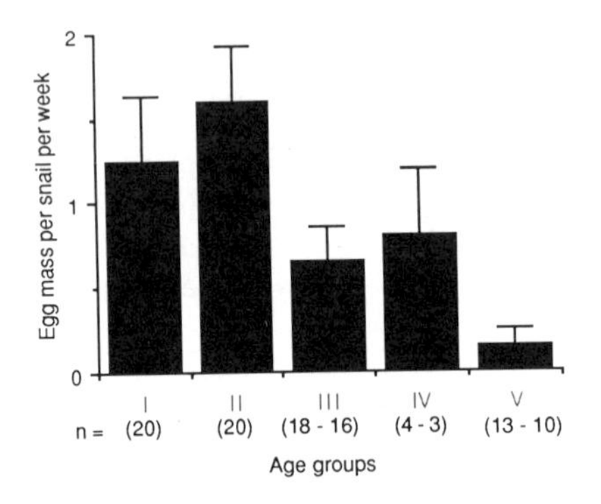

Fig. 4. Egg mass production of Lymnaea stagnalis at different age. Means and s.d. are given of measurements on 7 consecutive weeks. Between brackets the number of animals at the beginning and the end of the measurements are indicated. Median age; I: 4; II: 6; III: 12; IV: 16; V: 21 (months).

C. Age-related Changes in Identified Peptidergic Neurons

In *Aplysia* and in *Lymnaea*, peptidergic neurons are the object of extensive multi-disciplinary studies (Joosse, 1986; Boer and van Minnen, 1988; Geraerts *et al.*, 1988). Studies on age-related changes in identified peptidergic neurons in molluscs have sofar been done in *Lymnaea* . In these

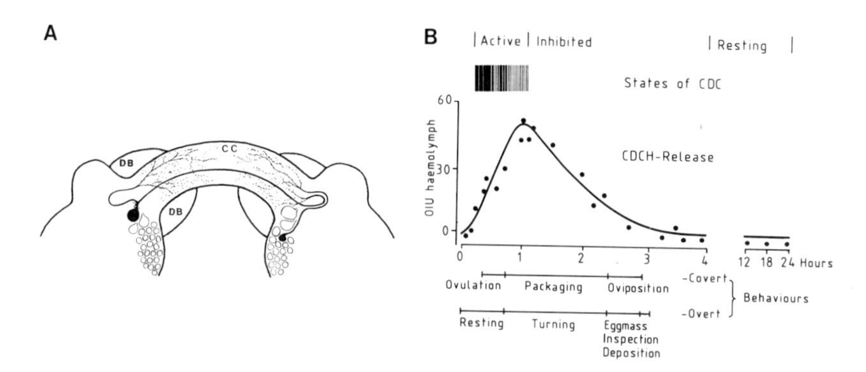

Fig. 5. A. Ventral view of the cerebral ganglia of Lymnaea with the morphology of CDC. The branching pattern of two different types of CDC is indicated, one with an axon at the ipsilateral side only and one with the second axon crossing to the contralateral side. Electrical activity in the clusters of CDC results in release of CDCH and other peptides at the periphery of the commissure. B. Diagram of interrelations of electrical activity of CDC, release of CDCH and overt and covert behaviours induced by an egg-laying stimulus (water change). DB: Dorsal Body; CC: Commissure; OIU: Ovulation Inducing Units as measured by bioassay (Geraerts et al., 1984) (After Kits, 1982 and Joosse, 1986).

animals this have been done on the CDC (see above) and on two giant identified peptidergic cells, VD1 and RPD2.

1. The CDC-system in *Lymnaea*

In *Lymnaea* the CDC control egg laying. The CDC are located in two bilateral clusters of about 50 electrotonically coupled cells each in the cerebral ganglia near the origin of the intercerebral commissure (Fig. 5A). Just before the start of ovulation and egg-laying the CDC become electrically active and secrete a hormone (CDCH) and multiple other peptides in the haemolymph which induce a stereotyped pattern of overt and covert behaviours which terminates in the deposition of an egg mass (Fig. 5B) (see also Geraerts *et al.*, 1988). As described above *Lymnaea* has a post-reproductive period. Experiments performed in our laboratory showed that CDC of senescent non-laying snails can still be electrically activated and still contain a substance which, upon injection, induces ovulation in young reproductive snails. Moreover, senescent non-layers still lay eggs upon injection with synthetic CDCH (Janse, ter Maat and Pieneman, in prep.). These observations indicate that in old non-laying snails the CNS still contains CDC which are in principle functionally intact. Probably activation of the CDC complex is hampered either by impairment of sensory input to the CDC or by impairment of communication between CDC.

Lymnaea can live several months after cessation of reproduction (unpublished observations). This implies that in old *Lymnaea* CDC can be inactive for a considerable time. It is known that neurons degenerate when they are not actively used. Morphological studies indeed showed that in old non-laying *Lymnaea* degeneration of CDC occurs (Joosse, 1964; Janse *et al.*, 1987b; Janse *et al.*, in prep.). During development a similar situation exists. In juvenile *Lymnaea* the CDC contain CDCH and are probably not activated (Dogterom *et al.*, 1983). In the developmental stage, however, no degeneration occurs in the CDC.

In addition to the changes reported above immunocytochemical studies suggested that peptide contents of CDC change with age. CDC of young *Lymnaea* showed enkephalin-like immunoreactivity. In *Lymnaea* of intermediate age the CDC showed enkephalin-like and gastrin/cholecystokinin-like immunoreactivity. Old *Lymnaea*, however, showed only gastrin/cholecystokinin-like immunoreactivity in their CDC (Gesser and Larsson, 1985). It is known that secretory products of the CDC are of importance for synchronization of electrical activity in the CDC system (ter Maat *et al.*, 1988). How age-related changes in peptide content in CDC relate to changes in reproductive activity of *Lymnaea* are unknown.

2. The VD1-RPD2 system in *Lymnaea*

VD1 and RPD2 are two giant electrotonically coupled peptidergic neurons located in the visceral and right parietal ganglion, respectively. With immunocytochemical methods it has been demonstrated that the cells contain an ACTH-like peptide (Boer *et al.*, 1979). The cells branch extensively in the CNS and the peripheral nerves. Moreover, they send many fine fibres in the connective tissue sheath of the CNS where they form a fine net work.

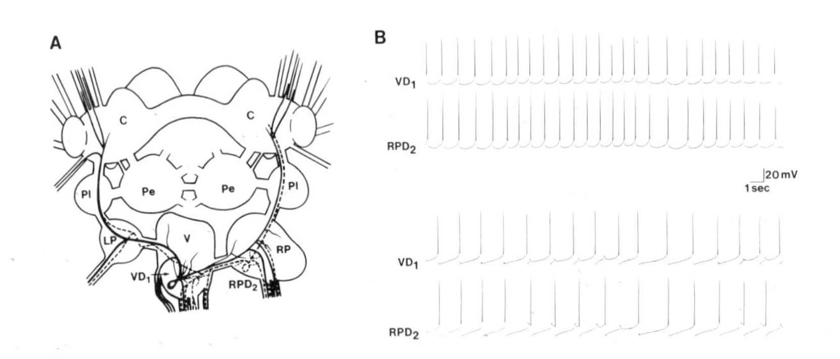

Fig. 6. A. Schematic diagram of the CNS of Lymnaea with branching pattern of VD1 and RPD2 (After Boer et al., 1979). B. Simultaneous intracellular recordings of VD1 and RPD2 in young (3 months) (upper two recordings) and old Lymnaea (12 months) (lower two recordings). Note the disturbances in firing synchrony of the two cells in the lower pair of recordings. C: cerebral ganglion, Pe: pedal ganglion, Pl: pleural ganglion, LP and RP: left and right parietal ganglion, V: visceral ganglion.

Fig. 6A shows the main branching pattern of the cells as revealed with immunocytochemical methods and with Lucifer Yellow fillings.

These neurons play a modulatory role in the control of respiratory behaviour (van der Wilt *et al.*, 1987;1988) and probably also of heart rate (Mojet and van der Wilt, unpublished observations). Neurophysiological studies show that both neurons receive input from oxygen sensitive receptors in the mantle and lung area (Janse *et al.*, 1985). Taken with the extensive branching pattern of the cells this led us to the idea that the neurons probably play an important more general role in the regulation of processes which are affected by the respiratory condition of the animal.

Recent observations revealed that, in addition to common synaptic input, VD1 and RPD2 each receive separate synaptic input (Mojet, Wildering and van der Wilt, unpublished observations). Integration of the different inputs of the two neurons is probably mediated by the tight electrical junction between the neurons.The junction is located in the left parieto-visceral connective (Benjamin and Pilkington, 1986). Normally the neurons indeed fire their action potentials in close synchrony (Janse *et al.*, 1986b).

With age VD1 and RPD2 increase their diameter (Janse *et al.*, 1986b). Such a neuronal growth has also been observed in other molluscs (Rattan and Peretz, 1981). In addition, however, a number of integrative properties change resulting in a change of the firing characteristics of the neurons (Janse *et al.*, 1986b;1987a). Fig. 6B shows the firing pattern of the neurons in young and old animals. One of the changes observed concerned the

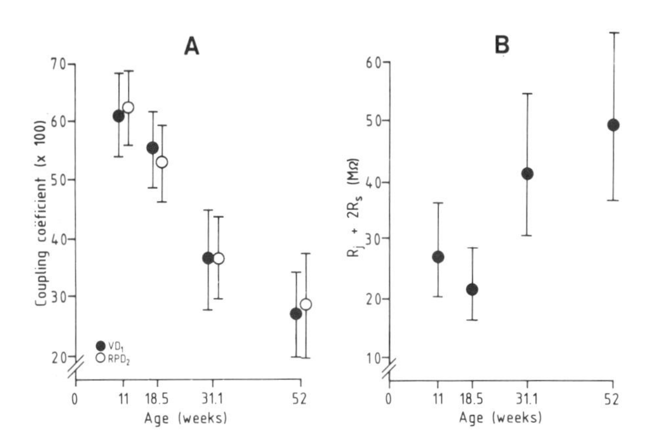

Fig. 7. A. Coupling coefficient of two giant, identified, peptidergic cells,VD1 and RPD2, in Lymnaea of different age. Coefficients are measured by injecting current in VD1 (VD_1) and by injecting current in RPD2 (RPD_2) and dividing the resulting voltage change in the non-injected cell by that in the injected cell. B. Resistance between VD1 and RPD2 (R_j + $2R_s$), this approximates the junction resistance. The results indicate that increase of junction resistance starts at an age of 18.5 weeks. Means and s.d. are indicated (A and B after Janse et al., 1986b).

effectiveness of the electrical junction between the cells. The coupling efficiency of the neurons decreases considerably with age, probably due to an increase of the junctional resistance between the neurons (Fig. 7). Recent studies confirmed this but also showed that in animals of about 19 months of age the junction between the two cells is as effective as in young animals. Moreover, in old animals individual variance is less, probably due to an increase of mortality in animals with a high junctional resistance (Wildering, van der Roest and Janse, in prep.). These data suggest that communication between VD1 and RPD2 is essential for survival. This is in line with the idea stated above that the cells are probably involved in regulation of processes dependent on the respiratory condition of the animal and that communication between the cells is necessary for integration of the different inputs on the two cells.

D. Ageing, Transmitters, Peptides and Receptors

In *Mytilus* ageing has been studied in a (gill) cilia-beat control system which involves intercellular and intracellular messengers in the brain. So far this is the only molluscan system in which coupling of receptors to intra-cellular messengers has been considered in ageing studies. Unfortunately the neurons which are involved in the system are not identified. Pharmacological studies showed that endogenous opioids, dopamine and serotonin are involved in central regulation of ciliary beat. Dopamine applied to the CNS inhibits ciliary beat and application of serotonin stimulates the beat. Dopamine release is presynaptically inhibited by the opioids (Leung and Stefano, 1987). In old *Mytilus* opioid levels increase considerably and simultaneously there is a decrease in the ability of opioids to influence dopamine levels in the CNS. This could be explained by the observation that opioid receptor density diminished, probably through down regulation of opioid receptor induction. The affinity of opioid receptors, however, did not change (Makman and Stefano, 1984; Leung and Stefano, 1987).

The dopamine release system becoming less sensitive to the inhibitory effects of opioids results in a higher release of dopamine. Dopamine stimulates adenylate cyclase activity in the neurons. With age the sensitivity of this adenylate cyclase system to dopamine also decreases (Stefano *et al.*, 1982). This probably also occurs through down regulation of the number of dopamine receptors in the CNS (Leung and Stefano, 1987). The general result of these different age-related changes is that no age-related changes in beating frequency of the cilia in *Mytilus* occurr. Probably the changes found in the CNS compensate for each others effects on ciliary beat. It is believed, however, that the age-related changes will diminish the regulatory capacity to respond to environmental changes (Leung and Stefano, 1987; Stefano *et al.*, 1987).

E. Ageing in Identified Non-peptidergic Giant Neurons

In *Lymnaea stagnalis* several aspects of identified giant neurons have been studied by Frolkis *et al.* (1984) and Nagy *et al.* (1985). For three different types of neurons the authors reported in older animals a decrease in excitability, a decrease in firing frequency and a decrease in velocity of repolarization of the action potential. Moreover they found an increase in the sensitivity of the neurons to acetylcholine, norepinephrine and serotonin. (Frolkis *et al.*, 1984). No changes in membrane potential were found in these neurons with age. In still another type of neuron the authors reported an increase of the membrane potential and an increase of the intracellular K^+-concentration (Nagy *et al.*, 1985). Obviously, not all neurons in *Lymnaea* show changes in membrane potential with age (see above). In our laboratory we also observed that age-related changes may differ in different neurons (Wildering *et al.*, 1987).

Frolkis *et al.* (1984) also studied ultrastructural aspects of *Lymnaea* neurons. They reported desintegration of cell organelles in neurons of old snails. In a more detailed study Papka *et al.* (1981) reported an increase in lipofuscinic material in a giant identified neuron of *Aplysia*. It is known that lipofuscin accumulation with age depends on feeding conditions (Winstanley and Pentreath, 1985). It is, however, not known whether the presence of lipofuscin affects the functioning of neurons.

F. Nervous Plasticity and Ageing

One of the conspicuous features of the ageing brain both in higher animals and in man is that there is a decrease of plasticity. This is expressed as a diminished ability of learning and memory capacity and a diminished capacity to compensate functionally for brain damage (Drachman, 1983). In molluscs several aspects of nervous plasticity have been studied in relation to age. In *Aplysia* it has been demonstrated that arousal behaviour and senzitization decrease considerably with age. Moreover, long-term retention of habituation is impaired (Bailey *et al.*, 1983).These findings are of special interest because sensitization and habituation are considered as simple forms of learning. In *Aplysia* much is known about the neuronal substrates of these functions (Bailey *et al.*, 1983; Mpitsos and Lukowiak, 1986; Hawkins 1987). Thus the *Aplysia* system might provide an opportunity to study fundamental mechanisms underlying age-related impairment of learning and memory.

In *Lymnaea* it has been shown that association of food with a chemostimulus is affected by age. Old animals need to be starved to learn the association whereas young animals can learn the association when well-fed, although they also need to be starved to express the learned ability (Audesirk *et al.*, 1982). In *Lymnaea* much is known about the neuronal mechanism of feeding behaviour (Benjamin, 1983). In molluscs intracellular messenger systems which are influenced by serotonin are thought to be of importance for food-induced motivation (Bailey *et al.*, 1983).

In *Aplysia* Peretz and coworkers showed that age-related changes occurred in the neuronal network involved in the gill-withdrawal reflex (Rattan and Peretz, 1981; Peretz *et al*., 1982; 1984). They found that the width of the synaptic cleft of the neuromuscular synapses changed concomitantly with the integrative functions of the synapses. In still other neurons innervating the same muscles these properties did not change with age. This also demonstrates the occurrence of differential ageing in molluscan neurons. In this case occurrence of age-related changes probably depends on the frequency at which the neurons are used; frequently used neurons showed less decrease in functioning (Peretz at al., 1984). In the intact aged animal restoration of functioning indeed occurrs in neurons after a period of extra sensory input . This indicates that external influences may modify processes at the neuronal level which are related to age (Peretz and Zolman, 1987; Zolman and Peretz, 1987).

Capacity of the molluscan nervous system to repair damage has been investigated in *Helisoma* (Bulloch, 1987) and *Lymnaea* (Janse *et al*., 1979; Benjamin and Allison, 1985). These studies showed that restoration of synaptic contacts takes place after disruption of nervous connections. In *Lymnaea* it has been shown that restoration of the tentacle-withdrawal reflex after damage of sensory input is impaired in old animals. Indications were obtained that this is due to impairment of restoration of synaptic connections in the CNS (Janse *et al*., 1986a). A similar loss of capacity of the brain in forming new synaptic connections after damage occurs also in vertebrates and men (Cotman and Schiff, 1979). In pulmonates it has been shown that somatostatin influences axon outgrowth and synapse formation (Bulloch, 1987) and that peptidergic cells regulating growth contain a somatostatin like peptide (Schot *et al*., 1981). It is conceivable that in old *Lymnaea* levels of peptides involved in control of synapse formation decrease.

IV. PERSPECTIVES: MOLLUSCS AS ANIMAL MODELS FOR AGEING STUDIES

The fundamental mechanisms which underly ageing processes are still poorly understood. This is illustrated by the great number of theories to explain ageing which exists at present (cf., Adelman and Roth, 1982). The use of lower animals, because of their relative simplicity, might provide an opportunity to unravel mechanisms of ageing. Several animal species are at present used in experimental studies of ageing processes (cf., Mitchell and Johnson, 1984). Among the lower animals molluscs seem to provide suitable model systems for ageing studies in the nervous and neuroendocrine systems. These animals have a relatively simply organized CNS and, especially many gastropods, possess a nervous system with giant, well identifiable, nerve cells. These include peptidergic cells which are involved in neuroendocrine regulatory processes. In several laboratories ageing studies on molluscan preparations are now in progress and age-related changes have been found at the level of the entire animal, at the cellular level and at the molecular level.

Lymnaea seems to offer suitable model systems to study changes in neuroendocrine control mechanisms. Age-related changes occur in peptidergic cells which control reproduction and cardio-respiratory functions. In the latter system there is evidence that impairment of intercellular communication might be lethal to the animal. Changes in the CDC system do not seem to introduce lethal changes in the animal but decrease reproductive effort. Both systems consist of peptidergic giant neurons which are relatively easy accessible for different techniques. It seems to be important to study both these neuroendocrine systems because they differ in complexity and in function; studies might thus be complementary. Interestingly in both systems impairment of functioning seems to be related to disorders in interneuronal communication. These disorders may be related to the changes in synaptic plasticity described above. Future studies will focuss on mechanisms underlying age-related changes in these neuroendocrine control systems.

In judging the relevance of molluscan studies for ageing studies in vertebrates and men two aspects should be kept in mind. On the one hand one should realize that the complexity of the organisation of the behavioural and regulatory systems is much higher in vertebrates as compared to molluscs. The impact for the animal of age-related changes in these systems might thus differ between these animal groups. On the other hand, however, from studies done sofar it is clear that principles of organisation of neuronal and neuroendocrine systems and fundamental mechanisms governing processes at the cellular and sub-cellular level in the vertebrate and molluscan nervous system, resemble each other to a high degree. It is therefore on the basis of these basic similarities that results of ageing studies in molluscs can be of direct relevance for ageing studies in higher animals.

ACKNOWLEDGEMENTS

The authors thank Prof. K. Lukowiak for reading a previous draft of the manuscript and W. C. Wildering and G. J. van der Wilt for comments.

V. REFERENCES

Adelman, R.C. and Roth, G.S. (1982). "Testing the Theories of Aging." CRC Press, Boca Raton, Fla.
Alkon, D.L. (1980). Biol. Bull. *159*, 505.

Alkon, D.L., Sakakibara, M., Naito, S., Heldman, E. and Lederhendler, I. (1986). Neurosc. Res. *3*, 487.

Audesirk, T.A., Alexander, J.E. jr., Audesirk, G.J. and Moyer, C.M. (1982). Behav. Neur. Biol. *36*, 379.

Bailey, C.H., Castellucci, V.F., Koester, J. and Chen, M (1983). Behav. Neur. Biol. *38*, 70.

Benjamin, P.R. (1983). *In* "Neural Origin of Rhythmic Movements" (A. Roberts and B. Roberts eds.), P. 159. Cambridge Univ. Press, Cambridge.

Benjamin, P.R. and Allison, P. (1985). Proc. R. Soc. B *226,* 159.

Benjamin, P.R. and Pilkington, J.B. (1986). J. Physiol. *370*, 111.

Boer, H.H., Schot, L.P.C., Roubos, E.W., Maat, A. ter, Lodder, J.C., Reichelt, D. and Swaab, D.F. (1979). Cell Tiss. Res. *202*, 231.

Boer, H.H., Schot, L.P.C., Reichelt, D., Brand, H. and Maat, A ter. (1984). Cell Tiss. Res. *238*, 197.

Boer, H.H. and Minnen, J van. (1988) *In* "Neurohormones in Invertebrates" (M. C. Thorndyke and G. J. Goldsworthy eds.) (in press).

Bulloch, A.G.M. (1987). Br. Res. *412*, 6.

Byrne, J., Castellucci, V. and Kandel, E. (1974). J. Neurophysiol. *37*, 1041.

Calow, P. (1983). *In* "The Mollusca. Ecology," Vol. 6. (W. D. Russell-Hunter ed.) p. 649. Academic Press, New York.

Capo, T. R., Perritt, S. E. and Berg, C. J. jr. (1979). Biol. Bull. *157*, 360.

Comfort, A. (1957). Proc. Mal. Soc. London *32*, 219.

Comfort, A. (1979). "The Biology of Senescence." 3rd ed. Churchill Livingstone, Edinburgh, London.

Cotman, C. W. and Schiff, S. W. (1979). *In* "Physiology and Cell Biology of Aging," Vol. 8. (A. Cherkin *et al*., eds.), p. 109. Raven Press, New York.

Cottrell, G. A. (1977). Neurosc. *2*, 1.

Cottrell, G. A., Schot, L. P. C. and Dockray, G. J. (1983). Nature *304*, 638.

Dogterom, G. E., Loenhout, H. van, Geraerts, W. P. M. (1983). Gen. Comp. Endocrinol. *52*, 242.

Drachman, D. A. (1983). *In* "Aging of the Brain," Vol. 22. (D. Samuel *et al*., eds.), p. 19. Raven Press, New York.

Frolkis, V. V., Stupina, A. S., Martinenko, O. A., Toth, S. and Timchenko, A. I. (1984). Mechs. Ageing Dev. *25*, 91.

Geraerts, W.P.M. and Joosse, J. (1975). Gen. Comp. Endocrinol. *27*, 450.

Geraerts, W. P. M., Tensen, C. P. and Hogenes, Th. M. (1983). Neurosci. Lett. *41*, 151.

Geraerts, W.P.M., Maat, A ter and Hogenes, Th. M. (1984). *In* "Biosynthesis, Metabolism and Mode of Action of Invertebrate Hormones" (J. Hoffmann and M. Porchet eds.), p. 44. Springer Verlag, Berlin.

Geraerts, W.P.M., Maat, A. ter and Vreugdenhil, E. (1988). *In* "Invertebrate Endocrinology," Vol. II. (H. Laufer ed.), p. 325. Alan R. Liss Inc. (in press).

Gesser, B. P. and Larsson, L. (1985). J. Neurosc. *5* , 1412.
Hirsch, H. R. and Peretz, B. (1984). Mechs. Ageing Dev. *27*, 43.
Hawkins. R. D. (1987). *In* "Neurobiology. Molluscan Models" (H. H. Boer, W. P. M. Geraerts and J. Joosse eds.), p. 311. North Holland Publ. Co., Amsterdam, New York.
Janse, C. (1974). Neth. J. Zool. *24*, 93.
Janse, C., Kits, K. S. and Lever, A. J. (1979). Mal. *18*, 485.
Janse, C., Wilt, G. J. van der, Plas, J. van der. and Roest, M. van der (1985). Comp. Biochem. Physiol. *82A*, 459.
Janse, C., Beek, A., Oorschot, I. van and Roest, M. van der (1986a). Mechs. Ageing Dev. *35*, 179.
Janse, C., Roest, M. van der and Slob, W. (1986b). Brain Res. *376*, 208.
Janse, C., Roest, M. van der, Bedaux, J.J.M. and Slob, W. (1987a). *In* "Neurobiology. Molluscan models" (H.H.Boer, W.P.M. Geraerts and J. Joosse eds.), p. 335. North Holland Publ. Co. Amsterdam, New York.
Janse, C., Wildering, W.C., Minnen, J. van, Roest, M. van der, Roubos, E.W. and Slob, W. (1987b). Proc. 28th Dutch Fed. Meeting 239.
Janse, C., Slob, W., Popelier, C. M. and Vogelaar, J. W. (1988a). Mechs. Ageing Dev. (in press).
Janse, C., Wilt, G. J. van der, Roest, M van der and Pieneman, A. W. (1988b). In " Neurobiology of Invertebrates. Transmitters, Modulators and Receptors" (J. Salánki ed.). (in press).
Janse, C., Wilt, G. J. van der, Roest, M van der and Pieneman, A. W. (1988c). Comp. Biochem. Physiol. (in press).
Joosse, J. (1964). Arch. Néerl. Zool. *16*, 1.
Joosse, J. (1979). *In* "Hormones and Evolution" (E. J. W. Barington ed.), p. 119. Academic Press, New York.
Joosse, J. (1986). *In* "Comparative Endocrinology: Developments and Directions" (Ch. Ralph ed.), p. 13. Alan R. Liss. Inc.
Joosse, J. (1988). *In* "Invertebrate Endocrinology," Vol. II (H. Laufer ed.), p. 325. Alan R. Liss Inc. (in press).
Joosse, J. and Geraerts, W. P. M. (1983). *In* "The Mollusca. Physiology," Vol. 4. (K. M. Wilbur and A. S. M. Yonge eds.), p. 317. Academic Press, New York.
Joosse, J., Vlieger, T. A. de and Roubos, E. W. (1982). *In* "Chemical Transmission in the Brain" (R. M. Buijs, P. Pévet and D. F. Swaab eds.), Progr. Br. Res. Vol. 55, 379.
Kandel, E. R. (1976). "Cellular Basis of Behavior." W. H. Freeman and Co., San Francisco.
Kandel, E. R. (1979). "Behavioral Biology of *Aplysia*.", W. H. Freeman and Co., San Francisco.
Kempf, S. C. and Willows, A. O. D. (1977). J. Exp. Mar. Biol. Ecol. *30*, 261.
Kirkwood, B. L. (1985). *In* "Handbook of the Biology of Aging" 2 nd ed., (C. E. Finch and E. L. Schneider eds.), p. 27. Van Nostrand Reinhold, New York.
Kits, K. S. (1982). Comp. Biochem. Physiol. *72*A, 91.

Kooijman, S. A. L. M. (1988). Proc. 2nd Symp. Math. Ecol. Trieste 1986. (in press).
Kriegstein, A. E., Castelllucci, V. and Kandel, E. R. (1974). Proc. Natl. Acad. Sc. USA *71*, 3654.
Levin, P., Janda, J. K., Joseph, J. A., Ingram, D. K. and Roth G. S. (1981). Science *214*, 561.
Leung, M. K. and Stefano, G. B. (1987). Progr. Neurobiol. *28*, 131.
Maat, A. ter, Geraerts, W. P. M., Jansen, R. F. and Bos, N. P. A. (1988). Brain Res. (in press).
Makman, M. and Stefano, G. (1984). *In* "Invertebrate Models in Aging Research" (D. H. Mitchell and T. E. Johnson eds.), p. 165 CRC Press, Boca Raton Fla.
Mitchell, D. H. and Johnson, T. E. (eds.) (1984). "Invertebrate Models in Aging Research.", CRC Press, Boca Raton, Fla.
Mpitsos, G. J. and Lukowiak, K. (1985). *In* "The Mollusca. Neurobiology and Behavior," Vol. 8. (A. O. D. Willows ed.), p. 95. Academic Press, New York.
Murphy, A. D., Lukowiak, K. L. and Stell, W. (1985). Proc. Natl. Acad. Sci. USA *82*, 7140.
Nagy, I. Zs., Tóth, S. and Lustyik, Gy. (1985). Arch. Gerontol. Geriatr. *4*, 53.
Newell, G. E. (1964). *In* "Physiology of Mollusca," Vol. I. p. 59. Academic Press, New York.
Papka, R., Peretz, B., Tudor, J. and Becker, J. (1981). J. Neurobiol. *12*, 455.
Peretz, B., Ringham, G. and Wilson, R. (1982). J. Neurobiol. *13*, 141.
Peretz, B., Romanenko, A. and Markesbery, W. (1984). Proc. Natl. Acad. Sci. USA *81*, 4332.
Peretz, B. and Zolman, J. F. (1987). *In* "Neurobiology. Molluscan Models" (H. H. Boer, W. P. M. Geraerts and J. Joosse eds.), p. 330. North. Holl. Publ. Co., Amsterdam, New York.
Rattan, K. S. and Peretz B,. (1981). J. Neurobiol. *12*, 469.
Rózsa, K. S. (1984). Progr. Neurobiol. *23*, 79.
Roubos, E. W. (1984). Int. Rev. Cytol. *89*, 295.
Russell-Hunter, W. (1964). *In* "Physiology of Mollusca," Vol. I. p. 83. Academic Press, New York.
Scheller, R. H., Jackson, J. F., McAllister, L. B., Rothman, B. S., Mayeri, E. and Axel, R. (1983). Cell *32*, 7.
Schot, L. P. C., Boer, H. H., Swaab, D. F. and Noorden, S. van (1981). Cell Tiss. Res. *216*, 273.
Schot, L. P. C., Boer, H. H. and Wijdenes, J. (1983). *In* "Molluscan Neuro-Endocrinology "(J. Lever and H. H. Boer eds.), p. 203. North Holland Publ. Co. , Amsterdam, New York.
Slob, W., Janse, C. (1988). Mechs. Ageing Dev. (in press).
Steen, W. J. van der, Hoven, N. P. van den, Jager J. C. (1969). Neth. J. Zool. *19*, 131.
Stefano, G. B., Stanec, A. and Catapane, E. J. (1982). Cell. Mol. Neurobiol. *2*, 249.

Stefano, G. B., Braham, E., Finn, P., Aiello, E. and Leung, M. K. (1987). Cell. Mol. Neurobiol. *7*, 209.
Van Heukelem, W. F. (1979). *In* "The Biology of Aging" (J. A. Behnke, C. E. Finch and G. B. Moment eds.), p. 211. Plenum Press. New York.
Villee, C. L., Walker, W. F. jr. and Barnes R. D. (1978). "General Zoology." W. B. Saunders Co., Philadelphia.
Weindruch, R. and Walford, R. L. (1982). Science *215*, 1415.
Wendelaar Bonga, S. E. (1970). Z. Zellforsch. *108*, 190.
Wildering, W. C., Roest, M van der and Janse, C. (1987). Eur. J. Cell Biol. *44*, 129.
Wilt, G. J. van der, Roest, M. van der and Janse, C. (1987). In "Neurobiology. Molluscan Models" (H. H. Boer, W. P. M. Geraerts and J. Joosse eds.) , p. 292. North Holland Publ. Co. , Amsterdam, New York.
Wilt, G. J. van der, Roest, M van der. and Janse, C. (1988). In " Neurobiology of Invertebrates. Transmitters, Modulators and Receptors" (J. Salánki ed.) (in press).
Winlow, W. and Haydon, P. G. (1985). Comp. Biochem. Physiol. *83*A, 13.
Winstanley, E. K. and Pentreath, V. W. (1985). Mechs. Ageing Dev. *29*, 299.
Wodinsky, J. (1977). Science *198*, 948.
Young, J. Z. (1971). "The Anatomy of the Nervous System of *Octopus vulgaris*." Clarendon Press, Oxford.
Zolman, J. F. and Peretz, B (1987). Behav. Neurosc. *101*, 524.

5

NEUROENDOCRINE ASPECTS OF AMPHIBIAN METAMORPHOSIS

David O. Norris

Department of Environmental, Population and Organismic Biology
University of Colorado, Boulder, CO 80309

James Norman Dent

Department of Biology, University of Virginia
Charlottesville, VA 22901

I. AMPHIBIAN LIFE CYCLES

Most amphibians are terrestrial or semiterrestrial animals that return to water to lay their eggs. Many species, however, are either completely terrestrial or are permanently aquatic (Dent, 1968; Hourdry and Beaumont, 1985). The unique aspect of the typical amphibian life history is the intercalation of a fish-like aquatic larval stage between the embryo and the terrestrial adult. The aquatic larva which usually bears little resemblance to the adult, represents an intermediate feeding stage. The length of the larval period varies greatly from only a few weeks in certain species to more than two years in others. At the end of the larval period, the larva undergoes a dramatic morphological, biochemical, behavioral and ecological transformation called metamorphosis (see Wald, 1981). This life history pattern decreases the amount of energy (as yolk) that the female must invest in each offspring and eliminates or at least reduces competition between the aquatic larvae and the terrestrial adults (for more discussion, see Dodd and Dodd, 1976; Just et al., 1981). In this review we shall focus on the possible roles of hypothalamic neurotransmitters and neurohormones in the regulation of metamorphic events.

Although amphibians generally seek water for reproduction, many exhibit modifications in reproduction and/or metamorphosis (see Lynn 1961; Dent, 1968, Hourdry and Beaumont, 1985). Some species retain their eggs so that they develop in oviducts, in skin pouches or embedded in the skin, etc. Retained eggs may develop into larvae which are then deposited in water, or the embryos may metamorphose directly into miniature adults prior to being released. A few species, such as the frog *Nectophrynoides occidentalis* and the salamander *Salamandra atra*, are truly viviparous and give birth to fully metamorphosed juvenile animals.

At the other extremes are species which emphasize the aquatic habitat and retain at least some of the characteristics of their larval form (e.g., gills, tail fins) in sexually mature animals. These species are termed paedogenetic, referring to the retention of larval characters. When paedogenesis is thought to occur because of delayed development of somatic tissues rather than via precocious sexual maturation, the animals are called neotenic. The latter term will be applied here to the paedogenetic salamanders as suggested by Gould (1977). Among the neotenic salamanders, some species are facultative neotenes in that under some conditions they may metamorphose to a terrestrial form prior to becoming sexually mature whereas under others they may become sexually mature without prior metamorphic change.

Although most investigators interested in metamorphosis have focused on those species which exhibit the standard aquatic tadpole larva, it may be instructive to learn more about neuroendocrine roles and about maturation of the neuroendocrine systems in direct development as well as in neotenic species, especially the facultative neotenes.

II. HORMONAL CONTROL OF METAMORPHIC EVENTS

Endocrine controls of amphibian metamorphosis have been reviewed frequently (see Dent, 1968, 1988; Etkin, 1968; Dodd and Dodd, 1976; White and Nicoll, 981; Galton, 1983; Norris, 1983, 1985; Rosenkilde, 1985), and only the major features will be summarized here. Current views suggest that low levels of thyroid hormones stimulate maturation of the hypothalamo-hypophysial system. Neurohormones produced in the hypothalamus regulate production and release of three adenohypophysial hormones: thyrotropin (TSH), corticotropin (ACTH) and prolactin (PRL). The resulting increase in thyrotropin secretion causes increased thyroid gland activity and a surge in thyroid hormone levels in the blood. This surge in thyroid hormones brings about the dramatic changes of metamorphosis. It appears that ACTH causes a surge in the release of corticosteroids from the interrenal gland, and there is also a surge in PRL that accompanies metamorphosis. The possible importance of these surges is discussed below.

A. The Hypothalamus in Metamorphosis

The importance of hypothalamic stimulation in the initiation of amphibian metamorphosis was demonstrated in two experiments. First, the transplantation of the pituitary gland to a site in the tail region of the tadpoles was sufficient to maintain growth and development but was insufficient to induce metamorphosis (Etkin, 1938). Second, the placement of a physical barrier between the hypothalamus and the adenophypophysis of of larval *Ambystoma maculatum* blocked metamorphosis except in animals where the blood vessels of the hypothalamo-hypophysial portal system had become re-established (Etkin and Sussman, 1961). Further evidence for thyroid hormone stimulation and/or differentiation of the hypothalamus and median eminence come from the observation that thyroxine (T4) injected into the brain was effective at inducing metamorphosis even though a similar dose administered intraperitoneally was ineffective (Norris and Gern, 1976).

B. The Hypothalamo-hypophysial-thyroid axis.

Metamorphosis is initiated by the hypothalamo-hypophysial-thyroid axis. In mammals, hypothalamic nuclei produce a tripeptide neurohormone known as thyrotropin-releasing hormone (TRH). This neurohormone is stored in the median eminence of the neurophypophysis until it is released into the hypothalamo-hypophysial portal system that carries TRH to the thyrotropes in the adenohypophysis. The thyrotropes respond to TRH by releasing TSH into the general circulation. Thyrotropin stimulates the thyroid to release thyroxine (T4) which is converted peripherally (primarily by liver cells) to triiodothyronine (T3), the more active form of thyroid hormone in mammals. All of these elements are present in amphibians, but as the following discussions will emphasize, there are major differences between their organization and function in mammals and amphibians.

A considerable amount of immunoreactive TRH (iTRH) has been identified in the amphibian hypothalamus (Jackson and Reichlin, 1974; King and Millar, 1981) and is distributed among numerous nuclei (Seki et al. 1983; Mimnagh et al., 1987). Although hypothalamic iTRH increases at the onset of metamorphosis (King and Millar, 1981) and appears in the median eminence (Mimnagh et al., 1987), attempts to stimulate metamorphosis with synthetic mammalian TRH have been disappointing (see Section IV.D.), suggesting that mammalian TRH may not be the amphibian thyrotropin-releasing factor (aTRF).

1. Thyroid hormone surge.

During early larval life the thyroid gland secretes low levels of T4 (Piotrowski and Kaltenbach, 1985) that stimulate growth and maturation of the nervous system (Kollros, 1981). The major circulating thyroid hormone at this time is probably T4 since the capacity for peripheral conversion of T4 to T3 is either low (Buscaglia et al., 1985) or does not develop until metamorphosis itself has begun (Galton, 1983). One of the maturational actions of T4 is believed to be stimulation of the differentiation of hypothalamic centers that release aTRF, of the median eminence, and of the portal vessel system between the median eminence and the adenohypophysis. Subsequently, the resultant release of aTRF stimulates the secretion of TSH which in turn causes a surge in the levels of thyroid hormones that bring about the changes constituting metamorphosis (Etkin, 1968). It is important to note that not only does the hypothalamus-pituitary system change, but the sensitivity of the thyroid gland to TSH stimulation may increase during metamorphosis (Norman and Norris, 1987).

The surge in thyroid hormones has been documented in both anurans (Leloup and Buscaglia, 1977; Miyauchi et al., 1977; Regard et al., 1978; Mondou and Kaltenbach, 1979; Suzuki and Suzuki, 1981; Weil, 1986) and urodeles (Eagleson and McKeown, 1978; Larras-Regard et al., 1981; Larras-Regard, 1985; Albrech et al., 1986; Norman et al., 1987). A surge in T4 is accompanied by an increase in the proportion of T3 present in the circulation (Galton, 1983; Buscaglia et al., 1985).

C. The Hypothalamo-hypophysial-interrenal Axis.

Neurosecretory neurons in the mammalian hypothalamus produce a corticotropin-releasing hormone (CRH) which, like TRH, is stored in the median eminence until it is released to the adenohypophysis where it stimulates the release of ACTH from the corticotropes. In response to circulating ACTH levels, the adrenocortical cells (homologous to the interrenal cells of amphibians) release corticosteroids. A similar system appears to be operational in amphibians except that the cells homologous to the cells of the mammalian adrenal cortex are embedded in the ventral surface of the kidneys and constitute what is termed the interrenal gland.

1. Corticosteroid Surge.

Circulating corticosterone exhibits a surge during metamorphosis corresponding to the timing of the T4 surge in *Rana catesbeiana* (Jaffe, 1981; Krug et al., 1983), *X. laevis* (Jolivet-Jaudet and Leloup-Hatey, 1984) and tiger salamanders (Carr and Norris, 1988). Increased secretion of aldosterone also has been reported during metamorphosis of *R. catesbeiana* (Kikuyama et al., 1986; Krug et al., 1983) and *X. laevis* (Jolivet-Jaudet and Leloup-Hatey, 1984).

Early studies demonstrated an enhancement of metamorphosis by treatments with corticosteroids (see Dent, 1988, for a review). Corticosteroids enhance tail resorption in the bullfrog (Kikuyama et al., 1982; Suzuki and Kikuyama, 1983) and may interact with insulin to effect changes in the intestine associated with metamorphosis (El Maraghi-Ater et al., 1986). Corticotropin stimulates an abrupt water loss in the newt, *Taricha torosa* (Brown et al., 1984), and corticosteroids may be responsible at least in part for the marked water losses that occur during metamorphosis (Etkin, 1932).

Yu et al. (1985) proposed that the increase in interrenal activity during metamorphosis is a consequence of a surge in ACTH during metamorphic climax, presumably as a consequence of the maturational actions of T4 on the hypothalamo-hypophysial pathway (see Mimnagh et al., 1987). Unpublished experiments by J.A. Carr have demonstrated no increase in interrenal sensitivity to ACTH (tested *in vitro*) during spontaneous or T4-induced metamorphosis of bullfrog tadpoles, unlike the change in sensitivity reported for the tiger salamander thyroids by Norman and Norris (1987).

D. Prolactin (PRL).

When it was discovered in 1932 that the pars distalis of the pituitary gland secretes a hormone that regulates the synthesis of milk in the mammal and of crop milk in the bird, it seemed reasonable to name the hormone "prolactin." That name is now well-established but is no longer appropriate because it has become apparent that among the various vertebrate species PRL performs a multitude of actions that bear little relation to the production of milk (see De Vlaming, 1980). One such action was inferred by Etkin and Lehrer in 1960 when they found that tadpoles with ectopically positioned grafts of pituitary glands failed to metamorphose and grew more rapidly than control animals. They suggested that PRL might be responsible for those responses. Their suggestion has been confirmed in a wide variety of studies (see Clarke and Bern, 1980; Dent, 1985).

1. Antagonism with thyroid hormones.

The nature of the antagonistic action between thyroid hormones and PRL is a matter of considerable interest. Conflicting results have been obtained in studies designed to test whether interaction occurs at the level of the thyroid gland. Some findings indicate that PRL inhibits the functioning of the thyroid gland whereas others suggest that PRL has no effect on the larval thyroid (White and Nicoll, 1981; Dent, 1988). On the other hand, inhibition of the regressive action of thyroid hormones by PRL at the tissue level has been demonstrated clearly in studies conducted *in vitro* (Derby and Etkin, 1968; Derby, 1975; Platt et al., 1978).

2. Mechanisms of action.

Prolactin may act to inhibit metamorphic change by one or more of several mechanisms. For example, injection of tadpoles with PRL diminishes the increase in activity of several of the degradative enzymes that are induced within regressing tissues by thyroid hormone (Blatt et al., 1969; Jaffe and Geschwind, 1974a; Derby, 1975). The increase in some liver enzymes following treatment with thyroid hormones, however, is not affected by PRL (Blatt et al., 1968; Jaffe and Geschwind,

1974a, b). Regression of tadpole tail fin cultured *in vitro* is enhanced by thyroid hormones (Tata, 1966; Hickey, 1971; Greenfield and Derby, 1972; Yamamoto et al., 1979). The addition of purified PRL to the medium inhibits regression (Yamamoto et al., 1979; Ray and Dent, 1986a) but does not affect the specific activity of the thyroid-induced enzymes, hexosaminidase and alkaline phosphatase (Ray and Dent, 1986a). Similarly, PRL blocks regression of tail fins in larvae of *Ambystoma tigrinum* that have been injected with T4 but not an induced rise in alkaline phosphatase (Platt et al., 1986). It was suggested that the shrinkage observed was an osmoregulatory response induced by PRL and that PRL failed to inhibit the induction of enzymes *in vitro* because the *in-vitro* preparation lacked synlactin, a factor which is produced in the liver of the intact animal and which mediates the action of PRL (Delidow et al., 1988).

Although there is little evidence to support the view that cAMP serves to mediate the actions of either thyroid hormone or PRL, we (Ray and Dent, 1986b) have confirmed observations of other workers indicating that dibutyryl cAMP (DBcAMP) mimics the antagonistic effect of thyroid hormone on the shrinkage of tail fin in culture. Like PRL, DBcAMP does not affect levels of hexosaminidase during the antagonistic response. An interesting parallel in the antagonism of DBcAMP to mullerian-inhibiting substance in the developing male rat (Ikawa et al., 1984) suggests that additional study of the sensitivity of these degenerative processes to cAMP may help elucidate the presently unknown mechanism by which the hormones involved induce cell death in their respective target tissues.

The osmoregulatory action of PRL may be basic to many of its numerous effects (Bern, 1975; Nicoll, 1980) including its metamorphic antagonism to thyroid hormone. Prolactin reverses thyroid-induced regression of fin and gills of the tiger salamander both *in vivo* and *in vitro*, the reversal being accomplished by reduction of water loss (Platt and Christopher, 1977; Platt et al., 1978). Its antimetamorphic effects are antagonized in interaction with such established osmoregulatory adrenocorticoids as aldosterone (Platt and Hill, 1982) and deoxycorticosterone (Kikuyama et al., 1983). Further evidence of osmoregulatory participation of PRL in metamorphic control comes from the finding of White and Nicoll (1981) to the effect that renal binding of PRL is low during premetamorphosis in the bullfrog but increases during climax, signaling a shift in participation of PRL in hydromineral control.

3. The prolactin surge.

Accumulated information regarding the antagonistic interaction between thyroid hormones and PRL during metamorphosis led to the view that metamorphic events are regulated by a balance between those two endocrine factors and the expectation that the well-documented increase of thyroid hormone levels during climax would be accompanied by a corresponding decrease in circulating levels of PRL (Bern et al., 1967; Etkin and Gona, 1967; Dodd and Dodd, 1976).

However, Clemons and Nicoll (1976b) showed that in bullfrog larvae PRL levels surged during climax. This finding was confirmed and, additionally, a corresponding increase in PRL synthesis was demonstrated (Yamamoto and Kikuyama, 1982b; Yamamoto et al., 1986). It was speculated that perhaps the PRL surge fails to produce an antagonistic response because by the time it occurs the climactic events already have been induced by the thyroid. In the support of that view Kawamura et al. (1986) have reported that although metamorphosis is halted if the thyroid is removed prior to Taylor and Kollros (TK) stage XVIII, metamorphosis is unaffected if the gland is extirpated after TK stage XXII when the surge is first fully developed (Yamamoto et al., 1986). Although the PRL surge may provide no antimetamorphic function, it may relate to the significant alteration in osmoregulatory activity that occurs during climax. Prior to TK stage XX, the skin of the tadpole does not have the adult capacity for active transport of sodium. That capacity is acquired and accelerated between TK stages XXII and XXV (Takada, 1986) concurrently with the PRL surge (Yamamoto et al., 1986).

The reduction in pituitary PRL content reported for metamorphosing tiger salamander larvae (Norris et al., 1973) could be interpreted as evidence for release of stored PRL during metamorphosis in urodeles. Treatment with ovine PRL at doses insufficient to block metamorphosis enhances survival of T4-treated tiger salamander larvae (Choun and Norris, 1974) and suggests a possible role for a PRL surge in those animals similar to that proposed for anurans.

E. Growth Hormone.

There is some evidence to indicate that PRL is the primary stimulator of growth in the amphibian larva and that growth hormone acts in a subsidiary role until the completion of metamorphosis when it becomes the predominant somatotropic factor, but numerous contradictory findings cast doubt on that position (White and Nicoll, 1981; Rosenkilde, 1985). It is reasonably evident that growth hormone does not act to inhibit metamorphosis. Both heterologous (Clemons and Nicoll, 1977a) and homologous (Yamamoto and Kikuyama, 1982a, 1982b) antisera to PRL accelerate metamorphosis in bullfrog larvae, whereas similar antisera to growth hormone do not.

Both PRL and growth hormone have marked osmoregulatory capacity among some fishes (Nicoll, 1981). Growth hormone enhances survival and reduces plasma sodium concentrations when freshwater-adapted salmonids are transferred to salt-water (Bolton et al., 1987). During the course of evolution, growth hormone, unlike PRL, has failed to retain its osmoregulatory function in the amphibians and the higher vertebrates.

In chickens, growth hormone increases levels of 5'-monodeiodinase activity in the liver (Kühn et al., 1987). This enzyme is responsible for conversion of T4 to T3. These observations suggest another possible role for growth hormone (or even PRL) during amphibian metamorphosis that should be examined.

III. ORGANIZATION OF THE AMPHIBIAN HYPOTHALAMUS

The organization of the amphibian hypothalamus and the distribution of various peptides and monoamines has been the subject of several reviews (Fasolo and Franzoni, 1976; Hanke, 1976; Goos, 1978; Peter, 1986). The following generalized account is derived from those reviews with some supplemental information. The approximate locations of the hypothalamic nuclei described below are provided in Figure 1, and a summary of the nuclear localization of monoamines and neuropeptides is provided in Table 1.

A. Anterior Hypothalamus.

The major peptidergic and aminergic centers of the anterior hypothalamus are the preoptic nucleus (NPO), the preoptic recess organ (PRO) and the organum vasculosum laminae terminalis (OVLT). The NPO is subdivided into magnocellular (mNPO) and parvocellular (pNPO) regions.

The mNPO cells produce arginine vasotocin (AVT) and mesotocin (MST) which are carried via axons to the median eminence and to the pars nervosa. Immunoreactive TRH also is present in mNPO cells. Recently, a mammalian growth hormone-releasing hormone-like peptide also has been demonstrated in the mNPO (Marivoet et al., 1987). The pNPO cells produce the catecholamines norepinephrine (NE) and dopamine (DA) as well as immunoreactive somatostatin (iSS), iCRH, and iTRH. The presence of immunoreactive alpha-melanotropin (iMSH, Anderson et al., 1987), immunoreactive melanin-concentrating hormone (iMCH, Anderson et al., 1987) and immunoreactive atrial natriuretic factor (iANF, Netchitailo et al., 1987) have been indicated in the NPO. In addition, ACTH-like activity has been reported for the anterior hypothalamus using an *in-vitro* interrenal bioassay (Thurmond et al., 1986). This corticotropic activity has not been assigned to any particular nucleus, but likely occurs in the pNPO with iMSH.

The PRO and OVLT are strictly aminergic nuclei. Both DA and NE are produced in the PRO and are transported via axons to the median eminence. The importance of catecholamine synthesis in the OVLT is not known.

B. The Posterior Hypothalamus.

The posterior hypothalamus contains the paraventricular organ (PVO), the nucleus infundibularis dorsalis (NID) and the nucleus infundibularis ventralis (NIV). The NIV is equivalent to the pars ventralis of the tuber cinereum, a term used by some authors. Both the PVO and NID contain catecholaminergic and serotinergic neurons. In

Table 1. Distribution of monoamines and neuropeptides in amphibian hypothalamus.

	Hypothalamic Nuclei							
	pNPO	mNPO	NID	NIV	PRO	PVO	OVLT	ME
Monoamines Catechols	+		+		+	+	+	+
Indoles			+			+		
Peptides								
TRH	+	+	+	+				+
CRH	+							+
AVT		+						+
MST		+						+
SS	+		+					+
GHRH		+						
GnRH	+							+
MSH	+			+				
ANF	+		+					
NPY				+				

TRH = thyrotropin-releasing hormone; CRH = corticotropin-releasing hormone; AVT = arginine vasotocin; MST = mesotocin; SS = somatostatin; GHRH = growth hormone-releasing hormone; GnRH = gonadotropin-releasing hormone; MSH =α melanotropin; ANF = atrial natriuretic factor; NPY = neuropeptide Y.

pNPO = parvocellular cells of preoptic nucleus; mNPO = magnocellular cells of preoptic nucleus; NID = dorsal infundibular nucleus; NIV = ventral infundibular nucleus; PRO = preoptic recess organ; PVO = paraventricular organ; OVLT = organum ventralis laminae terminalis, ME = median eminence.

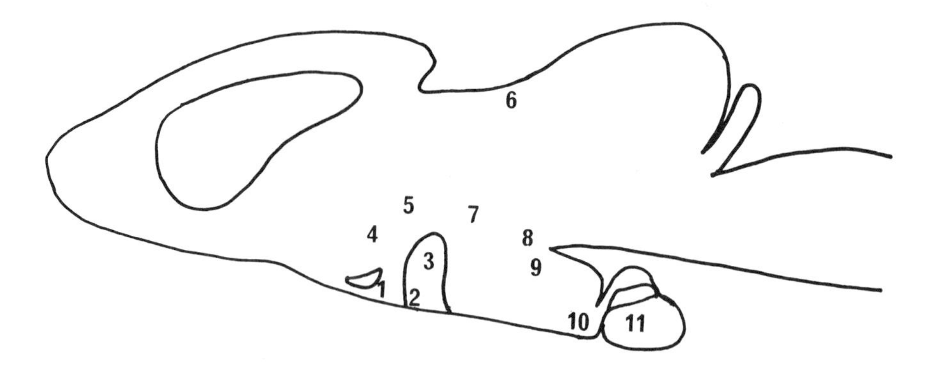

Figure 1. Approximate location of major hypothalamic nuclei in amphibian brain. 1 = PRO; 2 = OVLT; 3 = optic chiasma; 4 = pNPO; 5 = mNPO, 6 = pineal region; 7 = PVO; 8 = NID; 9 = NIV; 10 = ME; 11 = Pars distalis. See Table 1 or text for explanation of abbreviations.

addition, the NID contains iTRH (Seki and Kikuyama, 1983; Mimnagh et al., 1987), iSS (Leroy et al., 1987), iMCH (Anderson et al., 1987) and iANF (Netchitailo et al., 1987). The NIV sends peptidergic fibers to the median eminence. This nucleus has been shown to contain iTRH (Mimnagh et al., 1987), iMSH (Anderson et al., 1987) and immunoreactive neureopeptide Y (iNPY, Danger et al., 1987). Bioassayable ACTH-like material has been measured in the posterior hypothalamus (Thurmond et al., 1986), but the precise location has not been ascertained.

The structure and distribution of monoamines and peptides in the mammalian hypothalamus has been reviewed extensively and will not be duplicated here (see Reichlin, 1986; Cooper et al., 1986; Ganten and Pfaff, 1986; Halasz, 1986; Peter, 1986). A major difference between the mammalian and amphibian hypothalami is the origin of monoamines found there. In mammals only some of the DA is made in the hypothalamus, and all of the NE and serotonin (5-HT) are synthesized extrahypothalamically as is most of the DA. In amphibians all three monoamines are synthesized within hypothalamic nuclei.

IV. THE ROLE OF BRAIN MONOAMINES AND NEUROPEPTIDES

A. Catecholamines and Serotonin.

The effects of monoamines on release of hormones from the mammalian adenophypophysis have been reviewed extensively (see Collu,

1977; Reichlin, 1981, 1986; Nakai et al., 1986; Norman and Litwack, 1987). In mammals NE and DA stimulate release of TRH whereas 5-HT inhibits TRH release. Release of CRH is stimulated by 5-HT but can be blocked by NE, presumably at the pituitary level. In addition, release of CRH can be altered by gamma amino butyric acid (GABA) and arginine vasopressin (AVP). Dopamine functions as a PRL release-inhibiting hormone and probably inhibits release of MSH from the pars intermedia as well. Release of both TSH and PRL by TRH has been established, and it is known that CRH stimulates ACTH release.

The roles of monoamines in the release of TSH and ACTH are not known for amphibians. However, DA is thought to be the physiological inhibitor of PRL release as evidenced by studies employing DA and its known agonists and antagonists (Platt, 1976; Dubin and Brick, 1978; Seki and Kikuyama, 1979, 1982; Kikuyama and Seki, 1980). Furthermore, Seki and Kikuyama (1982) showed that PRL release was enhanced by a direct action of 5-HT at the pituitary gland, suggesting that 5-HT might be a prolactin-releasing neurohormone.

In anuran larvae, the only aminergic nuclei active in the hypothalamus prior to metamorphosis are the PVO and NID (see Hanke, 1976). McKenna and Rosenbluth (1975) provided the first link between thyroid hormones and hypothalamic monoamines by showing an increase in the intensity of fluorescing monoamines in the PRO of metamorphosing *Bufo marinus* tadpoles. Treatment of *Bufo japonicus* tadpoles with T4 caused an increase in catecholamine fluorescence in the PRO but not in either the PVO or NID (Kikuyama et al., 1979).

We (Carr et al., 1987) have made some preliminary observations on catecholamine and indoleamine turnover in the midbrain (including the hypothalamus) during metamorphosis of the tiger salamander, *A. tigrinum.* Larvae were placed under photoperiodic and temperature conditions that induce metamorphosis in the laboratory (Norris et al., 1977; Norman et al., 1987; Norman and Norris, 1987). Extracts of the midbrain region prepared from premetamorphic and metamorphosing larvae were subjected to separation by high performance liquid chromatography (HPLC) and quantitiation by use of coulometric electrochemical detection. This procedure allows for the simultaneous quantitative determination of DA, NE, and 5-HT as well as of their major metabolites, 3,4-dihydroxyphenyl acetate (DOPAC), 3-methoxy-4-hydroxyphenyl glycol (MHPG), and 5-hydroxyindole acetic acid (5-HIAA), respectively. Comparison of larvae, mid-metamorphic animals, and animals at the end of metamorphosis indicated a significant increase in the metabolism of NE as well as a significant reduction in the turnover of 5-HT. Both of these observations are consistent with the hypothesis that NE activates TRH and subsequent TSH and PRL secretion and the hypothesis that 5-HT can inhibit TRH release. The metabolism of DA showed a slight but not significant reduction during metamorphosis. It is important that future studies examine more stages of metamorphosis and look for changes in specific hypothalamic nuclei. Furthermore, studies of brain monoamines and their relationship to neuropeptide production and release in amphibians during the development of hypothalamic activity and metamorphosis are sorely needed.

B. Melatonin.

Some studies suggest melatonin or some other pineal factor can accelerate thyroid function and/or metamorphosis (Platt and Norris, 1973; Norris et al., 1981). In contrast are other studies showing that melatonin (and/or another pineal factor) depresses or delays metamorphic events (Remy and Disclos, 1970; Delgado et al., 1987). Still other experiments suggest no influence of the pineal system on thyroid function and/or metamorphosis (Kelly, 1958; Norris and Platt, 1973). In general circulating melatonin levels are elevated during periods of darkness (see Gern and Karn, 1983; Gern et al., 1978; Gern and Norris, 1979) although this response can be influenced strongly by environmental temperatures (Gern et al., 1983).

Observations on the effect of light on metamorphosis are more consistent, although they have not allowed assignment of any definite role to melatonin. Long photoperiod activates thyroid function (Norris and Platt, 1973) and accelerates metamorphosis in tiger salamander larvae (Norris et al., 1977; Norman et al., 1987; Norman and Norris, 1987). Diurnal fluctuations in circulating T4 and melatonin are related positively (Norris et al., 1981) suggesting a possible functional relationship. Similarly, Wright et al. (1988) report that injections of T4 into *Rana pipiens* tadpoles are less effective if given during darkness, and even a short pulse of light during the scotophase enhances the metamorphic action of injected T4.

Studies employing continuous light or dark are difficult to interpret because both are unnatural stimuli. Continuous light was not as effective as long photophases in the stimulation of thyroid activity in tiger salamander larvae, but the effect was similar to that observed under continuous darkness (Norris et al., 1977). Continuous light delayed metamorphosis in *Rana ridibunda* (Delgado et al., 1984) and both continuous light and darkness delayed metamorphosis in *X. laevis* (Delgado et al., 1987). However, darkness accelerated metamorphosis in *Discoglossus pictus* (Guiterrez et al., 1984). The duality of photic effects on metamorphosis suggests that either photic stimuli may influence more than one neuroendocrine pathway or that other factors are influencing the pineal-hypothalamic pathway.

C. Octapeptide neurohormones.

Little information is available regarding the participation of neurohypophysial hormones in the regulation of metamorphic events although some potentially significant findings have been published. For example, the onset of molting is a part of preparation for life on land and occurs during the terminal phase of metamorphosis. In urodeles, molting is controlled primarily by the thyroid. In hypophysectomized toads, on the other hand, it is induced by adrenal corticoids. That action of corticosteroids is enhanced by oxytocin (see Larsen, 1976).

In another series of studies, it was demonstrated that the antimetamorphic action of PRL is blocked in larval tiger salamanders by oxytocin (Platt and LiCause, 1980), by lysine vasopressin (LVP) and by AVT (Platt and Hill, 1982), presumably by interference with an osmoregulatory action of PRL.

In addition, it is suggested by Wells and Moser (1982) that among urodeles neoteny may result from reduced levels of neurohypophysial hormones. They found that both LVP and oxytocin induced significant metamorphic changes in adult Mexican axolotls, *Ambystoma mexicanum*. Wells and Moser (1982) point to an investigation conducted by Norris and Gern (1976) in which neotenic tiger salamanders metamorphosed in response to a single intrahypothalamic injection of T4, whereas the same dose given intraperitoneally was ineffective. It was reasoned that localized elevation of T4 levels brought about maturation of the hypothalamo-hypophysial axis giving the pituitary access to hypothalamic aTRF and bringing about the release of sufficient TSH to activate the thyroid. Wells and Moser (1982) propose that neoteny in permanent larvae occurs because PRL inhibits the differentiative action of thyroid hormones and that elevated levels of neurohypophysial hormones block effects of PRL in animals that metamorphose.

D. Thyrotropin-releasing hormone.

It is generally agreed that among birds and mammals the tripeptide, now known as TRH exercises control over secretion of TSH (see Griffiths and Bennett, 1982). The initial demonstrations of the thyrotropin-releasing action of TRH (see Vale et al., 1977) led to the expectation that the mammalian TRH might be aTRF which, as discussed in a preceding section (II.A.), is essential to completion of metamorphosis. However, although iTRH is present in a wide variety of ectothermic vertebrates, most recent reviewers have concluded that TRH does not play a prominent role in the regulation of the pituitary-thyroid axis below the level of birds (Crim et al., 1978; Ball, 1981; Jackson and Bolaffi, 1983; Rosenkilde, 1985). Indeed, the broad-spread distribution of stimulatory effects of TRH on the secretion of PRL and growth hormone (to be discussed below) among fishes and amphibians (see Hall et al., 1986) leads to the suggestion that those effects represent the most primitive function of the TRH (Ball, 1982; Preece and Licht, 1987). Even though most of the findings regarding the release of TSH in response to mammalian TRH among the ectotherms have been negative, TRH does have that capacity, at least in some ectothermic species and under some conditions. Convincing evidence supporting that limited position comes from observations of Preece and Licht (1987) to the effect that the thyrotropes of the turtle *Chysemys picta*, have functional receptors for mammalian TRH and that mammalian TRH stimulates the secretion of TSH in that animal. Also, Darras and Kühn (1982) reported that mammalian TRH administered in low doses by a perfusion technique to adults of the frog, *R. ridibunda*, elevated levels of T3 and T4 at intervals of 1, 2, and 4 hours with a peak at 2 hours. A thyrotropic role for mammalian TRH during metamorphosis is supported somewhat by the observed increase of iTRH during metamorphosis (King and Millar, 1982; Mimnagh et al., 1987), but this increase might be related to other actions of TRH. Using immunocytochemical techniques, Malogon-Poyato et al. (1987) report that mammalian TRH caused a reduction of pituitary TSH and PRL content in *R. ridibunda.*

Among urodeles, evidence that the effectiveness of mammalian TRH as the aTRF is limited comes from work done on the neotenic Mexican axolotl, *A. mexicanum.* Although perfusion of axolotls with bovine TSH or extracts of axolotl pars distalis induced peaks of both T3 and T4, infusion with mammalian TRH brought no alteration in levels of thyroidal hormone (Darras and Kühn, 1983). Conversely, mammalian TRH elevated levels of T4 in metamorphosed axolotls, suggesting that the responsiveness of the axolotl hypothalamus requires previous exposure to T4 (Jacobs and Kühn, 1987). These results may have important implications for earlier demonstrations of unresponsiveness to mammalian TRH in larval amphibians.

1. TRH as a prolactin-releasing hormone.

Even though mammalian TRH probably exercises little or no metamorphic action through the thyroid gland, there is ample evidence that it can act as a prolactin-releasing hormone (PRH) in all vertebrate classes (Jackson and Bolaffi, 1982). In that action TRH may affect the metamorphic sequence. *In-vitro* studies have demonstrated that mammalian TRH can enhance the synthesis and release of PRL in three different genera of Anura (Clemons et al. 1979; Seki and Kikuyama, 1982; Hall and Chadwick, 1982). In one anuran *R. ridibunda*, increased plasma concentrations of PRL were detected following injections *in vivo* of mammalian TRH (Kühn et al., 1982). Also, antibodies to mammalian TRH reduced PRL release from the bullfrog pituitary *in vitro* (Kikuyama et al., 1982). The inhibitory effects of catecholamines on the secretion of PRL in amphibians were discussed in a preceding section (IV.A.). In a recent study Seki and Kikuyama (1986) examined the release and synthesis of PRL by pituitary glands taken from bullfrogs and cultured with mammalian TRH and/or DA. Their results indicated that, under the conditions of their experiments, TRH induced the release but not the synthesis of PRL and that DA counteracted the stimulatory action of TRH, confirming earlier findings of Hall and Chadwick (1982) and suggesting that the secretion of PRL is controlled primarily by catecholamines.

There is as yet no clear understanding of the significance of the PRL surge at metamorphic climax. Equally mysterious is the mechanism by which it is activated. Some enlightenment comes from a series of experiments in which the pituitary glands of some bullfrog tadpoles were stalk-sectioned at TK stage XXII and those of others were transplanted to ectopic positions (Kawamura et al., 1986). Although severing hypothalamic connections at earlier stages prevents climax (Etkin and Lehrer, 1961), both of these operated groups completed climax in unison with control groups. Enhanced levels of PRL were seen in none of the operated animals, even in stalk-sectioned tadpoles that were treated with T4. These findings show that the PRL surge has little or no effect on metamorphic events. Both Kawamura et al. (1986) and Etkin and Lehrer (1961) indicate, that some sort of hypothalamic factor is essential to its occurrence.

White and Nicoll (1981) have hypothesized that the disappearance of aminergic nerve fibers from the pars distalis at climax (Aronsson, 1976) might release thyrotropes from tonic inhibition and initiate the PRL surge. That explanation must be discarded since isolation of the pars distalis from the hypothalamus did not permit the surge to take place (Kawamura et al., 1986). One must conclude that the isolation prevents passage to the pars distalis of some stimulatory control factor, possibly TRH (Kawamura et al., 1986). That possibility seems plausible in the light of results obtained by Seki and Kikuyama (1986). They demonstrated that the prolactin-releasing effect of homologous hypothalamic extract on partes distales from bullfrogs was severally attenuated in culture when exposed to IgG from anti=TRH serum (Seki and Kikuyama, 1986).

2. Other prolactin releasing factors.

Although mammalian TRH may be the major prolactin-releasing factor, other substances can also stimulate the release of PRL. Both vasoactive intestinal polypeptide (VIP) and peptide histidine isoleucine (PHI) stimulated the synthesis and release of PRL in bullfrog pituitary glands maintained in culture (Koiwai et al., 1986), and in the rat, a potent prolactin-releasing factor has been found in the pars nervosa (Hyde et al., 1986).

3. Other actions of TRH.

Morphological localizations of iTRH have been followed in the brains and pituitary glands of larval and adult bullfrogs (Mimnagh et al., 1987). Distributional patterns of iTRH-positive fibers gave indication of the passage of TRH into the portal capillaries of the median eminence from which it might move to carry out neuroendocrine roles in the pars distalis. Other TRH-positive fibers extended from hypothalamic nuclei to regions other than the pituitary with the implication that, as in the mammal (Morley, 1979), there is reason to suspect that TRH can serve as a neurotransmitter.

Earlier observations (Seki et al., 1983; Bolaffi, 1983) to the effect that iTRH-positive fibers enter the amphibian pars intermedia were confirmed (Mimnagh et al., 1987) in agreement with studies which show that mammalian TRH antagonizes the dopamine-induced inhibition of MSH release in the frog (Adjeroud et al., 1986). Changes in coloration and color patterns are usually an important aspect of metamorphosis.

One of the many sets of radical changes that constitute the cataclysmic events of metamorphosis takes place when the immune system of the tadpole is reformed as the immune system of the adult (Ruben et al., 1985). A part of that reformation consists of the regression of larval lymphatic organs in response to thyroidal stimulation (Riviere and Cooper, 1973). The thymus glands that provide cells for the immune restructuring are themselves reformed. The restructuring is accompanied by a continued increase in levels of thymic TRH which, perhaps, may perform some regulating action with respect to the immunologic alterations (Balls et al., 1985).

Growth hormone, as we have indicated in an earlier section, may have some somatotropic effects on amphibian larvae, but there is no indication that it acts to regulate metamorphic events (see Rosenkilde, 1985). It is worthy of note, however, that mammalian TRH stimulates secretion of growth hormone as well as PRL in larvae of the anurans, *R. pipiens* and *X. laevis* (Hall and Chadwick, 1983). That stimulatory action of mammalian TRH on the release of growth hormone is inhibited by somatostatin but not by DA (Hall and Chadwick, 1982). Also, as mentioned earlier, growth hormone does cause elevation of 5'-monodeiodinase in chickens (Kühn et al., 1987) and could be a factor affecting the increased conversion of T4 to T3 seen during metamorphosis (Galton, 1983).

E. Corticotropin-releasing Hormone.

The participation of corticosteroids in the regulatory action of the thyroid is a matter of great importance in the control of metamorphic changes. Only limited information is available regarding the regulation of the release of corticosteroids in the Amphibia. Neurophypophysial octapeptides can release ACTH in *Bufo* (Jorgensen, 1976); however, lesioning experiments have demonstrated the presence of a CRH-like factor in the ventral portion of the preoptic nucleus in larvae of *Xenopus* (Notenboom et al., 1976), and iCRH is present in the pNPO cells of several urodelan and anuran species (see section III.A.).

V. SUMMARY

We are now beginning to understand what hypothalamic actions might occur during the postembryonic events involved in the initiation and completion of metamorphosis. New and sensitive techniques make it possible to localize and monitor changes in monoamine metabolism. Yet, studies are needed to ascertain roles of monoamines on release of hypothalamic neuropeptides. The potential roles for mammalian TRH in this system and nature of aTRF are still unresolved, and more information on other hypothalamic peptides are needed. Immunocytochemical approaches to the study of brain peptides and the potential uses of recombinant DNA techniques to develop molecular probes for the examination of hypothalamic neuropeptides will undoubtedly shed new light on this area. Through such approaches we will be able to disclose the timed sequence of ontogenetic hypothalamic events necessary to the initiation of metamorphosis and to the identification of how the external (e.g., photoperiod, temperature) and internal (e.g., premetamorphic neural and endocrine factors) events influence hypothalamic activity.

VI. REFERENCES

Adjeroud, S., Tonon, M.-C., Gouteux, L., Leneveu, E., Lamacz, M., Cazin, L. and Vaudry, H. (1986). *In vitro* study of frog (*Rana ridibunda* Pallas) Neurointermediate lobe secretion by use of a simplified perifusion system. IV. Interaction between dopomine and thyrotropin-releasing hormone on α-melanocyte-stimulating hormone secretion. *Gen. comp. Endocrinol.* **64**, 428-434.

Alberch, P., Gale, E. A., and Larsen, P. R. (1986). Plasma T_4 and T_3 levels in naturally metamorphosing *Eurycea bislineata* (Amphibia; Plethodontidae). *Gen. Comp. Endocrinol.* **61**, 153-161.

Anderson, A.C., Leroux, P., Eberle, A. N., Jegou, S. and Vaudry, H. (1987). Immunohistochemical localization of melanin-concentrating hormone (MCH) in the frog brain. *Gen. Comp. Endocrinol.* **66**, 17.

Aronsson, S. (1976). The ontogenesis of monoaminergic nerve fibers in the hypophysis of *Rana temporaria* with special reference to the pars distalis. *Cell Tissue Res.* **171**, 437-448.

Ball, J. N. (1981). Hypothalamic control of the pars distalis in fishes, amphibians, and reptiles. *Gen. Comp. Endocrinol.* **44**, 135-170.

Balls, M., Clothier, R. H., Rowles, J. M., Kiteley, N. A. and Bennett, G. W. (1985). TRH distribution, levels, and significance during the development of *Xenopus laevis.* In "Metamorphosis" (M. Balls and M. Bownes, eds.) pp. 260-272. Clarendon Press, Oxford.

Bern, H. A. (1975). On two possible primary activities of prolactin: Osmoregulatory and developmental *Verh. Dtsch. Zool. Ges.* **1975**, 40-46.

Bern, H. A., Nicoll, C. S., and Strohman, R. C. (1967). Prolactin and tadpole growth. *Proc. Soc. Exp. Biol. Med.* **126**, 518-520.

Blatt, L.M., Slickers, A., and Kim, K.H. (1969). Effect of prolactin on thyroxine-induced metamorphosis. *Endocrinology* **85**, 1213-1215.

Bolaffi, J. L. (1983). Immunohistochemical localization of thyrotropin-releasing hormone in amphibian brain. In "65th Ann. Meeting Endocrine Society" *Abst.* 274.

Bolton, K. P., Collie, N. L., Kawauchi, H. and Hirano, T. (1987). Osmoregulatory actions of growth hormone in rainbow trout (*Salmo gairdneri*). *J. Endocrinol.,* **112**, 63-68.

Brown, P. S., Brown, S. C., and Specker, J. L. (1984). Osmoregulatory changes during the aquatic-to-terrestrial transition in the rough-skinned newt, *Taricha granulosa*: The roles of temperature and ACTH. *Gen. Comp. Endocrinol.* **56**, 130-139.

Buscaglia, M., Leloup, J., and deLuze, A. (1985). The role and regulation of monoiodination of thyroxine to 3,5,3'-triiodothyronine during amphibian metamorphosis. In "Metamorphosis" (M. Balls and M. Brownes, eds.) pp. 273-293. Clarendon Press, Oxford.

Carr, J. A. and Norris, D. O. (1988). Interrenal activity during metamorphosis of the tiger salamander *Ambystoma tigrinum. Gen. Comp. Endocrinol.*, in press.

Carr, J. A., Norris, D. O., Desan, P. H., Smock, T., and Norman, M. F. (1987). Quantitative measurement of brain amines and metabolites during metamorphosis of tiger salamander larvae. *The Physiologist* **30**, 142.

Choun, J. L. and Norris, D. O. (1974). The effects of prolactin dose on rate of TSH-induced metamorphosis, weight loss and feeding frequency in larval tiger salamanders, *Ambystoma tigrinum. J. Colo.-Wyo. Acad. Sci.,* **7(4)**, 68.

Clarke, W. C., and Bern, H. A. (1980). Comparative endocrinology of prolactin. In "Hormonal Proteins and Peptides", (C. H. Li, ed.) V8, pp. 106-197. Academic Press, New York.

Clemons, G. K. and Nicoll, C. S. (1977a). Effects of antisera to bullfrog prolactin and growth hormone on metamorphosis of *Rana catesbeiana* tadpoles. *Gen. Comp. Endocrinol.* **31**, 495-497.

Clemons, G. K. and Nicoll, C. S. (1977b). Development and preliminary application of a homologous radioimmunoassay for bullfrog prolactin. *Gen. Comp. Endocrinol.* **32**, 531-535.

Clemons, G. K., Russell, S. M., and Nicoll, C. S. (1979). Effect of mammalian thyrotropin releasing hormone on prolactin secretion by bullfrog adenohypophyses *in vitro. Gen. Comp. Endocrinol.* **38**, 62-67.

Collu, R. (1977). Role of central cholinergic and aminergic neurotransmitters in the control of anterior pituitary hormone secretion. In "Clinical Neuroendocrinology". (L. Martini and G. Besser, eds.) pp. 43-65. Academic Press, New York.

Cooper, J.R., Bloom, F. E., and Roth, R. H. (1986). "The biochemical Basis of Neuropharmacology". Oxford University Press, New York.

Crim, J. W., Dickhoff, W. W., and Gorbman, A. (1978) Comparative endocrinology of piscine hypothalamic hypophysiotopic peptides and distribution and activity. *Amer. Zool.* **18**, 441-424.

Danger, J. M., Caillez, D., Guy, J., Benyamina, M., Polak, J. M., Pelletier, G., and Vaudry, H. (1987). NPY in amphibians: localization in the frog brain and inhibitory action on pars intermedia secretion. *Gen. Comp. Endocrinol.* **66**, 13.

Darras, V. M. and Kühn, E. R. (1982). Increased plasma levels of thyroid hormones in a frog *Rana ridibunda* following intravenous administration of TRH. *Gen. Comp. Endocrinol.* **48**, 469-475.

Darras, V. M. and Kühn, E. R. (1983). Effects of TRH, bovine TSH and pituitary extracts on thyroidal T_4 release in *Ambystoma mexicanum. Gen. Comp. Endocrinol.* **51**, 286-291.

Delgado, M. J. Gutierrez, P., and Alonso-Bedate, M. (1984). Growth response of premetamorphic *Rana ridibunda* and *Discoglossus pictus* tadpoles to melatonin injections and photoperiod. *Acta Embryol. Morph. Exper.; n.s.,* **5**, 23-39.

Delgado, M. J. Gutierrez, P., and Alonso-Bedate, M. (1987). Melatonin and photoperiod alter growth and larval development in *Xenopus laevis* tadpoles. *Comp. Biochem. Physiol.* **86A**, 417-421.

Delidow, B. C., Baldocchi, R. A., and Nicoll, C. S. (1988). Evidence for hepatic involvement in the regulation of amphibian development by prolactin. *Gen. Comp. Endocrinol.* in press.

Dent, J. N. (1968). Survey of amphibian metamorphosis. In "Metamorphosis" (W. Etkin and L. I. Gilbert eds.) pp. 271-311. Appleton-Century-Crafts, New York.

Dent, J. N. (1985). Interaction between prolactin and thyroidal hormones in amphibian systems. In "Current Trends in Comparative Endocrinology" (B. Lofts and W. N. Holmes, eds.) pp. 541-544. Hong Kong University Press, Hong Kong.

Dent, J. N. (1988). Hormonal interaction in amphibian metamorphosis. Amer. Zool., in press.

Derby, A. (1975). The effect of prolactin and thyroxine on tail resorption of *R pipiens: In vivo* and *in vitro. J. Exp. Zool.* **193**, 15-20.

Derby A., and Etkin, W. (1968). Thyroxine induced tail resorption *in vitro* as affected by anterior pituitary hormones. *J. Exp. Zool.* **169**, 1-8.

De Vlaming, V. L. (1980). Actions of prolactin among the vertebrates. In "Hormones and Evolution" (E. J. W. Barrington, ed.) pp. 561-642. Academic Press, New York

Dodd, M. H. I., and Dodd, J. M. (1976). The biology of metamorphosis. In "Physiology of the Amphibia" (B. Lofts, ed). Vol. III, pp. 467-599. Academic Press, New York.

Dubin, E., and Brick, I. (1978). Acceleration of T_4-induced metamorphosis by ergocornine-induced prolactin depletion. *Amer. Zool.* **18**, 582.

Eagleson, G. W., and McKeown, B. A. (1978). Changes in thyroid activity of *Ambystoma gracile* (Baird) during larval, transforming, and postmetamorphic phases. *Can. J. Zool.* **56**, 1377-1381.

El Maraghi-Ater, H., Mesnard, J., and Hourdry, J. (1986). Hormonal control of the intestinal brush border enzyme activities in developing anuran amphibians. I. Effects of hydrocortisone and insulin during and after spontaneous metamorphosis. *Gen. Comp. Endocrinol.* **61**, 53-63.

Etkin, W. (1932). Growth and resorption phenomena in anuran metamorphosis I. *Physiol. Zool.* **5**, 275-300.

Etkin, W. (1938). The development of the thyrotropic function in pituitary grafts in the tadpole. *J. Exp. Zool.* **77**, 347-377.

Etkin, W. (1968). Hormonal control of amphibian metamorphosis. In "Metamorphosis" (W. Etkin and L. I. Gilbert, eds.) pp. 313-348. Appleton-Century-Crofts, New York.

Etkin, W., and Gona, A. G. (1967). Antagonism between prolactin and thyroid hormone in amphibian development *J. Exp. Zool.* **165**, 149-258.

Etkin, W., and Lehrer, R. (1960). Excess growth in tadpoles after transplantation of the adenohypophysis. *Endocrinology*, **67**, 457-466.

Etkin, W., and Sussman, W. (1961). Hypothalamo-pituitary relations of *Ambystoma. Gen. Comp. Endocrinol.* **1**, 70-79.

Fasolo, A. and Franzoni, M. F. (1976). The neurohypophysis of urodela: a comparative analysis. *Monitore Zool. Ital. (N.S.)* **10**, 149-189.

Fellmann, D., Bugnon, C., Bresson, J. L., Gouget, A., Cardot, J., Clavequin, M. C. and Hadjiyiassemis, M. (1984). The CRF neuron: immunocytochemical study. *Peptides* **5**, *Suppl.* 19-33.

Galton, V. A. (1983). Thyroid hormone action in amphibian metamorphosis. In "Molecular Basis of Thyroid Hormone Action" (J. H. Oppenheimer and H. H. Samuels, eds.) pp. 445-483. Academic Press, Inc., New York.

Ganten, D. and Pfaff, D. (1986). "Morphology of Hypothalamus and Its Connections". Springer-Verlag, Berlin.

Gern, W. A. and Karn, C.M. (1983). Evolution of melatonin's functions and effects. *Pineal Res. Revs.* **1**, 49-90.

Gern, W. A. and Norris, D. O. (1979). Plasma melatonin in the neotenic tiger salamander (*Ambystoma tigrinum*): effects of photoperiod and pinealectomy. *Gen. Comp. Endocrinol.* **38**, 393-398.

Gern, W. A., Owens, D. W., and Ralph, C. L. (1978). Plasma melatonin in the trout: day-night change demonstrated by radioimmunoassay. *Gen. Comp. Endocrinol.* **34**, 453-458.

Gern, W. A., Norris, D. O., and Duvall, D. (1983). The effect of light and temperature on plasma melatonin in neotenic tiger salamanders (*Ambystoma tigrinum*). *J. Herpetol.* **17**, 228-234.

Goos, H. J. Th. (1978). Hypophysiotropic centers in the brain of amphibians and fish. *Amer. Zool.* **18**, 401-410.

Gould, S. J. (1972). "Ontogeny and Phylogeny". Belknap Press, Cambridge, MA.

Greenfield, P. and Derby, A. (1972). Activity and localization of acid hydrolases in the dorsal tail fin of *Rana pipiens* during metamorphosis. *J. Exp. Zool.* **179**, 129-142.

Griffiths, E. C. and Bennett, G. W. (1983). "Thyrotropin-Releasing Hormone". Raven Press, New York.

Gutierrez, P., Delgado, M. J., and Alonso-Bedate, M. (1984). Influence of photoperiod and melatonin administration on growth and metamorphosis in *Discoglossus pictus* larvae. *Comp. Biochem. Physiol..* **79A**, 255-260.

Halasz, B. (1986). A 1985 view of the hypothalamic control of the anterior pituitary. In "Neuroendocrine Perspectives" (E. E. Muller and R. M. MacLean, eds.), Vol. 5, pp. 1-12, Elsevier. Amsterdam.

Hall, T. R. and Chadwick, A. (1983). Hypothalamic control of prolactin and growth hormone secretion in the pituitary gland of the pigeon and the chicken: *in vitro* studies. *Gen. Comp. Endocrinol.* **49**, 135-143.

Hall, T. R., and Chadwick, A. (1984). Effects of synthetic mammalian thyrotropin releasing hormone, somatostatin and dopamine on secretion of prolactin and growth hormone from amphibian and reptilian pituitary glands incubated *in vitro. J. Endocrinol.* **102**, 175-180.

Hall, T. R., Harvey S., and Chadwick, A. (1986). Control of prolactin secretion in birds: a review *Gen. Comp. Endocrinol.* **62**, 171-184.

Hanke, W. (1976). Neuroendocrinology. In "Frog Neurobiology" (R. Llinas and W. Frecht, eds.) pp. 975-1020. Springer-Verlag, Berlin.

Hickey, E. D. (1971). Behavior of DNA, protein and acid hydrolases in response to thyroxine in isolated tail tips of *Xenopus*-larvae. *Wilhelm Roux Archiv.* **166**, 303-330.

Hourdry J., and Beaumont, A. (1985). "Les Métamorphoses des Amphibiens". Masson, Paris.

Hyde, J. F., Murai, I., and Ben-Jonathan, N. (1987). The rat posterior pituitary contains a potent prolactin-releasing factor: studies with perifused anterior pituitary cells. *Endocrinology* **121**, 1531-1539.

Ikawa, H., Hutson, J. M., Budzik, G. P., and Donahoe, P. K. (1984). Cyclic adenosine 3',5'-monophosphate modulation of Mullerian duct regression. *Endocrinology* **114**, 1686-1691.

Jackson, I. M. D. and Bolaffi, J. L. (1983). Phylogenetic distribution of TRH: Significance and function. In "Thyrotropin-Releasing Hormone" (E.C. Griffiths and G. W. Bennett, eds.). pp. 191-202. *Raven Press*, New York.

Jackson, I. M. D., and Reichlin, S. (1974). Thyrotropin-releasing hormone (TRH): Distribution in hypothalamic and extrahypothalamic brain tissues of mammalian and submammalian chordates. *Endocrinology* **95**, 854-862.

Jacobs, G. F. M. and Kühn, E. R. (1987). TRH injection induces thyroxine release in the metamorphosed but not in the neotenic axolotl, *Ambystoma mexicanum. Gen. Comp. Endocrinol.* **66**, 40.

Jaffe, R. C. 1981. Plasma concentration of corticosterone during *Rana catesbeiana* tadpole metamorphosis. *Gen. Comp. Endocrinol.* **44**, 314-318.

Jaffe, R. C. and Geschwind, I. I. (1974a). Studies on prolactin inhibition of thyroxine-induced metamorphosis in *Rana catesbeiana* tadpoles. *Gen. Comp. Endocrinol.* **22**, 289-295.

Jaffe, R. C. and Geschwind, I. I. (1974b). Influence of prolactin on thyroxine-induced changes in hepatic and tail enzymes and nitrogen metabolism in *Rana catesbeiana* tadpoles. *Proc. Soc. Exp. Biol. Med.* **146**, 961-966.

Jolivet-Jaudet, G. and Leloup-Hatey, J. (1984). Variations in aldosterone and corticosterone plasma levels during metamorphosis in *Xenopus laevis* tadpoles. *Gen. Comp. Endocrinol.* **56**, 59-65.

Jorgensen, C. B. (1976). Sub-mammalian vertebrate hypothalamic-pituitary-adrenal interrelationships. In "General, Comparative and Clinical Endocrinology of the Adrenal Cortex" (I. C. Jones and I. W. Henderson, eds.), Vol. 1, pp. 153-206. Academic Press, New York.

Just, J. J., Kraus-Just, J. and Check, D. A. (1981). Survey of Chordate Metamorphosis. In "Metamorphosis", (L. I. Gilbert and E. Frieden, eds.), pp. 265-326, Plenum Press, New York.

Kawamura, K., Yamamoto, K., and Kikuyama, S. (1986). Effects of thyroid hormone, stalk section, and transplantation of the pituitary gland on plasma prolactin levels at metamorphic climax in *Rana catesbeiana. Gen. Comp. Endocrinol.* **64**, 129-135.

Kelly, D. E. (1958). Embryonic and larval epiphyectomy in the salamander *Taricha torosa*, and observations on scoliosis. *J. Morph.* **103**, 503-538.

Kikuyama, S. and Seki, T. (1980). Possible involvement of dopamine in the release of prolactin-like hormone from the bullfrog pituitary gland. *Gen. Comp. Endocrinol.* **41**, 173-179.

Kikuyama, S., Miyakawa, M., and Arai, Y. (1979). Influence of thyroid hormone on the development of preoptic-hypothalamic monoaminergic neurons in the tadpoles of *Bufo bufo japonicus. Cell Tissue Res.* **198**, 27-33.

Kikuyama, S., Niki, K. Mayumi, M., and Kawamura, K. (1982). Retardation of thyroxine-induced metamorphosis by Amphenone B in toad tadpoles. *Endocrinol. Japon.* **29**, 659-662.

Kikuyama, S., Niki, K., Mayum, M., Shibayama, R., Nishikawa, M., and Shintake, N. (1983). Studies on corticoid action on the toad tadpole tail *in vitro. Gen. Comp. Endocrinol.* **52**, 395-399.

Kikuyama, S., Suzuki, M. R., Iwamuro, S. (1986). Elevation of plasma aldosterone levels of tadpoles at metamorphic climax. *Gen. Comp. Endocrinol.* **63**, 186-190.

Kikuyama, S., Koiwai, K., Seki, T., and Sakai, M (1987). Prolactin-releasing activities of the bullfrog hypothalamus. *Gen. Comp. Endocrinol.* **66**, 6.

King, J.A. and Millar, R. P. (1981). TRH, GH-RIH, and LH-RH in metamorphosing *Xenopus laevis. Gen. Comp. Endocrinol.* **44**, 20-27.

Koiwai, K., Kikuyama, S., Seki, T., and Yanaihara, N. (1986). *In vitro* effect of vasoactive intestinal polypeptide and peptide histidine isoleucine on prolactin secretion by the bullfrog pituitary gland. *Gen. Comp. Endocrinol.* **64**, 254-259.

Kollros, J. J. (1981). Transitions in the nervous system during amphibian metamorphosis. In "Metamorphosis", (L.I. Gilbert and E. Frieden, eds.). pp. 445-459, Plenum Press, New York.

Krug, E. C., Honn, K. V., Battista, J. and Nicoll, C. S. (1983). Corticosteroids in serum of *Rana catesbeiana* during development and metamorphosis. *Gen. Comp. Endocrinol.* **52**, 232-241.

Kühn, E. R., Darras, V. M., and Verlinden, T. M. (1985). Annual variations of thyroid reactivity following thyrotropin stimulation and circulating levels of thyroid hormones in the frog *Rana ridibunda. Gen. Comp. Endocrinol.* **57**, 226-273.

Kühn, E. R., Verheyen, G., Chaisson, R. B., Huts, C., Huybrechts, L., Vanden Steen, P., and Decuypere, E. (1987). Growth hormone stimulates the peripheral conversion of thyroxine into triidothyronine by increasing the liver 5'-monodeiodinase activity in the fasted and normal fed chicken. *Horm. Metabol. Res.* **19**, 304-308.

Larras-Regard, E. (1985). Hormonal determination of neoteny in facultative neotenic urodeles. In "Metamorphosis", (M. Balls and M. Bowes, eds.), pp. 294-312, Clarendon Press, Oxford.

Larras-Regard, E., Taurog, A., and Dorris, M. (1981). Plasma T_4 and T_3 levels in *Ambystoma tigrinum* at various stages of metamorphosis. *Gen. Comp. Endocrinol.* **43**, 443-450.

Larsen, L. O. (1976). Physiology of Molting. In "Physiology of Amphibia" (B. Lofts, ed.), Vol. 2, pp. 54-100. Academic Press, New York.

Leloup, J., and M. Buscaglia. (1977). La triiodothyronine, hormone de la metamorphose des Amphibiens. *C. R. Acad. Paris Ser. D.* **284**, 2261-2263.

Leroy A., Leroux, P., Benoit, R. and Vaudry, H. (1987). Coincidence of somatostatin (SS)-related peptide-like-immunoreactivity (S-LI) and somatostatin-binding sites (S-BS) in the frog (*Rana ridibunda*) brain. *Gen. Comp. Endocrinol.* **66**, 44.

Lynn. W. G. (1961). Types of amphibian metamorphosis. *Amer. Zool.* **1**, 151-161.

Malagón-Poyato, M. M., Garcia-Navarro, S., and Garcia-Navarro, F. (1987). Immunohistochemical study of PRL and TSH cells in normal and TRH-injected frogs. *Gen. Comp. Endocrinol.* **66**, 5.

Marivoet, S., Eilen, C., and Vandesande, F. (1987). Growth hormone-releasing factor (GRF)-like immunoreactivity in the hypothalamo-hypophysial system of the frog *Rana temporaria*) and a possible coexistence with vasotocine (VT). *Gen. Comp. Endocrinol.* **66**, 6.

McKenna, O. C. and Rosenbluth, J. (1975). Ontogenetic studies of a catecholamine-containing nucleus of the toad hypothalamus in relation to metamorphosis. *Exper. Neurol.* **46**, 496-505.

Mimnagh, K. M., Bolaffi, J. L., Montgomery, N. M., and Kaltenbach, J. C. (1987). Thyrotropin-releasing hormone (TRH): immuno-histochemical distribution in tadpole and frog brain. *Gen. Comp. Endocrinol.* **66**, 394-404.

Miyauchi, H., LaRochelle, F. T., Jr., Suzuki, M., Freeman, M. and Frieden, E. (1977). Studies on thyroid hormones and their binding in bullfrog tadpole plasma during metamorphosis. *Gen. Comp. Endocrinol.* **39**, 343-349.

Mondou, P. M. and Kaltenbach, J. C. (1979). Thyroxine concentrations in blood serum and pericardial fluid of metamorphosing tadpoles and of adult frogs. *Gen. Comp. Endocrinol.* **39**, 343-349.

Morley, J.E. (1979). Extrahypothalamic thyrotropin releasing hormone (TRH)-its distribution and its functions. *Life Sci.* **25**, 1539-1550.

Nakai, Y., Shioda, S., Ochiai, H., and Kozasa, K. (1986). Catecholamine-peptide interactions in the hypothalamus. In "Morphology of Hypothalamus and Its Connections", (D. Ganten and D. Pfaff, eds.), pp. 135-160. Springer-Verlag, Berlin.

Netchitailo, P., Feuilloley, M., Pelletier, G., Cantin, M., Andersen, A., Leboulenger, F., and Vaudry, H. (1987). Immunocytochemical localization of atrial natriuretic factor (ANF) in the frog brain. *Gen. Comp. Endocrinol.* **66**, 43-44.

Nicoll, C. S. (1980). Ontogeny and evolution of prolactin functions. *Fed. Proc.* **30**, 2563-2566.

Nicoll, C. S. (1981). Role of prolactin in water and electrolyte balance in vertebrates. In "Current Endocrinology" (R. Jaffe, ed.), pp. 127-166. Elsevier North-Holland Press, New York.

Norman, A. W., and Litwack, G. (1987). "Hormones", pp. 133. Academic Press, New York.

Norman, M. F. and Norris, D. O. (1987). Effects of metamorphosis and captivity on the *in vitro* sensitivity of thyroid glands from the tiger salamander, *Ambystoma tigrinum*, to bovine thyrotropin. *Gen. Comp. Endocrinol.* **67**, 77-84.

Norman, M. F., Carr, J. A. and Norris, D. O. (1987). Adenohypophysial-thyroid activity of the tiger salamander *Ambystoma tigrinum*, as a function of metamorphosis and captivity. *J. Exp. Zool.* **242**, 55-66.

Norris, D. O. (1983). Evolution of endocrine regulation of metamorphosis in lower vertebrates. *Amer. Zool.* **23**, 709-718.

Norris, D. O. (1985). Endocrine regulation of amphibian metamorphosis. In "Vertebrate Endocrinology" (D. O. Norris), pp. 425-443, Lea and Febiger, Philadelphia.

Norris, D. O. and Gern, W. (1976). Thyroxine-induced activation of hypothalamo-hypophysial axis in neotenic salamander larvae. *Science* **194**, 525-527.

Norris, D. O. and Platt, J. E. (1973). Effects of pituitary hormones, melatonin, and thyroidal inhibitors on radioiodine uptake by the thyroid glands of larval and adult tiger salamanders, *Ambystoma tigrinum* (Amphibia: Caudata). *Gen. Comp. Endocrinol.* **21**, 368-376.

Norris, D. O., Jones, R. E., and Criley, B. B. (1973). Pituitary prolactin levels in larval, neotenic and metamorphosed salamanders (*Ambystoma tigrinum*). *Gen. Comp. Endocrinol.* **20**, 437-442.

Norris, D. O., Duvall, D., Greendale, K., and Gern, W. A. (1977). Thyroid function in pre- and postspawning neotenic tiger salamanders (*Ambystoma tigrinum*). *Gen. Comp. Endocrinol.* **35**, 512-517.

Norris, D. O., Gern, W. A., and Greendale, K. (1981). Diurnal and seasonal variations in thyroid function of neotenic tiger salamanders (*Ambystoma tigrinum*). *Gen. Comp. Endocrinol.* **45**, 134-137.

Notenboom, C. D., Terlou, M., and Maten, M. L. (1976). Evidence for corticotropin releasing factor (CRF) synthesis in the preoptic nucleus of *Xenopus laevis* tadpoles. *Cell Tissue Res.* **169**, 23-31.

Olivereau, M., Vandesande, F., Boucique, E., Ollevier, F. and Olivereau, J. M. (1987). Immunocytochemical localization and spatial relation to the adenohypophysis of a somatostatin-like and a corticotropin-releasing factor-like peptide in the brain of four amphibian species. *Cell Tiss. Res.* **247**, 317-324.

Peter, E. E. (1986). Vertebrate neurohormonal systems. In "Vertebrate Endocrinology: Fundamentals and Biochemical implications". Vol 1, Morphological Considerations (P. K. T. Pang and M. P. Schreibman, eds.,) pp. 57-104, Academic Press. New York.

Piotrowski, D.C. and Kaltenbach, J. C. (1985). Immunofluorescent detection and localization of thyroxine in blood of *Rana catesbeiana* from early larval through metamorphic stages. *Gen. Comp. Endocrinol.* **59**, 82-90.

Platt, J. E. (1976). The effects of ergocornine on tail height, spontaneous and T_4-induced metamorphosis and thyroidal uptake of radioiodine in neotenic *Ambystoma tigrinum*. *Gen. Comp. Endocrinol.* **28**, 71-81.

Platt, J. E., and Christopher, M. A. (1977). Effects of prolactin on the water and sodium content of larval tissues from neotenic and metamorphosing *Ambystoma tigrinum*. *Gen. Comp. Endocrinol.* **31**, 243-248.

Platt, J. E. and P. L. Hill. (1982). Inhibition of the antimetamorphic action of prolactin in larval *Ambystoma tigrinum* by using vasopressin, arginine vasotocin, and aldosterone. *Gen. Comp. Endocrinol.* **48**, 355-361.

Platt, J. E., and LiCause, M. J. (1980). Effects of oxytocin in larval *Ambystoma tigrinum*: Acceleration of induced metamorphosis and inhibition of the antimetamorphic action of prolactin. *Gen. Comp. Endocrinol.* **41**, 84-91.

Platt, J.E. and Norris, D. O. (1973). The effects of melatonin, bovine pineal extract and pinealectomy on spontaneous and induced metamorphosis and thyroidal uptake of ^{131}I in larval *Ambystoma tigrinum*. *J. Colo.-Wyo. Acad. Sci.* **7**(3), 40.

Platt, J. E., Christopher, M. A., and Sullivan, C.A. (1978). The role of prolactin in blocking thyroxine-induced differentiation of tail tissue in larval and neotenic *Ambystoma tigrinum*. *Gen. Comp. Endocrinol.* **35**, 402-408.

Platt, J. E., Brown, G. B., Erwin, S. A., and McKinley, K. T. (1986). Antagonistic effects of prolactin and oxytocin on tail fin regression and acid phosphatase activity in metamorphosing *Ambystoma tigrinum*. *Gen. Comp. Endocrinol.* **61**, 376-382.

Preece, H. and Licht, P. (1987). Effects of thyrotropin-releasing hormone *in vitro* on thyrotropin and prolactin release from the turtle pituitary. *Gen. Comp. Endocrinol.* **67**, 247-255.

Ray, L. B., and J. N. Dent. (1986a). Observations on the interaction of prolactin and thyroxine in the tail of the bullfrog tadpole. *Gen. Comp. Endocrinol.* **64**, 36-43.

Ray, L. B. and Dent, J. N. (1986b). Investigations on the role of cAMP in regulating the resorption of the tail fin from tadpoles of *Rana catesbeiana*. *Gen. Comp. Endocrinol.* **64**, 44-51.

Regard, E., Taurog, A. and Nakashima, T. (1978). Plasma thyroxine and triiodothyronine levels in spontaneously metamorphosing *Rana catesbeiana* tadpoles and in adult anuran amphibia. *Endocrinology* **102**, 674-684.

Reichlin, S. (1981). Neuroendocrinology. In "Textbook of Endocrinology" (J. D. Wilson and D. W. Foster, eds.), pp. 492-567, *W. B. Saunders*, Philadelphia.

Reichlin, S. (1986). Neural functions of TRH. Acta Endocrinol. **112** (Suppl. 276), 21-33.

Remy, C. and Disclos, P. (1970). Influences de l'epiphysectomie sur le developpement de la thyroide et des gonades chez le têtards d'Alytes obstetricans. *C. R. Soc. Biol.* **164**, 1989-1993.

Riviere, H. B. and Cooper E. L. (1973). Thyroxine-induced regression of tadpole lymph glands. *Proc. Soc. Exp. Biol. Med.* **143**, 320-322.

Rosenkilde, P. (1985). The role of hormones in the regulation of amphibian metamorphosis. In "Metamorphosis" (M. Balls and M. Bownes, eds.), pp. 221-259, Clarendon Press, Oxford.

Ruben, L. N., Clothier, R. H., Jones, S. E., and Bonyhadi, M. L. (1985). The effects of metamorphosis on the regulation of humoral immunity in *Xenopus laevis*, the South African clawed toad. In "Metamorphosis" (M. Balls and M. Bownes, eds.), pp. 360-398, Clarendon Press, Oxford.

Seki, T. and Kikuyama, S. (1979). Effects of ergocornine and reserpine on metamorphosis in *Bufo bufo japonius* tadpoles. *Endocrinol. Japon.* **26**, 675-678.

Seki, T. and Kikuyama, S. (1982). *In vitro* studies on the regulation of prolactin secretion in the bullfrog pituitary gland. *Gen. Comp. Endocrinol.* **46**, 473-479.

Seki, T. and Kikuyama, S. (1986). Effects of thyrotropin-releasing hormone and dopamine on the *in vitro* secretion of prolactin by the bullfrog pituitary gland. *Gen. Comp. Endocrinol.* **61**, 197-202.

Seki, T., Nakai, Y., Shioda, S., Mitsuma, T., and Kikuyama, S. (1983). Distribution of immunoreactive thyrotropin-releasing hormone in the forebrain and hypophysis of the bullfrog, *Rana catesbeiana*. *Cell Tissue Res.* **223**, 507-516.

Suzuki, S. and Kikuyama, S. (1983). Corticoids augment nuclear binding capacity for triiodothyronine in bullfrog tadpole tail fins. *Gen. Comp. Endocrinol.* **52**, 272-278.

Suzuki, S. and Suzuki, M. (1981). Changes in thyroidal and plasma iodine compounds during and after metamorphosis of the bullfrog, *Rana catesbeiana. Gen. Comp. Endocrinol.* **45**, 74-81.

Takada, M. (1986). The short-term effect of prolactin on the active Na transport system of the tadpole skin during metamorphosis. *Comp. Biochem. Physiol.*, **85A**, 755-759.

Tata, J. R. (1966). Requirement for RNA and protein synthesis for induced regression of the tadpole tail in organ culture. *Develop. Biol.* **13**: 77-94.

Thurmond, W., Kloas, W. and Hanke, W. (1986). The distribution of interrenal stimulating activity in the brain of *Xenopus laevis. Gen. Comp. Endocrinol.* **63**, 117-124.

Tonon, M.-C., Burlet, A. Lauber, M., Cuet, P., Jegou, S., Gouteux, L., Ling, N., and Vaudry, H. (1985). Immunohistochemical localization and radioimmunoassay of corticotropin-releasing factor in the forebrain and hypophysis of the frog *Rana ridibunda. Neuroendocrinology* **40**, 109-119.

Vale, W., Rivier, C., and Brown, M. 1977. Regulatory peptides of the hypothalamus. *Annu. Rev. Physiol.* **39**, 473-527.

Verhaert, P., Marivuet, S., Vandesande, F., and De Loof, A. (1984). Localization of CRF immunoreactivity in the central nervous system of three vertebrate and one insect species. *Cell Tissue Res.* **238**, 49-53.

Wald, G. (1981). Metamorphosis: An overview. In "Metamorphosis" (L.I. Gilbert and E. Frieden, eds.), pp. 1-39. Plenum Press, New York.

Wells, D. E. and Moser, C. R. (1982). Influence of vasopressin and oxytocin on neoteny in the adult Mexican axolotl *Ambystoma mexicanum. J. Exp. Zool.* **221**, 173-179.

Weil, M. R. 1986. Changes in plasma thyroxine levels during and after spontaneous metamorphosis in a natural population of the green frog, *Rana clamitans. Gen. Comp. Endocrinol.* **62**, 8-12.

White, B. A. and Nicoll, C. S. (1981). Hormonal control of amphibian metamorphosis In "Metamorphosis", (L. I. Gilbert and E. Frieden, eds.), pp. 363-396. Plenum Press, New York.

White, B. A. Lebovic, G. S., and Nicoll, C. S. (1981). Prolactin inhibits the induction of its own renal receptors in *Rana catesbeiana* tadpoles. *Gen. Comp. Endocrinol.* **43**, 30-38.

Wright, M. L., Jorey, S. T., Blanchard, L. S., and Baso, C. A. (1988). Effect of a light pulse during the dark on photoperiodic regulation of the rate of thyroxine-induced, spontaneous, and prolactin-inhibited metamorphosis in *Rana pipiens* tadpoles. *J. Exp. Zool.* in press.

Yamamoto, K., and Kikuyama, S. (1982a). Radioimmunnoassay of prolactin in plasma of bullfrog tadpoles. *Endocrinol. Japon.* **29**, 159-167.

Yamamato, K. and Kikuyama, S. (1982b). Effects of prolactin antiserum on growth and resorption of tadpole tail. *Endocrinol. Japon.* **29**,81-85.

Yamamoto, K., Kikuyama, S., and Yasumasu, I. (1979). Inhibition of thyroxine-induced resorption of tadpole tail by adenosine 3'5'-cyclic monophosphate. *Dev. Growth Differ.* **21**, 255-261.

Yamamoto K., Niinuma, K., and Kikuyama, S. (1986). Synthesis and storage of prolactin in the pituitary gland of bullfrog tadpoles during metamorphosis. *Gen. Comp. Endocrinol.* **62**, 247-253.

Yu, N. W., Hsu, C. Y., Ku, H. H., and Wang, H.C. (1985). The development of ACTH-like substance during tadpole metamorphosis. *Gen. Comp. Endocrinol.* **57**, 72-76.

II
Development and Aging of Reproductive Systems and Functions

6

AN INTRODUCTION TO DEVELOPMENT AND AGING OF REPRODUCTIVE SYSTEMS AND FUNCTIONS

Martin P. Schreibman and Colin G. Scanes

Department of Biology
Brooklyn College of the
City University of New York
Brooklyn, NY 11210

Department of Animal Sciences
Bartlett Hall - Cook College
Rutgers - The State University of New Jersey
New Brunswick, NJ 08903

Aging is a continuum that spans birth (or even earlier, as for example, during gestation) and death and consists of development, maturation and senescence as its major components. Biological aging is a common denominator for all living organisms. Its various morphological and physiological manifestations occur at every level of organization in plants and animals. Yet despite increasing interest and study, we know relatively little about this basic universal phenomenon. Many theories of aging have been presented and although all are attractive and worthy of consideration, not one has gained general acceptance. Essentially, the theories that have been advanced are related to cellular and molecular mechanisms. They include such phenomena as: - cell 'wear and tear', inability to grow and proliferate, autointoxication, malfunctional changes in basic support systems such as vascular, respiratory, nervous, immune, endocrine and connective, and the break down of genetic information transfer and impairment of synthetic processes (for detailed discussion see Finch and Schneider, 1985). In Section Two, we examine these phenomena as they are specifically related to the 'aging' of the reproductive system and associated processes.

From birth to death the nervous and endocrine systems have essential roles in the maintenance of most physiological processes, including reproduction. In recent years an increasing number of investigators have found it desirable to view aging as a biological 'master plan' with a clock for senescence that is genetically encoded and based in the brain. The clock relays its signals, by way of the endocrine and autonomic nervous systems. According to this concept, neurotransmitters in the brain begin to exert regulatory action on hypothalamic neurosecretory cells during development. A chain of regulatory events (a cascade phenomenon?) is initiated in the pituitary and target glands which ultimately leads to the elaboration and release of target gland hormones that cause further developmental changes in tissues and processes. It is further suggested that with aging, the control mechanisms of the cortex, hypothalamus, pituitary and target glands, and the various feedback mechanisms that regulate them, become structurally and physiologically impaired. The neurons in the brain centers, the postulated 'pace makers' of the biological clock for aging, begin to falter in their neurotransmitter/neuroendocrine activity. The resulting repercussions on nervous, endocrine, muscular and secretomotor functions could cause the basic characteristics of senescence, that is, the inability to maintain homeostasis or to adapt to stress situations (see discussion by Timiras in this volume). The exponental increase of mortality that occurs with age may indicate an additive effect from the demise of physiologically-based systems that exceeds the essential minimum of support systems required for survival.

The concept that aging occurs according to a genetically determined time-table that is based in, and expressed through, neuroendocrine systems is attractive in its simplicity. It is also physiologically sound, based on aging studies of the brain, neurotransmission and endocrine gland interaction.

Programs for development and senescence undoubtedly vary phylogenetically and therefore it is not conceivable at this point to draw major generalizations. As related to reproductive processes, this diversity can be illustrated by examples such as the relationship between duration of reproductive competence and total life span, number of matings per unit time, age at puberty, age at reproductive senescence, association of geographical location and reproduction (i.e., migration patterns and mating), specific substrate upon which mating occurs (i.e., aquatic or terrestrial), number of offspring generated, and the relationship between reproductive ability, gender and longevity. The extreme of the variation among animals is illustrated by the neuroendocrine-mediated post-spawning deaths of the octopus (Wodinsky, 1977) and the Pacific salmon (see discussion by Dickhoff in this volume). The initial studies which form the basis for neuroendocrinology concentrated on the structure and function of the neurosecretory cell. The field of neuroendocrinology although conceived in molluscs, has spread rapidly throughout the animal kingdom from insects to vertebrates and onto humans. Although the base of information has grown exponentially in recent years, we are still far from a true understanding of anything but the simplest processes. Consider the fundamental question, 'how is aging *per se* related to reproductive senescence, and vice versa'?

The section, which follows, concentrates on the development and aging of structural and functional relationships between the neuroendocrine and reproductive systems. It begins with an examination of the structure and function of the brain-pituitary-gonadal axis in cold- and warm-blooded vertebrates. It continues with the neuroendocrine regulation of the fetus and maternal-fetal interaction in mammals. A consideration of puberty and reproductive senescence precedes an analysis of the dramatic relationship between reproduction and death in certain fishes. The section is concluded by addressing the question, 'is senescence intrinsic to eukaryotic somatic cells'? In keeping with the theme of this book, the chapters which follow present the state of the art and hopefully, by integrating the information, present us new interpretations, new concepts and point to new directions for study.

Literature Cited

1. Finch, C.E. and Schneider, E.L. (1985) Editors. "Handbook of the Biology of Aging". Second Edition. 1025 pp. Van Nostrand Rheinhold Co., New York.
2. Wodinsky, J. (1977) Hormonal inhibition of feeding and death in *Octopus*: Control by optic gland secretion. Science. 198, 948-951.

7

THE BRAIN-PITUITARY-GONAD AXIS IN POIKILOTHERMS

Martin P. Schreibman
Henrietta Margolis-Nunno

Department of Biology
Brooklyn College of the City University of New York
Brooklyn, New York 11210

I. INTRODUCTION

There is a patent paucity of fundamental information available on the specific mechanisms that regulate the aging of neuroendocrine systems, and on basic physiological phenomena in general, that lead to the subsequent demise of the individual. The majority of studies related to the brain-pituitary-gonadal (BPG) axis have been restricted to mammals (see Adelman and Roth, 1982; Cristofalo et al., 1984; Harman and Talbert, 1985; Meites, 1983; vom Saal and Finch, 1988). Detailed discussions of the reproductive endocrinology of fishes, amphibians and reptiles have been presented recently (Norris and Jones, 1987; Lance, 1984; Licht, 1984; Lofts, 1984). However, little information is available on age-related changes in the BPG axis of poikilotherms.

Our investigations of this axis in the platyfish, *Xiphophorus maculatus* , probably constitute the most extensive longitudinal study of the BPG axis in poikilotherms. We feel compelled,

Development, Maturation, and Senescence of
Neuroendocrine Systems: A Comparative Approach

therefore, to call attention to this special model in order to stimulate thought and to provoke similar comparative studies that might provide information for extrapolation to other species. It is not the intent of this discourse to present the specific details of age-related changes in the BPG axis of cold-blooded vertebrates (although specific data will be cited; see also Schreibman, et al., 1987). Rather, we will:

1) present general trends that we have observed in our studies, and provide the so-called, "take home messages", (see section III)

2) provide an interpretation/analysis where we can, and

3) suggest directions for possible fruitful future research in other animal models.

The platyfish, a small viviparous freshwater teleost, has been studied since 1927 (Gordon, 1927). Variation in the time of onset of sexual maturation of male and female platyfish is due to genetic factors (Kallman and Schreibman, 1973; Kallman, et al., 1973; Kallman, 1975; Schreibman and Kallman, 1977). A sex-linked gene, *P*, controls the age at which sexual maturation occurs as a result of maturational changes in the BPG axis. Five *P* alleles (P^1 through P^5) have been identified in natural populations and laboratory stocks (Kallman and Borkoski, 1978), and depending upon genotype, puberty occurs at a specific predictable age between 8 and 104 weeks. There is no overlap in the timing of maturation when different genotypes are raised under identical conditions (Kallman and Schreibman, 1973; Kallman et al., 1973; Schreibman and Kallman, 1977). The various *P* factors are carried on the sex chromosomes (X and Y) and are closely linked to a number of genes for body pigmentation that serve as their phenotypic markers (Table I)

Table I The Association of P Alleles and Pigment Genes and Their Relationship to Age At Maturation

P ALLELE-PIGMENT GENE ASSOCIATION	AGE AT MATURATION (WKS)
NSp (P^5 P^1)	11
NN (P^5 P^5)	34 to 104
*SpSr (P^1 P^2)	10
*NSr (P^5 P^2)	30
SpSp (P^1 P^1)	8

* = male fish, others are females; Sp is spot-sided (accumulation of melanocytes in a spot arrangement on the body), Sr is stripe-sided (melanocytes arranged to give effect of "stripes" on the body) and N is Nigra [heavy accumulations large "blotches" of melanocytes over the surface of the body; N is inherited with CPo, the gene responsible for orange color of the caudal peduncle ("tail")].

Unlike other vertebrates, the platyfish presents us with a unique model system with which to study the interaction of genetic and environmental factors, and the effect of these factors on the senescence of integral parts of the nervous and endocrine systems, and on the general phenomenon of aging. By using known genetic stocks, some with large degrees of homozygosity, we can eliminate the variability of animal populations that plague the analysis of so many aging studies.

II. METHODS OF ANALYSIS

All fish studied are of known genotype and are derived from the stocks maintained in the Genetics Section (Dr. Klaus D. Kallman, Director) of the Osborn Laboratories of Marine Sciences of the New York Aquarium and from their descendants reared in the laboratories at Brooklyn College, CUNY.

Essentially, the experimental design involves an examination of early and late maturing males and females at specific periods between birth and 30 months of age, the average lifespan of platyfish. The two genetic crosses shown below yield both early and late maturing brothers and sisters in a single mating and they provided the majority of the animals used in our studies:

Cross I:

	Females		Males
P1	X-N X-Sp (P^5P^1)	x	X-N Y-Sr (P^5P^2)
F1	X-N X-Sp (P^5P^1)		X-Sp Y-Sr (P^1P^2)
	X-N X-N (P^5P^5)		X-N Y-Sr (P^5P^2)

Cross II:

	Females		Males
P1	X-N X-Sp (P^5P^1)	x	X-Sp Y-Sr (P^1P^2)
F1	X-Sp X-Sp (P^1P^1)		X-Sp Y-Sr (P^1P^2)
	X-N X-Sp (P^5P^1)		X-N Y-Sr (P^5P^2)

In crosses I and II the progeny are easily distinguished because of differences in their sex and body pigment patterns. The various P factors are closely linked to specific color genes that serve as phenotypic markers for the P locus and, thus, the time at which the fish become sexually mature. Table I lists P allele - Pigment gene associations and the age at maturation for the fish studied.

The specific procedures of the histological (Schreibman, 1964), cytochemical (Schreibman *et al*.; 1982), immunocytochemical (ICC) (Margolis-

Kazan and Schreibman, 1981) and morphometric (Leatherland, 1970; Schwanzel-Fekuda *et al.*, 1981; Halpern-Sebold *et al*. 1986) methods have been presented previously.

Metamorphosis of the anal fin into a gonopodium (the intromittent organ) occurs in six stages that are easily recognizable and therefore invaluable for ascertaining the state of gonadal maturation by simple gross inspection (Grobstein, 1948; Kallman and Schreibman, 1973). All male platyfish, regardless of genotype, enter into stage 1 of anal fin transformation at 5 weeks of age. Stage 2 coincides with the initiation of sexual maturation. At stage 6, several weeks later, fish have functional testes and are fully mature (cf. Kallman and Schreibman, 1973). Females follow a time table for maturation that is comparable to males of similar *P* allele constitution (Schreibman and Kallman, 1977).

III. THE "TAKE HOME MESSAGES"

ONE. THERE ARE DISTINCT CHANGES IN THE DISTRIBUTION AND RELATIVE QUANTITY OF IMMUNOREACTIVE (IR-) NEUROPEPTIDES, NEUROTRANSMITTERS, AND PITUITARY AND GONADAL HORMONES, AS WELL AS IN THE STRUCTURE OF THE CELLS THAT MANUFACTURE OR ARE THE TARGET ORGANS FOR THEM DURING THE LIFESPAN OF PLATYFISH. THE CHANGES ARE AGE-(PHYSIOLOGICAL AND/OR CHRONOLOGICAL), GENOTYPE- AND SEX-RELATED.

A brief description of the BPG axis in neonatal animals is presented here in order for the reader to better comprehend the neuroendocrinological events that occur at puberty.

Ir-gonadotropin releasing hormone (GnRH;reactive with antibody generated against synthetic mammalian hormone) can already be localized in the brains of five week old early maturing fish. (See discussion below.) Virtually nothing is known about the distribution of neurotransmitters in this age group. These studies are now in progress.

The pituitary gland of neonatal fish is characterized by equal areas of neurohypophysial and adenohypophysial tissue. With the exception of the ventral gonadotropic zone in the caudal pars distalis (vCPD), all cell types characteristic of the mature gland are present in neonatal fish (Schreibman, 1964). Cells containing both ir-gonadotropin (GTH; reactive with antibody generated against the beta subunit of carp GTH) and ir-GnRH can be identified in one week old fish in the periodic acid-Schiff positive cells of the pars intermedia (PI) and in clusters of cells in the lateral pockets of the CPD; gonadotropes of the vCPD are represented by a few scattered isolated ir-cells (Schreibman *et al.*, 1982). The gonadotropes of the PI (GTH^p cells) are numerous at one week of age while those of the vCPD begin to appear just prior to the onset of sexual maturation and soon increase markedly in number, size and activity Halpern-Sebold *et al.*, 1983, 1984, 1986). There is a direct relation between the number of ir-GnRH containing neurons in the brain and the number of ir-GTH containing CPD and PI cells in the pituitary in both early and late maturers from one week to adulthood (Halpern-Sebold *et al.*, 1983, 1984, 1986). In adult early and late maturers, the number of ir-cells in the PI are similar, but late maturers have significantly fewer ir-cells in the CPD. In both genotypes, the initial appearance of ir-GnRH in the nucleus preopticus periventricularis (NPP) and nucleus lateralis tuberis (NLT) of the brain occurs concomitantly with the proliferation of gonadotropes in the CPD.

In the gonads, both 3-beta hydroxysteroid dehydrogenase (3BHSD) and glucose 6-phosphate dehydrogenase (G6PD) were found in the Leydig cells of the youngest males examined (3 days old). At a similar age in females, 3BHSD and G6PD were localized in the stroma of the ovaries which were comprised essentially of previtellogenic oogonia.

We believe that the early stages of testicular (up to the formation of spermatogonia) and ovarian (up to the oil droplet stage) development are regulated by gonadotropins which originate from the

GTH^P cells of the PI and, perhaps, from the cells in the lateral CPD. The gonadotropes which appear in the vCPD are presumably essential for the completion of gonadal development and for subsequent maintenance of gonadal structure and function.

TWO. GENETIC FACTORS DETERMINE THE TIMING AND THE SEQUENCE OF EVENTS IN THE BPG AXIS THAT CULMINATE IN PUBERTY. (See Introduction).

THREE. THE MATURATION OF THE REPRODUCTIVE SYSTEM IS CHARACTERIZED BY THE APPEARANCE OF IR-GnRH CONTAINING CELLS OR FIBERS IN THREE CENTERS OF THE BRAIN IN A SPECIFIC SEQUENCE WHOSE TIMING IS DETERMINED BY P ALLELE COMPOSITION. THE APPEARANCE OF IR-GnRH IN THESE CENTERS PRECEDES, AND IS ESSENTIAL FOR, THE ONTOGENY OF GONADOTROPES IN THE CPD AND THE SUBSEQUENT MATURATION OF THE GONADS.

The relationship of the three ir-GnRH containing brain centers and the pituitary gland and gonad is illustrated in Fig. 3. In addition, numerous ir-fibers emanate in all directions from the nucleus olfactoretinalis (NOR) and processes can be clearly delineated between this nucleus and the other nuclei, the pineal gland (habenular region), the olfactory tract and the optic nerve (Schreibman, et al., 1982; Halpern-Sebold and Schreibman, 1983; Munz, et al., 1981; Demski, 1984). Ir-GnRH appears in the NOR early in development in both early (at 5 wks) and late (9 wks) maturers. At the initiation of the maturation process, ir-GnRH appears in the NPP and very soon after in the NLT. This sequence of developmental changes in the brain which are related to the appearance of ir-GnRH in these three brain centers has been referred to by us as the "cascade effect". It is essential for the appearance of the gonadotropes in the vCPD and the subsequent maturation of the gonads. The attainment of sexual maturation (Stage 6) in males is revealed by the final transformation of the anal fin into the gonopodium. In addition, the P allele genotype, and thus the age of maturation, is indicated by the pigment patterns carried by the fish. The sequential development of the three ir-GnRH containing areas described above is directly related to the stage of sexual development and not to chronological age.

In the NOR of both early and late maturers, cellular and nuclear areas increase from one week (neonatal) to five weeks (stage 1 gonopodial development) of age. Ir-GnRH first appears in the NOR perikarya at 5 weeks of age in early maturers and 4 to 6 weeks later (still in stage 1), in late maturers. Ir-NOR perikarya are at their maximum number when they first appear, and this number, which is similar in both genotypes, remains constant into adulthood. There are, however, significant increases in the dimensions of NOR perikarya in both genotypes from stage 1 to stage 2, a period that is characterized by the appearance of ir-fibers between the NOR and the NPP and NLT, and a slight increase in the immunoreactive intensity of NOR perikarya in early maturers. These observations suggest increased cellular activity, a concept supported by cytological observations of increased basophilia, paler nuclei and more prominent nucleoli.

In the NPP of both genotypes, the dimensions of perikarya show substantial increases in size subsequent to those shown by the perikarya of the NOR. The NPP enlargement seen at stage 2 corresponds with the initial appearance of ir-NPP perikarya. By stage 6 the NPP of both genotypes attains its maximum number of ir-perikarya, with early maturers containing approximately fifty percent more cells than late maturing sibs.

In both genotypes, the dimensions of NLT perikarya follow a similar pattern of change, increasing up to gonopodial stage 2 and then decreasing from stage 2 to 6. Ir-GnRH fibers first appear in the NLT at stage 2 in both genotypes. Although ir-perikarya begin to appear shortly thereafter (at their maximum number) in early maturers, ir-GnRH cell bodies are never seen in the NLT of late maturers, whose NLT perikarya also display less basophilia than early maturers at stage 2. The intensity and number of ir-cell bodies in the NLT of early maturers decrease dramatically from stage 2 to stage 6. Although late maturers continue to lack NLT ir-perikarya, ir-fibers persist at stage 6.

Other essential differences also exist between the genotypes studied (see Halpern-Sebold et al., 1986). For example, in late maturers specific steps of the 'cascade effect' take place at similar developmental stages as in early maturers, but in older animals they require more time to complete. This delay in time creates significant differences in the cytometric, cytologic, and immunologic characteristics of the three brain regions in fish of the same age but different genotype.

In the pituitary gland the transformation from the immature to the mature state is characterized by the appearance of a zone of gonadotropes on the ventral border of the CPD. In the testis, 3BHSD- and glucose G6PD-positive cells become more numerous and display a more intense histochemical response at the periphery of the testes and in the stroma around the efferent duct system. This pattern of enzyme distribution is also characteristic of the fully developed testes where, in addition, the epithelium of the testicular main duct shows a positive G6PD reaction. In mature females 3BHSD- and G6PD-positive cells which are found in the stroma, increase in number and activity with ovarian development (Schreibman et al., 1982; unpublished with van den Hurk).

Eight month old animals are the youngest for which information has been reported on the distribution of neurotransmitters [i.e., serotonin (5HT), tyrosine hydroxylase (TH), arginine vasotocin (AVT) and somatostatin (SRIH)]. However, studies are now in progress which will determine the distribution of these substances at earlier developmental stages beginning at birth. Preliminary experiments suggest the presence of relatively high contents of ir-5HT and ir-dopamine in the pituitary gland of 2 week old fish (unpublished).

FOUR. WITH INCREASING AGE, DISTINCT FLUCTUATIONS IN THE DISTRIBUTION AND INTENSITY OF IMMUNOREACTIVE NEUROPEPTIDES, NEUROTRANSMITTERS AND PITUITARY HORMONES OCCUR AT SPECIFIC TIMES.

In the paragraphs that follow, a survey will be presented for the distribution of ir-GTH, -GnRH, -5HT -TH, and -AVT in platyfish from 8 to 30 months old. A more detailed discussion of age-related changes in these substances appears in a recent review (Schreibman et al., 1987). Where applicable, background information on what is known of these substances in other poikilotherms will also be given.

GTH: Ir-GTH in the vCPD reaches its maximum intensity at 24 months of age in early maturing males. By comparison, similar high levels are maintained at 18, 24 and 30 months in late maturing males. In the PI, high levels are found at 18, 24 and 30 months in early maturing males whereas in late maturing males immunoreactivity is highest at 18 months and then is lowered at 24 and 30 months of age. It is interesting and important to note that aging changes in levels of GTH immunoreactivity closely follow or occur at the same time that morphometric and cytologic changes are noted in the cells that contain them.

Studies on age-related changes in GTH immunoreactivity are unavailable for other poikilotherms, however, it is becoming increasingly apparent that two forms of gonadotropin occur in each class of vertebrates. These gonadotropins differ in their chemistry (relative amounts of carbohydrate), the cells from which they emanate and their roles in reproductive physiology. We have long postulated that in platyfish there are two types of gonadotropes, in different regions of the gland (the CPD and PI), which may produce two types of gonadotropins (based on tinctorial and ICC observations) and function at different developmental stages (one, at least, early in development the other at puberty and adulthood)

(Schreibman and Margolis-Kazan, 1979; cf. Schreibman 1986). The observations of Idler and his associates that in flounder the two types of GTHs differ in carbohydrate content are now being extended to other species (see Idler, 1987 and Sherwood, 1987). Most recently Swanson and her colleagues (1987) have confirmed the presence of two distinct GTHs in salmonids which are homologous to the LH and FSH of tetrapods. It is now established that two GTHs are present in at least two orders of amphibians (anurans and urodeles) and in crocodilian and chelonian reptiles (Licht and Porter 1987).

GnRH: With increasing age there are distinct fluctuations in the distribution and intensity of ir-GnRH with major changes detected at 18 months. In early maturing males the most intense immunoreactive responses occur in the NOR, NPP and NLT at 18 months of age. These do not occur in late maturing males where NOR ir-GnRH is most intense at 8 months (while they are still in stage 2) or early maturing females where immunoreactivity is consistently low at all ages between 8 and 30 months.

In the light of the different GTHs discussed above, it is interesting and provocative to note that there may also be more than one form of GnRH in vertebrates (Peter, 1986). Chromatographic and immunological data in at least 12 species of fishes have demonstrated that more than one form of GnRH occurs within a species (Sherwood, 1987). The concept of multiple forms of GnRH fits well with the many functions (other than its main role of synthesis and release of GTH) that have been demonstrated for it (see Demski 1984, Schreibman et al., 1988). The importance of GnRH as a humoral agent which executes the purported and demonstrated roles of the terminal nerve system (discussed in EIGHT below) has been demonstrated for virtually every vertebrate class. The notion that the gonads may be innervated by GnRH- containing nerve tracts (a motor function) has been suggested for such diverse animals as Amphioxus (Schreibman et al., 1986), fish (cf. Demski, 1984; Schreibman et al., 1988; Subhedar et al., 1987), and rats (Marchetti et al., 1985).

SEROTONIN (5HT): In early maturing platyfish, the localization of ir-5HT in the brain and pituitary gland of 5 to 8 month old sexually mature fish shows no sexual dimorphism (Margolis-Kazan, et al., 1985). In early maturing males, ir-5HT first appears in the brain in perikarya of the nucleus preopticus (NPO) at 18 months of age and its activity in the paraventricular organ (PVO) is most intense at 24 months. In late maturing males the pattern of distribution is similar, but the ir-response is significantly lower than seen in early maturing males. The most obvious differences between males of early and late genotypes is that in late maturers no ir-5HT is found in the NPO, and PVO activity is unchanged at 24 months and decreased at 30 months of age. Early maturing females show changes in 5HT immunoreactivity with increasing age that are similar to those observed in early maturing males at all ages. The major exception is that ir-5HT containing NPO perikarya are not seen in females. This is also true for late maturing males.

In the pituitary of early maturing males ir-5HT is present in its greatest intensity in vCPD cells at 18, 24 and 30 months, and is found in the PI at all ages beginning at 5 months of age. In late maturing males, ir-5HT is elevated only at 24 months and is not found at 30 months of age. In early maturing females ir-5HT in the vCPD and the PI is always much paler than is seen in males. Pale ir-5HT is first seen in the vCPD at 18 months and its intensity remains unchanged to 30 months of age in early maturing females. It has been suggested that the age-related changes in ir-5HT distribution are somehow related to the reproductive senescence of the platyfish (Margolis-Nunno et al., 1986).

In most vertebrate classes the presence of serotonin in nervous tissue has been identified by histochemical and biochemical methods, however, there is no information available on age-related changes in ir-5HT distribution in poikilotherms. The most common characteristic of monoaminergic systems in the hypothalamus of nonmammalian vertebrates is the presence of a group of cerebrospinal fluid-contacting serotonergic cells which form the paraventricular organ. These cells are numerous in anamniotes and have been documented in several species (Rana temporaria, Rao and Hartwig, 1974; Xenopus laevis, Terlou and Ploemacher, 1973; Carassius auratus, Kah and Chambolle, 1983; Yoshida et al., 1983) as well as in platyfish. The PVO is believed to be involved in the regulation of various endocrine events, including ovulation (cf. Parent, 1984).

Tyrosine hydroxylase (TH): TH is an enzyme necessary for catecholamine synthesis. In 12 month old early maturing male platyfish, ir-TH is localized in perikarya and processes of the NOR and the NPO. Paler immunoreactivity is found in perikarya in the wall of the third ventricle and in the PVO (Halpern-Sebold, et al., 1985). In the NOR, TH immunoreactivity is greatest at 12 months and lowest at 8 and 30 months; in the NPO it is at its highest levels between 12 and 24 months. In the PVO it is at low levels between 8 and 24 months and is absent at 30 months. In late maturing males low levels of ir-TH are found in the PVO between 8 and 24 months only; no ir-TH perikarya are localized in the NOR or NPO (unpublished).

In male and female platyfish of early maturing genotype, prolactin cells contain ir-TH at all ages. In addition, paler ir-TH can be seen in some cells of the PI of some fish at 12 and 18 months of age (Halpern-Sebold, et al., 1985; unpublished). There are no other reports available for age-related changes in ir-TH distribution in the brain of poikilotherms, however, ir-TH has been localized in "adult" goldfish (Yoshida, et al., 1983; Kah, et al., 1984), and Parent (1984) has reviewed catecholamine localizations in all vertebrate classes. Similarly, age-related changes in ir-TH in the pituitary have not been investigated in other cold-blooded vertebrates or, for that matter, any vertebrates.

Arginine Vasotocin (AVT): Ir-AVT has been identified in sexually mature platyfish in perikarya of the NOR (but not the NPP and NLT) and in the pars parvocellularis and pars magnocellularis of the NPO and in the neurohypophysis. Ir-AVT containing tracts, are seen between the NOR and the olfactory bulb and NPO, and between the NPO and the pituitary gland (Schreibman and Halpern, 1980; and unpublished). Studies of age-related changes in ir-AVT are now in progress.

FIVE. THE PROCESS OF SENESCENCE IN THE BPG AXIS IS SEXUALLY DIMORPHIC.

This is demonstrated by the following observations:

a) The age-related changes in the distribution of gonadotropes [in females ir-GTH containing cells begin to appear among the prolactin cells in the rostral pars distalis at 8 months; in males this occurs almost a year later and to a lesser degree (Margolis-Kazan and Schreibman, 1984)].

b) The differences noted between early maturing males and females in GnRH immunoreactivity (discussed above).

c) The sexual dimorphism of 5-HT immunoreactivity: in females there is no ir-5HT in the NPO and they have less ir-material in the pituitary as compared to males (see above).

d) The sexual dimorphism of somatostatin (SRIH) immunoreactivity: There is a decrease in ir-intensity in the NLT and nucleus anterior tuberis (NAT) and in the pituitary (PI) in older females but not males (Margolis-Nunno et al., 1987). [This statement is included here because of its relevance to sexual dimorphism and not for any apparent involvement of SRIH in the functioning of the BPG axis.]

e) The dimorphic age-related changes that occur in gonad morphology (e.g., ovarian changes take place later and are less apparent than those that take place in the testis) suggest that there may be differential mechanisms of aging in the testis and ovary.

SIX. THE INTERDIGITATING CYCLICAL EVENTS RELATED TO HORMONE AND NEUROTRANSMITTER SYNTHESIS AND RELEASE, WHICH OCCUR AT SPECIFIC TIMES DURING (PLATYFISH) "AGING", SUGGEST THE EXCHANGE OF INFORMATION AMONG COMPONENTS OF THE BPG AXIS.

For example, at 18 months:

a) there is an increase in ir-GnRH in the brain of early maturing males.

b) in the pituitary gland of both early and late maturing platyfish ir-GTH and -GnRH are elevated (at 18 and/or 24 months).

c) 5-HT first appears in the NPO and in the vCPD gonadotropes, and increases in ir-5HT ir-intensity are noted in GTH^{p} cells.

Our observations in platyfish at 18 months suggest that changing interplays of neurotransmitters occur during aging. It is, therefore, conceivable that these changes may be related, or may contribute, to reproductive senescence. It is possible that they are in some way translated into the "degenerative" transformations seen in the testes of 18 month old early maturers.

It certainly would appear that some sort of control mechanism is operative at this age. The aging changes observed between 12 and 18 months of age need to be examined at shorter time intervals in order to determine the specific sequence of events and the direct cause and effect relationships that lead to senescence. There is an obvious need to examine, too, early (neonatal to puberty) stages of development.

SEVEN. IN THE OLDEST ANIMALS EXAMINED, GONADS RETAIN THE CAPACITY FOR GAMETOGENESIS AND STEROID

HORMONE PRODUCTION. IN ADDITION, IR-GTH AND IR-GNRH ARE STILL PRESENT IN THE PITUITARY GLAND.

1. Testes

Histologically, the testes of young, sexually mature fish are composed of cysts, each containing cells representative of a particular stage of spermatogenesis. Cysts representing the more advanced stages of gametogenesis are closest to the center of the testis, in proximity to the main testicular ducts (Schreibman et al., 1982; van den Hurk, 1974).

The testes from males aged 18 months and older can be recognized on dissection by the appearance of white lobules in a somewhat translucent matrix and by an abnormally large number of melanophores on the surface of the gonad. All stages of spermatogenesis can be recognized in the testes. However, with advancing age beginning at about 18 months, the pattern of organization becomes distorted, especially in the cysts furthest from the testis periphery, and there is an increase in collagenous fibers and adipose tissue. In the oldest fish examined (4.2 yrs), proportionately more than the normal complement of spermatozeugmata (packets of mature sperm in almost perfect spheres) appear to be present. However, the older testes lack the definitive organization characteristic of the younger gonad, and frequently large clusters of loose sperm are seen. The testes of aged platyfish show no apparent differences in efferent duct columnar or main duct cuboidal epithelia or in the histological appearance of Leydig cells (Schreibman et al., 1983). The spermatozeugmata-laden ducts suggest a decrease in sperm release which may be related either to an age-dependent decline in the reproductive capacity of live-bearing poeciliids (Gerking, 1959), or to celibacy imposed by our experimental design (Schreibman et al., 1983).

Our observations that testis in aging platyfish continue to produce sperm despite connective tissue infiltration, correlate with observations in other teleosts (Poecilia reticulata, Woodhead and Ellett, 1969; Astyanax mexicanus, Rasquin and Hafter, 1951; bitterlings,

Haranghy et al., 1977). In aged Betta (Woodhead, 1974), however, the observed increase in connective tissue infiltration is accompanied by an almost total regression of the gonads and an absence of germinal tissue similar to that seen in hypophysectomized fish (Pickford and Atz, 1957).

Histological studies of male platyfish indicate that there is a general decrease in 3BHSD, 3HSD, G6PD and lactic dehydrogenase and an increase in acid phosphatase activity, suggesting hampered cell metabolic and steroidogenic activity, and cellular degeneration with increasing age (unpublished with R. van den Hurk).

Little is known about the aging of the testis in other poikilotherms. The endocrinology of reproduction in male reptiles has been recently summarized (Lance, 1984), however, no age-related changes are discussed in this review.

2. Ovary

Age-related changes in the gross morphology of the platyfish ovary are less consistent than those observed in the testis. It frequently appears as if the ovary is atrophied and there is an increase in the brown or bright yellow "bodies" observed on its surface (unpublished). With increasing age, yolky oocytes become depleted and those remaining become very hard. Nevertheless, yolky oocytes may still be observed even in the oldest females indicating that oogenesis persists. There are marked decreases in 3BHSD, 3HSD, and G6PD activity with increasing age, suggesting, as in testes, that metabolic and steroidogenic activity is reduced with age (unpublished with R. van den Hurk).

In many teleosts, resorption of yolky eggs, thickening of the ovarian wall, increased connective tissue and general atrophy are characteristics of aging ovaries (cf. Woodhead and Setlow, 1979). However, these observations do not hold for all teleosts since few age-related changes occur in ovaries of guppies and bitterlings (cf. Haranghy et al., 1977). Thus, the age dependent decrease in reproductive capacity found in some teleosts (Gerking, 1959) need not be an obligatory phenomenon.

Studies in reptiles have shown that there are significant differences in the numbers and quality of eggs produced in young, mid-age and old female alligators that appear to be related to their reproductive and behavioral biology (Ferguson, 1984, 1985). In addition, "reproductively senescent" females have few growing oocytes, no corpora lutea, resorbing ovarian follicles, and eggs in the oviduct that are undergoing resorption (cf. Ferguson, 1985). Although there is a limit on the number of oocytes that a female can produce, it appears that oogonial proliferation continues through much of adulthood (Ferguson, 1984). Hormonal regulation of the reptilian reproductive cycle requires further study.

EIGHT. OUR STUDIES, WHICH CORRELATE CHANGES IN THE OLFACTORY SYSTEM WITH EVENTS OCCURRING IN THE BPG AXIS DURING THE LIFESPAN OF PLATYFISH DEMONSTRATE THAT AN IMPORTANT, PRIMARY AXIS OF NEUROENDOCRINE INTERACTION EXISTS BETWEEN THE OLFACTORY AND REPRODUCTIVE SYSTEM (NOR).

Based on morphological (including histology, immunocytochemistry, horseradish peroxidase (HRP) tracing, transmission and scanning electron microscopy) and physiological studies, we find profound structural and functional modifications in the olfactory epithelium at specific stages of sexual maturation , independent of chronological age (cf. Schreibman *et al*., 1986, 1988). There are also differences in the levels of GnRH immunoreactivity in the brain that occur at these distinct stages of maturation. For example, the level of ir-GnRH present in fibers between the NOR and the olfactory bulb is significantly increased at the time of sexual maturation. Additionally, the olfactory epithelium undergoes what appears to be "degenerative modifications" at times when there are age-related changes occurring in the structure and function of the reproductive system (*i.e.*, at 18 to 24 months of age).

We have observed at the light microscope level that HRP is transported anterogradely from the olfactory epithelium to the NOR (Fig. 1). Ultrastructurally we see that these HRP-labeled processes make synaptic contact with the cells of the NOR (unpublished with R. Schild, P. Rao and S. Jahangir). The patent structural and functional link between the olfactory and reproductive systems, as demonstrated by our studies and those of other investigators, needs to be clarified by further investigation. It is apparent that we need to learn more about the specific mechanisms of interaction between the NOR (nervus terminalis) and the rest of the neuroendocrine and genetic systems. One of several basic questions which demands further clarification concerns the nature of the direction of information transfer between the NOR and structures with which it is most closely associated. Fig. 1 presents a tentative model and suggests the following generalizations:

1. Information transfer may occur into and out of the NOR.

2. This information may be conveyed by several chemical means, that is, by neurotransmitters and/or neuropeptides.

3. Information transfer may occur using different structural pathways. The HRP-labeled processes may not be the same ones that contain LHRH. Thus, the nature and direction of information transfer may be accomplished by different neuronal pathways.

4. Fig. 1 also indicates that neuropeptide-containing processes may innervate the gonads and, thus, neurons of the NOR may have motor as well as sensory functions (for further discussion see Schreibman *et al.*, 1986, 1988).

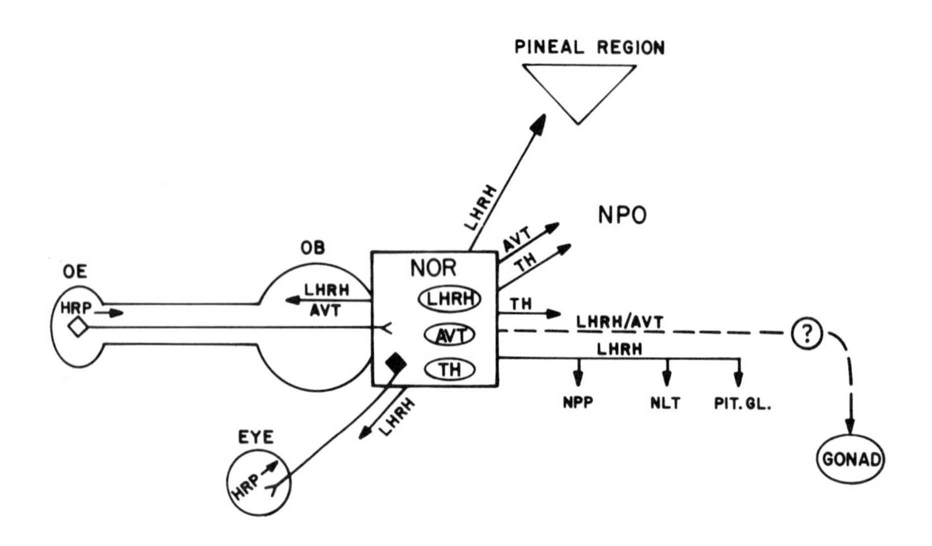

FIGURE 1. This scheme suggests directions for "information" transfer into and out of the NOR. It is based on data derived from histological, immunocytochemical, nerve tract tracing (HRP) and ultrastructural analyses of platyfish during normal development or experimental manipulation. The arrows for TH (tyrosine hydroxylase), AVT (arginine vasotocin) and LHRH (luteinizing hormone releasing hormone) which leave the NOR are based on our localizations of ir-processes projecting from ir-perikarya in the NOR. HRP applied to disrupted olfactory epithelium (OE) is transported anterogradely (open diamond represents cell bodies) to synapse on NOR cells. When applied into the eye, HRP is transported retrogradely and labels NOR cell bodies (solid diamond). The dashed line with LHRH and AVT represents our hypothesis that the gonad is innervated by the NOR. (From Schreibman et al, 1988, with permission of the publisher.)

NINE. THERE APPEARS TO BE A DISTINCT RELATIONSHIP BETWEEN P ALLELE CONSTITUTION AND LONGEVITY.

Although this statement is based on incomplete (and perhaps biased) data we feel that it merits attention and discussion. We have found that male and female platyfish which live the longest carry the P^1 allele marked by the gene Sp (spot-sided) when compared to the other genotypes examined. Based on an analysis of 103 fish (11 broods) resulting from matings which yielded early and late maturing brothers and sisters and which lived more than 20 months, we find the following association between P allele composition and longevity. The data summarized in Fig. 2 suggest that when siblings of different P genotype are grown together, those that carry the P^1 (Sp) allele (i.e., allele for early maturity) have the longest maximum lifespan. These observations suggest that:

a) the P alleles are associated with the genetic determinants of longevity,

b) the P^1 (Sp) allele is, or serves as a marker for, the genes responsible for maximum lifespan,

(c) the presence of a functional reproductive system at an early age may be related to longevity since P^1 is also the allele responsible for early maturation. Our data suggest that fish which mature earlier and, therefore have a reproductive system that is functional for a longer duration, also live longer. It remains to be determined whether it is the age at maturation or the duration of reproductive function that is (are) the determining factor(s) for longer life.

It has been noted that, in general, pure Jamapa stocks (which always carry a P^1 allele) live longer in the laboratory than pure Belize stocks which do not carry P^1 alleles (K.D. Kallman, personal communication). Male Belize-Jamapa hybrids

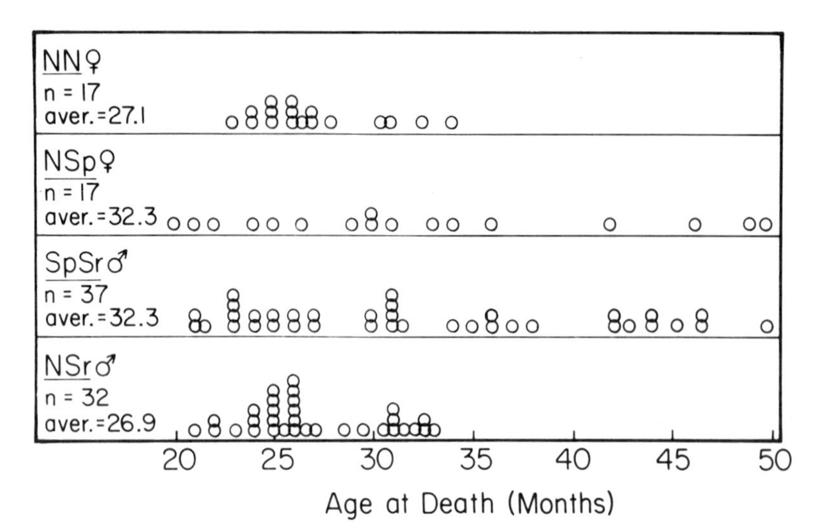

FIGURE 2. The data presented for the age at death is based on an analysis of 103 fish (11 broods) resulting from matings which yield early and late maturing siblings and which lived more than 20 months. The data suggest that siblings that carry the P_1 (Sp) allele have the longest maximum lifespan. See text for further analysis and discussion. (open circles, age at death for individual fish; n, number of fish; aver., average lifespan.)

(P^1P^2) do not differ in longevity from the males of pure Jamapa (P^1P^2) stocks (unpublished). It is also significant to note that in a study where fish were raised in isolation, those that lived the longest (up to 5 years) carried the P^1 (Sp) allele. It would appear that these observations support our hypothesis of a relationship between P^1 and longevity. Fig. 2 indicates that fish that live the longest possess P^1 or Sp alleles. P1 may also be marked by the gene for Sd (spotted dorsal; heavy concentration of melanocytes on the dorsal fin). Thus, it is essential to demonstrate whether P1 associated with Sp and P1 associated with Sd are equivalent in their relationships to longevity. The experimental approach that would be used to answer this question could also determine if the pigment genes themselves are associated with longevity.

TEN. GENETIC FACTORS DETERMINE THE TIMING AND SEQUENCE OF EVENTS IN THE BRAIN AND PITUITARY GLAND THAT CULMINATE IN PUBERTY. GENETIC REGULATION OF THE STRUCTURE AND FUNCTION OF THE BPG AXIS IS CARRIED INTO SENESCENCE.

The fundamental question persists; how, where, and in the case of senescence, when, is the language of the *P* gene for the timing of sexual maturation, and perhaps for longevity, translated into the neuroendocrine physiological actions that lead to the development, maturation and senescence of the reproductive system?

The role of genetic factors in determining the lifespan and the characteristics of the development and aging of the BPG axis of vertebrates is certainly important but remains to be clearly defined. That the patterns of aging and maximum lifespan are generally characteristic and consistent in most species from generation to generation lends additional credence to this notion. Suggestions as to the number of genes involved in the aging process has varied from the involvement of many genes to few (Cutter, 1975). Based on the discussions above, it is apparent that we suspect that few (perhaps even one) genes are associated with longevity in platyfish.

ELEVEN. EACH IR-GNRH-CONTAINING REGION OF THE BRAIN PLAYS A SIGNIFICANT AND PERHAPS UNIQUE ROLE DURING SPECIFIC PHASES OF THE PLATYFISH LIFESPAN.

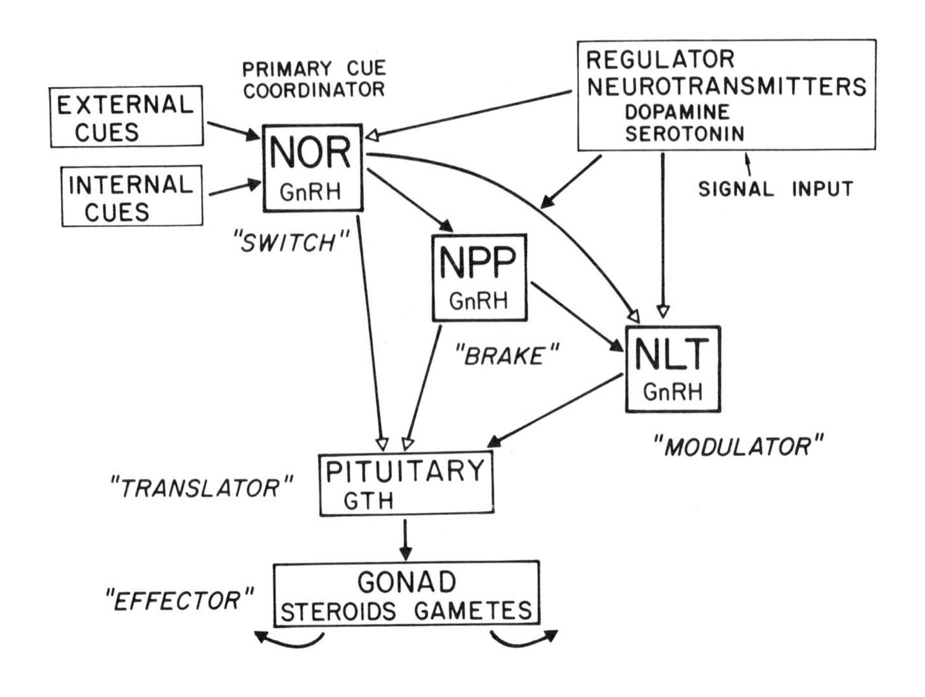

FIGURE 3. This diagram suggests a structural and functional relationship among the various components of the BPG axis, and select sites where a variety of cues may affect this axis. This diagram serves as a working model for investigating the sites and nature of P gene activity. The black arrowheads point to suspected primary, preferred pathways of interaction.

Fig. 3 provides a working model which indicates that it is possible for the P gene to exert its influence at any of the GnRH-containing centers of the brain and, indeed, at any level of the BPG axis.

The NOR is one of the most likely candidates for the location of the P gene "switch" for reproductive development. The NOR is the first region in the brain to contain ir-GnRH. The most significant role of the NOR may well be one of relaying a message to other brain centers which signals the beginning of a process of puberty, _i.e._, that it functions as an "initiator" of the cascade phenomenon. (Therefore, the terms "switch" and "initiator" may be interchangeable.) This suggestion is supported by observations that the NOR innervates the other two (NPP and NLT) ir-GnRH-containing brain centers as well as the pituitary, and that there are marked changes in NOR activity which are directly related to the initiation of activity in the NPP, NLT and pituitary. In addition, the association of the NOR with the eye, nose and pineal, and the positive correlation between the development of the nasal epithelium and the stage of sexual maturation, suggest a link between NOR-related sensory receptors and _P_ gene action. It is quite conceivable, too, that the ir-GnRH which is present in the NOR early in development may also serve to regulate the gonadotropic activity of the immature pituitary (Halpern-Sebold and Schreibman, 1983). Thus, in summary, the NOR:

(1) coordinates/translates the environmental cues that are fed into the BPG axis which are related to reproductive system structure and function,

(2) functions as (a) the initial source of GnRH (a "starter" for puberty?) (b) a "switch" to initiate the cascade effect for puberty, during development,

(3) functions when GTH levels are low, as in hypophysectomized and in immature animals,

(4) functions to regulate GTH activity in young platyfish by direct innervation of gonadotropes.

Another possibility is that the operational level of the _P_ gene lies in the NPP. The _P_ gene may serve to regulate inhibitors in the NPP which block the stimulatory action of the NOR. It has been suggested that the NPP is a center for a GTH release inhibitory factor (GRIF) (Peter and Paulencu, 1980, Chang and Peter, 1983a 1983b; Kah _et al_., 1984). It is believed that this inhibitor is dopamine (cf. Peter _et al_., 1986; Goos _et al_.,

1987). Dopamine acts as a GRIF in the spontaneous release of GTH as well as in modulating the action of GnRH at the level of the pituitary (cf. Peter, 1986). It may be that the *P* gene in platyfish is responsible for the regulation of this, and/or, other inhibitors of GnRH action which are present in the NPP. In this connection it is interesting to note that the NOR in late maturers is larger than in early maturing animals, suggesting that it may be working harder to overcome a "brake" factor located in the NPP. We have also postulated that the NPP may serve to initiate/regulate activity in the NLT. This is suggested by the observation that an anatomical link exists between them, and by the temporal sequence of activity in these two centers.

The NLT should, most certainly, not be excluded as a possible candidate for the site of *P* gene action. In our scheme of providing titles for these centers the NLT serves as a "modulator". The failure of ir-perikarya to appear in the NLT of late maturing fish between stage 2 and 6 may have considerable significance in the understanding of delayed maturation, and the lack of ir-GnRH-containing NLT perikarya may be related to the situation in homozygous Nigra fish (*NN*, P^5P^5) which frequently do not reach sexual maturity until they are more than one year old (Kallman and Borkoski, 1978) or to sterile hybrids of *Xiphophorus* which never become mature (Bao and Kallman, 1982). The lack of GnRH in the NLT perikarya of late maturers is the only striking difference between the brains of mature platyfish of the 2 genotypes, and therein may lie a clue to the delay in maturation and the expression of *P* gene activation. This lack of ir-GnRH in cells of the NLT, coupled with their cytologic and the cytometric profiles, suggests reduced GnRH production by these perikarya and may mean that the *P* gene expresses itself in a quantitative fashion. Late maturers also have fewer pars distalis gonadotropes and smaller gonads, and these observations may very well be a reflection of the postulated lower levels of GnRH within the NLT.

Thus, (1) the NLT is the ir-GnRH-containing brain nucleus which displays the greatest age-, sex-, and genotype-related variability in ir-LHRH; i.e., early maturing males have ir-material in the NLT at all ages but levels in perikarya are highest at stage 2 and at 18 months, early maturing females have ir-material only in fibers and not in the perikarya after an age equivalent to stage 2, and ir-GnRH is never seen in the NLT of late maturing males and females. (2) The NLT is innervated by the PVO; it is, therefore, entirely reasonable to suggest that the NLT is directly under 5HT and dopamine control.

Finally, it is possible that the P gene also operates in a quantitative manner at the level of the pituitary gland by regulating its sensitivity to GnRH. This could be accomplished by genetic control of either the number of gonadotropes or the number of active GnRH receptor sites on GTH cells. This idea supports the suggestion of Bao and Kallman (1982) that the P locus appears to regulate the physiological function and fate of GnRH and not its production. In fact, the term "P" gene was originally suggested because its action was thought to be at the level of the pituitary gland (Kallman and Schreibman, 1973) based on the observation that the abrupt development of the gonadotropic zone in the CPD coincided with the initiation of puberty. Indeed a direct correlation has been found between the number of ir-GnRH containing neurons in the brain (NPP and NLT) and the number of ir-GTH containing gonadotropes in both early and late maturers from one week to adulthood (Halpern-Sebold et al., 1986). In our scheme (Fig. 1), the pituitary gland has been labeled the "translator" and the gonad, in its production of steroids, as the "effector". Thus, there is a chain of communication; humoral and structural, that regulates development, maturation, and senescence of the BPG axis. Into this chain are fed a variety of messages. We see that regulator neurotransmitters (e.g., dopamine, serotonin,) serve to influence the activity at these various control centers. In addition, a variety of cues, both external (e.g., light, color, temperature) and internal, are suspected of playing very important roles in affecting the interaction among the various components of the BPG axis. Indeed, Fig. 3

presents a working hypothesis with a great deal of flexibility. The model must be manipulated and subjected to change as newer information is gathered and assimilated. Once we know how the _P_ gene functions in regulating development, we may have a better idea of its role in the aging process. We may then be able to determine if the "puberty" and "longevity" factors are one in the same. The newer methods of dissecting, analyzing, and manipulating genetic material hold great promise for these studies.

IV. SUMMARY

We have discussed the changes that occur from birth to senescence in the neuroendocrine axis which regulates reproductive function in platyfish of known genetic makeup. The platyfish is a unique research model in that it permits the analysis of the interplay of environmental, genetic, and neuroendocrine components and their concomitant influences on the aging of the reproductive system. Whittier and Crews (1987) have stated that "short-term studies cannot incorporate long-term variation and thus may lead to incorrect interpretations of control processes", and that in longer-lived species, the effects of specific cues on reproduction may vary from year to year just as they vary between sexes, from season to season and among individuals. These statements are entirely justified by our observations of age-related changes in the structure of, and in the humoral agents produced by, the BPG axis.

The timing of reproduction of poikilotherms in nature, especially in seasonal breeders, is clearly determined by an assortment of environmental stimuli (temperature, light, water, food, social conditions, etc.). "Fishes respond to a wider variety of cues than any other group of vertebrates" (Whittier and Crews, 1987). The specific cues, whatever they may be, must be received, interpreted and converted into neuroendocrine language that will insure gamete production and sexual encounter. The platyfish model lends itself to the study to these associations.

The concept that aging occurs according to a genetically determined timetable that is expressed through the nervous and endocrine systems is attractive in its simplicity. The platyfish appears to have been custom-designed to study the exciting question proposed by Timiras (1978) "Does a genetically programmed brain endocrine master plan code for aging processes?". Future research will help us to determine whether a clock for aging resides in the genotype of the organism in the same way that one exists for puberty.

ACKNOWLEDGEMENTS

The research of our laboratory, which provided much of the basic information on platyfish reported in this paper, has been supported over the years by grants from the NSF (PCM77-15981), the NIH-NIA (AGO-1938), The City University of New York (PSC-CUNY), NATO (333/84) and BARD (I-772-84). We thank Dr. Seymour Holtzman, Brooklyn College, for his constructive criticism of the manuscript.

REFERENCES

Adelman, R. C., and Roth, G. S. (1982). "Endocrine and Neuroendocrine Mechanisms of Aging." (R. C. Adelman and G. S. Roth, eds.), CRC Press, Fla.

Bao, I. Y., and Kallman, K. D. (1982). Genetic control of the hypothalamopituitary axis and the effect of hybridization on sexual maturation (Xiphophorus, Pisces, Poeciliidae). J. Exper. Zool. 220, 297-309.

Chang, J. P., and Peter, R. E. (1983a). Effects of pimozide and des Gly10, [D-Ala6] luteinizing hormone-releasing hormone ethylamide on serum gonadotropin concentrations, germinal vesicle migration, and ovulation in female goldfish, Carassius auratus. Gen. Comp. Endocrinol. 52, 30-37.

Chang, J. P., and Peter, R. E. (1983b). Effects of dopamine on gonadotropin release in female goldfish, Carassius auratus. Neuroendocrinology 36, 351-357.

Cristofalo, V. J., Baker, G. T., Adelman, R. C., and Roberts, J. (1984). In "Altered Endocrine Status During Aging." (V. J. Cristofalo, G. T. Baker, R. C. Adelman and J. Roberts, eds.), 216pp. Alan R. Liss, Inc., New York.

Cutter, R. G. (1975). Evolution of human longevity and the genetic complexity governing aging rate. Proc. Nat. Acad. Sci. USA 72, 4664-4668.

Demski, L. S. (1984). The evolution of neuroanatomical substrates of reproductive behavior: sex steriod and LHRH-specific pathways including the terminal nerve. Amer. Zool. 24, 809-830.

Ferguson, M. W. J. (1984). Craniofacial Development in Alligator mississippiensis. In "The Structure, Development and Evolution of the Reptiles." (M. W.J. Ferguson, ed.), pp. 223-273. Academic Press, Orlando, Fla.

Ferguson, M. W. J. (1985) Reproductive biology and embryology of the crocodilians. In "Biology of the Reptilia." (C. Gans, F. Billet and P. F. A. Maderson, eds.), Vol. 14, pp. 329-491. John Wiley & Sons, New York.

Gerking, S. D. (1959). Physiological changes

accompanying ageing in fishes. *In* "The Lifespan of Animals (Ciba Foundation Colloquia on Ageing)." (G.E W. Wolstenholme and M. O´Connor, eds.), pp. 181-208. Little, Brown & Co., Boston.

Goos, H. J. Th., Joy, K. P., DeLeeuw, R., van Oordt, P. G. W. J., van Delft, A. M. L., and Gielen, J. Th. (1987). The effect of luteinizing hormone-releasing hormone analogue (LHRHa) in combination with different drugs with anti-dopamine and anti-serotonin properties on gonadotropin release in the African Catfish, *Clarius gariepinus*. Aquaculture 63, 143-150.

Gordon, M. (1927). The genetics of a viviparous top-minnow *Platypoecilus*; the inheritance of two kinds of melanophores. Genetics 12, 253-283.

Grobstein, D. (1948). Endocrine and developmental studies of gonopod differentiation in certain poeciliid fishes. I. The structure and development of the gonopod in *Platypoecilus maculatus*. Univ. of Calif. Publ. Zool. 47, 1-22.

Halpern-Sebold, L.R. (1984). The ontogeny and functional significance of luteinizing hormone releasing hormone (LHRH) containing centers of the brain of the freshwater teleost, *Xiphophorus maculatus*. 105pp. Doctoral Dissertation. City University of New York, New York.

Halpern-Sebold, L. R., Margolis-Kazan, H., Schreibman, M. P., and Joh, T. (1985). Immunoreactive tyrosine hydroxylase in the brain and pituitary gland of the platyfish. Proc. Soc. Exper. Biol. Med. 178, 486-489.

Halpern-Sebold, L. R., and Schreibman, M. P. (1983). Ontogeny of centers containing luteinizing hormone-releasing hormone in the brain of platyfish (*Xiphophorus maculatus*) as determined by immunocytochemistry. Cell Tissue Res. 229, 75-84.

Halpern-Sebold, L.R., Schreibman, M.P., and Margolis-Nunno, H. (1986). Differences between early- and late-maturing genotypes of the platyfish (*Xiphophorus maculatus*) in the morphometry of their immunoreactive luteinizing hormone releasing hormone-containing cells: A developmental study. J. Exper. Zool. 240, 245-257.

Haranghy, L., Penzes, L., Kerenyi, T., and Penzes, B. (1977). Sexual activity and ageing in the teleost fish, Rhodeus sericeus amarus. Exp. Gerontol. 12, 17-25.

Harman, S. M., and Talbert, G. B. (1985). Reproductive aging. In "Handbook of the Biology of Aging." (C. E. Finch and E. L. Schneider, eds.), pp. 457-510. Second Edition, Van Nostrand Reinhold, New York.

Idler, D. R., and So, Y. P. (1987). Carbohydrate-poor gonadotropins. Proc. Third Intl. Symp. on Reproductive Physiology of Fish. Memorial Univ. of Newfoundland, St. John´s., pp. 57-60.

Kah, O., and Chambolle, P. (1983). Serotonin in the brain of the goldfish, Carassius auratus: An immunocytochemical study. Cell Tissue Res. 234, 319-333.

Kah, O., Chambolle, P., Thibault, J., and Geffard, M. (1984). Existence of dopaminergic neurons in the preoptic region of the goldfish. Neuroscience Letters 48, 293-298.

Kallman, K. D. (1985). The Platyfish Xiphophorus maculatus. In "Handbook of Genetics." (R. C. King, ed.), pp. 81-132. Plenum Press, New York.

Kallman, K. D., and Borkoski, V. (1978). A sex-linked gene controlling the onset of sexual maturation in female and male platyfish (Xiphophorus maculatus), fecundity in females and adult size in males. Genetics 89, 79-119.

Kallman, K. D., and Schreibman, M. P. (1973). A sex-linked gene controlling gonadotrop differentiation and its significance in determining the age of sexual maturation and size of the platyfish, Xiphophorus maculatus. Gen. Comp. Endocrinol. 21, 287-304.

Kallman, K. D., Schreibman, M. P., and Borkoski, V. (1973). Genetic control of gonadotrop differentiation in the platyfish, Xiphophorus maculatus (Poeciliidae). Science 58, 678-680.

Lance, V. (1984). Endocrinology of reproduction in male reptiles. In "The Structure, Development and Evolution of Reptiles." (M. W. J. Ferguson, ed.), pp. 357-383. Academic Press, Orlando, Fla.

Leatherland, J. F. (1970). Histological investigation of pituitary homotransplants in the marine form (Trachurus) of the three-spine stickleback, Gasterosteus aculeatus L. Z.

Zellforsch. 104, 337-344.
Licht, P. (1984). Reptiles. In "Physiology of Reproduction." (G.E. Lamming, ed.), pp. 206-282. Churchill Livingston, London.
Licht, P., and Porter, D. (1987). Role of gonadotropin-releasing hormone in regulation of gonadotropin secretion from amphibian and reptilian pituitaries. In "Hormones and Reproduction in Fishes, Amphibians, and Reptiles." (D. O. Norris and R. E. Jones, eds.), pp. 62-85. Plenum, New York.
Lofts, B. (1984). Amphibians. In "Physiology of Reproduction." (G.E. Lamming, ed.), pp. 127-205. Churchill Livingston, London.
Marchetti, B., Cioni, M., and Scapagnini, U. (1985). Ovarian LHRH receptors increase following lesions of the major LHRH structures in the rat brain: Involvement of a direct neural pathway. Neuroendocrinology 41, 321-331.
Margolis-Kazan, H., and Schreibman, M. P. (1981). Cross-reactivity between human and fish pituitary hormones as demonstrated by immunocytochemistry. Cell Tissue Res. 221, 257-267.
Margolis-Kazan, H., and Schreibman, M. P. (1984). Sexually dimorphic, age-related changes in pituitary gonadotrop distribution. Mech. Ageing Devel. 24, 325-333.
Margolis-Nunno, H., Halpern-Sebold, L., and Schreibman, M. P. (1986). Immunocytochemical changes in serotonin in the forebrain and pituitary of aging fish. Neurobiol. Aging 7, 17-21.
Margolis-Nunno, H., Schreibman, M. P., and Halpern-Sebold, L. (1987). Sexually dimorphic age-related differences in the immunocytochemical distribution of somatostatin in the platyfish. Mech. Ageing and Devel. 41, 139-148.
Meites, J." (1983). "Neuroendocrinology of Aging." Plenum Press, New York.
Munz, H., Stumpf, W. E., and Jennes, L. (1981). LHRH systems in the brain of platyfish. Brain Res. 221, 1-13.
Norris, D. O., and Jones, R. E. (1987). "Hormones and Reproduction in Fishes, Amphibians and Reptiles." 613pp. Plenum Press, New York.
Parent, A. (1984). Functional anatomy and evolution

of monoaminergic systems. Amer. Zool. 24, 783-790.

Peter, R. E. (1986) Structure-activity studies on gonadotropin-releasing hormone in teleosts, amphibians, reptiles and mammals. In "Comparative Endocrinology: Developments and Directions." (C. L. Ralph, ed.), pp. 75-93. A.R. Liss, Inc., New York.

Peter, R. E., Chang, J. P., Nahorniak, C. S., Omeljaniuk, R. J., Sokolowska, M., Shih, S. H., and Billard, R. (1986). Interactions of catecholamines and GnRH in regulation of gonadotrophin secretion in teleost fish. Rec. Prog. Horm. Res. 42, 513-548.

Peter, R. E., and Paulencu, C. R. (1980). Involvement of the preoptic region in gonadotropin release-inhibition in goldfish, Carassius auratus. Neuroendocrinology 31, 133-141.

Pickford, G. E., and Atz, J. W. (1957). "The Physiology of the Pituitary Gland in Fishes." New York Zoological Society, New York.

Prasada Rao, P. D., and Hartwig, H. G. (1974). Monoaminergic tracts of the diencephalon and innervation of the pars intermedia in Rana temporaria: A fluorescence and microspectrofluorometric study. Cell Tissue Res. 151, 1-26.

Rasquin, P., and Hafter, E. (1951). Age changes in the testes in the teleost, Astyanax mexicanus. J. Morphol. 89, 397-408.

Samuel, D., Algeri, S., Gershon, S., and Grimm, V. (1983). "Aging of the Brain." Raven Press, New York.

Schreibman, M. P. (1964). Studies on the pituitary gland of Xiphophorus maculatus (the platyfish). Zoologica 49, 217-243.

Schreibman, M. P. (1986). Pituitary gland morphology. In "Vertebrate Endocrinology: Fundamentals and Biomedical Implications." (P. K. T. Pang and M. P. Schreibman, eds.), pp. 11-55. Academic Press, New York.

Schreibman, M. P., Demski, L. S., and Margolis-Nunno, H. (1986). Immunoreactive LHRH in the "brain" of Amphioxus. Amer. Zool. 26, 30A.

Schreibman, M.P. and Halpern L.R. (1980). The demonstration of neurophysin and arginine vasotocin by immunocytochemical methods in the brain and pituitary gland of a teleost fish. Gen. Comp. Endocrinol. 40, 1-7.

Schreibman, M. P., and Kallman, K. D. (1977). The genetic control of the pituitary-gonadal axis in the platyfish, _Xiphophorus maculatus_. J. Exper. Zool. 200, 277-294.

Schreibman, M. P., and Margolis-Kazan, H. (1979). Immunocytochemical localization of gonadotropin, its subunits and thyrotropin in the teleost, _Xiphophorus maculatus_. Gen. Comp. Endocrinol. 39, 467-74.

Schreibman, M. P., Margolis-Kazan, H., Bloom, J. L., and Kallman, K. D. (1983). Continued reproductive potential in aging platyfish as demonstrated by the persistence of gonadotropin, luteinizing hormone releasing hormone and spermatogenesis. Mech. Ageing Devel. 22, 105-112.

Schreibman, M. P., and Margolis-Nunno, H. (1987). Reproductive biology of the terminal nerve (nucleus olfactoretinalis) and other LHRH pathways in teleost fish. _In_ "The Terminal Nerve (Nervus terminals) Structure, Function and Evolution." (L. S. Demski and M. Schwanzel-Fukuda, eds.), 519:60-68. N.Y. Acad. Sci., New York.

Schreibman, M. P., Margolis-Nunno, H., and Halpern-Sebold, L. (1986). The structural and functional relationships between olfactory and reproductive systems from birth to old age in fish. _In_ "Chemical Signals in Vertebrates." (D. Duval, D. Muller-Schwarze and R. M. Silverstein, eds.), pp. 155-172. Plenum Press, New York.

Schreibman, M. P., Margolis-Nunno, H., and Halpern- Sebold, L. (1987). Aging in the neuroendocrine system. _In_ "Hormones and Reproduction in Fishes, Amphibians, and Reptiles." (D. O. Norris and R. E. Jones, eds.), pp. 563-84. Plenum Press, New York.

Schwanzel-Fukuda, M., Robinson, J. A., and Silverman, A. J. (1981). The fetal development of the luteinizing hormone-releasing hormone (LHRH) neuronal systems of the guinea pig brain. Brain Res. Bull. 7, 293-315.

Sherwood, N. (1987). Gonadotropin-releasing hormone in fishes. _In_ "Hormones and Reproduction in

Fishes, Amphibians, and Reptiles." (D. O. Norris and R. E. Jones, eds.), pp. 31-60. Plenum Press, New York.

Subhedar, N., Rama Krishna, N. S., and Deshmukh, M. K. (1987). The response of nucleus preopticus neurosecretory cells to ovarian pressure in the teleost, Clarias batrachus (Linn.). Gen. Comp. Endocrinol. 68, 357-368.

Swanson, P., Suzuki, K. and Kawauchi, H. (1987). Isolation and biochemical characterization of two distinct pituitary gonadotropins from Coho Salmon, Oncorhyncus kisutch. Amer. Zool. 27, 79A.

Terlou, M., and Ploemacher, R. E. (1973). The distribution of monoamines in the tel-, di- and mesencephalon of Xenopus laevis tadpoles, with special reference to the hypothalamo-hypophysial system. Z. Zellforsch. 137, 521-540.

Timiras, P. (1978). Biological perspectives in aging: In search of a masterplan. Amer. Sci. 66, 605-613.

van den Hurk, R. (1974). Steroidogenesis in the testes and gonadotropic activity in the pituitary during postnatal development of the black molly, (Molliensia latipinna). Proc. Kon. Ned. Akad. Wetensch. Ser. C 77, 193-200.

vom Saal, F. S., and Finch, C. E. (1988). Reproductive senescence: Phenomena and mechanisms in mammals and selected mammals. In "The Physiology of Reproduction." (E. Knobil and J. D. Neill, eds.), pp. 2351-2413. Raven Press, New York.

Whittier, J. M., and Crews, D. (1987). Seasonal reproduction; patterns and control. In "Hormones and reproduction in fishes, amphibians, and reptiles." (D. O. Norris and R. E. Jones, eds.), pp. 385-410. Plenum Press, New York.

Woodhead, A. D. (1974). Ageing changes in the siamese fighting fish, Betta Splendens I. The testis. Exp. Geront. 9, 75-81.

Woodhead, A. D., and Ellett, S. (1969). Endocrine aspects of ageing in the guppy, Lebistes reticulatus (Peters). III. The testis. Exp. Geront. 4,17-25.

Woodhead, A. D., and Setlow, R. B. (1979). Aging changes in the ovary of the amazon molly,

Poecilia formosa. Exper. Geront. 14, 205-209.
Yoshida, M., Nagatsu, I., Kawakami-Kondo, Y., Karasawa, N., Spatz, M., and Nagatsu, T. (1983). Monoaminergic neurons in the brain of goldfish as observed by immunohistochemical techniques. Experientia 39, 1171-1174.

8

THE BRAIN-PITUITARY-GONAD AXIS IN HOMEOTHERMS

MARY ANN OTTINGER

University of Maryland
College Park, Maryland 20742

I. INTRODUCTION

The reproductive endocrine system is composed of a myriad of neural and cellular elements that constantly change in a dynamic and interactive manner. Major constituents of the system are the hypothalamus, pituitary gland, and gonads which represent an axis about which the rest of the system is regulated. Taken individually, each part of the system has specific functions; together, they comprise an interactive system capable of responding to a variety of external and internal variables. The status and nature of the response varies with the age and condition of the individual. Once organized during embryonic development, the system remains relatively quiescent until the initiation of sexual maturation. At this time, the brain-pituitary-gonad axis (BPG axis) shifts in response and increases hormone synthesis and secretion. Concentrations of gonadal hormones rise and other gonadal functions involved with gamete formation increase to adult levels. Many extra-hypothalamic areas of the brain, such as the pineal gland and limbic system modulate the reproductive system. With age, this whole system becomes altered in the nature, timing and level of response. The basis for and sequence of events in aging are unclear and there is considerable species variability. Even with favorable environmental circumstances, there are essential elements of the system that begin to break down with age. Among them, receptor systems, neuroendocrine and neuropeptide regulators, cellular responses, and enzyme systems. Another parameter critical to reproductive success is the expression of reproductive behavior. Courtship and mating behaviors are hormone

Development, Maturation, and Senescence of
Neuroendocrine Systems: A Comparative Approach

dependent and will not occur unless there are sufficient circulating concentrations of gonadal steroids. Steroids induce the perinatal sexual differentiation of the brain, which then predisposes the animal for later activation of male or female behavior. During sexual maturation, increasing testosterone levels stimulate the expression of reproductive behavior. Sex specific endocrine and morphological characters of reproduction change in synchrony with behavioral changes. Social interactions and environmental conditions alter behavioral response via neuroendocrine systems and with age behavioral responses falter, presumably due to neuroendocrine changes.

It is the purpose of this paper to review research relevant to the function of the BPG axis during ontogeny and aging in homeotherms. This review is not intended to be exhaustive, but rather will focus on neuroendocrine regulation of gonadotropin release and gonadal response with emphasis on the dynamics of the system. Successful reproduction requires appropriate behavioral responses which are regulated via mechanisms that also control endocrine processes. The BPG axis will be considered at the cellular and systemic levels, with hope that some of the dynamics dictating alterations in the BPG axis in mammals and birds will be clarified in spite of the functional complexity of the system. Basic function of the BPG axis and specific differences between mammals and birds will be discussed.

II. DYNAMICS OF THE SYSTEM

A. The Hypothalamus

Areas of the hypothalamus regulate a multitude of physiological functions including reproduction, water balance, appetite control, body temperature, metabolic level, response, aggressive and reproductive behavior and growth. Luteinizing hormone releasing hormone (LHRH) and follicle stimulating hormone releasing hormone (FSHRH) are produced by the hypothalamus and are critical for pituitary secretion of luteinizing hormone (LH) and follicle stimulating hormone (FSH) with subsequent stimulation of gonadal function. In mammals, hypothalamic areas important for secretion of LHRH and reproductive behavior, include the medial preoptic area, suprachiasmatic nucleus, median eminence and ventromedial nucleus (Gray et al., 1978; Selmanoff et al., 1980). These areas are responsive to steroid feedback and to a number of other hormonal and peptide factors, such as prolactin, thyroid hormone, adrenal hormones, opiates, and endorphin. The preovulatory LH surge in mammalian females is stimulated via a positive feedback from high concentrations of estradiol secreted by the tertiary follicle. This sort of cyclic change is not seen in the male, although both sexes have cyclic changes in hormone levels on a daily and often seasonal basis. In mammals, both LHRH and FSHRH are released in the median eminence and transported to their target cells in the adenohypophysis via long or short blood vessels to stimulate synthesis and secretion of LH and FSH.

The system operates in a similar manner in birds. However, there are 2 chemically distinct forms of LHRH, (Gln^8)-LHRH, termed LHRH I (King and Millar, 1982) and (His^5, Try^7, Tyr^8)-LHRH, termed LHRH II (Miyamoto et al., 1984), both different from mammalian LHRH. Both forms are active, with LHRH II more potent in a dispersed pituitary cell preparation (Connally and Callard, 1987). In addition, there are other peptides, notably vasoactive intestinal polypeptide (VIP), that are extremely potent in releasing LH from pituitary cells in vitro (Fehrer et al., 1987). Therefore, the actual control of LH secretion in the bird appears to depend on several hormonal regulators rather than a single LHRH(for review, see Scanes et al., 1984). The localization of the LHRH neurons have been carefully mapped in the domestic hen (Sterling and Sharp, 1982). The LHRH neurons were found to be localized in areas which sequester steroids (Arnold et al., 1976; Kim et al., 1978). Implantation of steroids or lesions in areas such as the preoptic , confirm their role in reproductive endocrine function (for review, see Scanes et al., 1984). Further, the LHRH content of specific hypothalamic areas changes with the ovulatory cycle (Johnson and Advis, 1985).

1. Modulation of the Hypothalamic Response

Feedback regulation of the BPG axis is generally thought of as a long loop negative feedback system between the gonadal steroids, testosterone, estradiol and progesterone with specific areas of the hypothalamus responsive to steroids. There is evidence that testosterone is metabolized to estradiol or 5Á-dihydrotestosterone (5Á-DHT), its active metabolites, for much of its action (Adkins-Regan et al., 1982; Balthazart et al., 1982). As mentioned earlier, LHRH neurons, localization of steroids and monoamine systems overlap in these hypothalamic areas (Heritage et al., 1980; Fuxe and Ljunggren, 1965; Oksche and Hartwig, 1980). Integration of internal and external information takes place here or at areas of the brain with projections to the hypothalamus, such as the limbic system, pineal or hippocampus.

Steroid and monoamine receptors have been studied

extensively because of their importance in feedback regulation and stimulation of behavior. Distribution of estrogen target cells are similar across species and overlap with androgen target cells (Stumpf and Sar, 1978). The response is closely related to receptor number and availability and varies with endocrine status (Singh and Muldoon, 1986), which is important for mediation of behavioral and endocrine effects (for review, see Etgen, 1984). Androgen receptors are concentrated in areas of the hypothalamus of the finch that are known to mediate behavioral responses (Harding et al., 1984). There is evidence for modulation of ß-adrenergic receptors by 5-HT (Stockmeier et al ., 1985). This has important ramifications for regulation of LHRH and for the interactions between neurochemical systems - i.e. does a physiological state in which the 5-HT system is activated then impinge on the adrenergic system? There is also noradrenergic modulation of estradiol receptors, as well as the converse, regulation of noradrenergic receptors by estradiol (Blaustein and Letcher, 1987; Weiland and Wise, 1987). This provides an elegant means by which catecholamines could affect estrogen target cells that function in LHRH secretion and/or behavioral response. The complexity of these interactions becomes almost indecipherable because small areas of the hypothalamus, often subareas within larger ones, show specific responses dependent on the stimulus. Therefore not only is it important to understand the endocrine status of the animal and to measure as precisely as possible, it is also essential to be aware of technical limitations when comparing receptor systems or neuroendocrine response.

It has been well established that monoamines regulate gonadotropin secretion. The turnover rate of NE decreased with testosterone implants and increased at puberty (Raum et al., 1980; Simpkins et al., 1980). Norepinephrine was found to stimulate pulsatile LH release via Á-ạdrenergic receptors (Weick, 1978; Gallo, 1980). Dopamine has had documented negative and positive effects on LH release, possibly due to endocrine status (Naumenko and Serova, 1976; Vijayan and McCann, 1978). Norepinephrine turnover decrease with exogenous estradiol (Renner et al., 1986). These data show not only a regulatory function of the monoamines, but also a modulatory effect, which may vary with endocrine status. Subtle effects of monoamines are possible through the diversity of receptor subtypes, such as the Á and the ß noradrenergic receptors (Snyder, 1984). There are also complex intercellular relationships between monoamines and peptides in the hypothalamus (Calas, 1985) which would permit integration of the effects of these systems.

Monoamines also regulate gonadotropins in birds. The preoptic area, in particular has catecholamine containing neurons as well as those containing LHRH and accumulate testosterone (Barfield et al., 1978; Davies and Follett, 1980). The critical role of the adrenergic systems for reproduction has been shown through pharmacology and turnover studies (Sharp, 1975; El Halawani et al., 1980), with LHRH stimulation via the Á-adrenergic receptors (Scanes et al., 1984).

B. The Pituitary Gland

There are anatomical differences in the mammalian and avian pituitary gland. The avian pituitary gland has caudal and cephalic portions, which synthesize and secrete hormones similar to those secreted by the mammalian pituitary (for review, see Scanes et al., 1984). In birds, isohormones exist for LH (Proudman, 1987). Hattori and coworkers (1986) have provided evidence for relatively autonomous regulation of LH and FSH in birds. There is species specificity that occurs for the protein and glycoprotein hormones across the species.

Feedback by the gonadal steroids has been shown to alter pituitary responsiveness. This appears to be through up or down regulation of the receptors (Savoy-Moore et al., 1980). Although this review will concentrate on the gonadotropins and the gonadal hormones, it should be noted that prolactin has powerful effects on the responses of the BPG axis in regulation of reproductive function. Further, the adrenal hormones and thyroid hormones also impact upon the reproductive endocrine system. Therefore if there are situations induced by stress, environment or other factors that involve these systems, the reproductive system will be affected, presumably through interactions of the regulatory systems in the hypothalamus that control all these systems.

C. The Gonads

Endocrine control of testicular function in mammalian and avian species is relatively similar. In the male, the most obvious differences between most mammals and birds is the internal location of the gonads in birds and the lack of many of the accessory sex glands found in mammals. The primary gonadal steroid is testosterone, which is generally metabolized to active forms at the cellular level with some of the metabolism occurring in the peripheral circulation (Ottinger and Mahlke, 1984). The active metabolites are estradiol, which is generally the biologically effective form in the central nervous system, and dihydrotestosterone, which has an active (5-Á) form and an inactive (5-ß) form and is effective primarily in peripheral target tissues. Leydig cells produce testicular steroids and are stimulated by LH; Sertoli cells produce inhibin and support the developing spermatozoa under the stimulation of FSH. There is a difference in the tubule arrangement, in that the mammalian testis is much more organized than the avian gonad (for more information, see Hafez, 1976; Nalbandov, 1976).

Ovarian function differs widely over species. In most mammals, estradiol is the most biologically potent estrogen and stimulates the preovulatory LHRH surge. The mammalian female may be an induced (by mating) or spontaneous ovulator, regulated by environmental factors particular to the species. There are two phases to the non-primate ovarian cycle: the follicular phase followed by ovulation and the luteal phase. The follicle produces the high concentration of estradiol necessary for the LH surge and progesterone is produced mainly in the luteal phase of the cycle by the corpora lutea (Nalbandov, 1976).

The avian female has one active ovary which develops in the embryo. Species may have one or more laying cycles a year, generally regulated by environment and lay either a set number of eggs (determinate) or until there are a number of eggs in the nest (indeterminate). In the latter case, removal of the egg will prevent the bird from initiating incubation behavior and stimulate continued egg production. There is no luteal phase of the cycle; both estradiol and progesterone are produced by the follicle. The source of estradiol is primarily from thecal cells; progesterone is produced by the granulosa cells (Huang et al., 1979; Bahr et al., 1983). There is a change in relative production of these steroids as the follicle matures in that the primary follicle produces primarily estradiol and the tertiary follicle yields progesterone (Huang et al., 1979). The preovulatory LH surge is induced by feedback effects of progesterone, rather than by estradiol (Johnson and van Tienhoven, 1984). The ovum is contained under one of the membranes surrounding the yolk, added in the ovary. Albumin and other proteins are added and into the uterus, termed the shell gland (Nalbandov, 1976).

D. Reproductive Behavior: Interactions With BPG Axis

There is a large literature on reproductive behavior and it regulation, which will not be reviewed here (see Balthazart et al., 1983). Generally, female behavior is stimulated by estrogen and progesterone and male behavior is stimulated by testosterone. Areas of the brain that mediate behavioral responses are mainly in the hypothalamus with input from other areas of the brain. In songbirds, neuroendocrine and morphological changes with behavior have been carefully assessed and neural pathways have been mapped, linking regulation of behavioral, neural and endocrine components of reproduction

(DeVoogd and Nottebohm, 1981; Williams and Nottebohm, 1985). The complexities of the neural basis of behavior still remain unclear. There is evidence for direct effects of monoaminergic systems on behavior. Male copulatory behavior in rats is modulated via Á adrenoceptors and other monoamine systems (Meyerson et al., 1979; Joyce et al., 1984; Clark et al., 1985). In birds, Barclay (1984) reported direct effects of pharmacological agents that affect the noradrenergic system on male courtship behavior in doves. We have found male courtship and mating in castrate male Japanese quail with stimulation of the Á and ß adrenergic receptors (Ottinger et al., 1987a; Rawlings and Ottinger, 1987). Conversely, antagonists to these same receptor systems suppressed behavior in castrate, testosterone implanted male quail.

Most reproductive behaviors will not occur without availability of sufficient circulating concentrations of the gonadal steroids. The hormone dependency of courtship and mating behavior has been shown experimentally many times by the loss of behavior with castration and the restoration of behavior with testosterone replacement therapy (for review, see Davidson, 1972; Ottinger, 1983; Ottinger et al., 1984). In the male, estradiol is thought to be the active metabolite of testosterone responsible for many of the behaviors (Adkins et al., 1980; Adkins-Regan and Garcia, 1986). There are specific behaviors known to be modulated by DHT, the other major testosterone metabolite (Erskine et al., 1985; Deviche et al., 1987).

Peptides such as LHRH, opiates and vasopressin have been linked to courtship and mating behavior (Boyd and Moore, 1985; Deviche, 1985; Dudley and Moss, 1985). The role of LHRH in behavior is an intriguing idea as it provide an additional mechanism for the synchronization of behavior with endocrine state. Dudley and coworkers (1983) found that behavior was activated by one half of the LHRH molecule. This implies that a duality in function may exist directed through distinct portions of the LHRH molecule. Therefore, experimental evidence does implicate the monoaminergic, LHRH and peptide systems in the modulation of behavior. Additional data with emerging techniques will help to verify the regulatory mechanisms and unravel intricacies of the interactions between the systems.

There is evidence for behavioral effects on the BPG axis. Brain estradiol and estradiol receptor binding increase in female voles exposed to males (Cohen-Parsons and Carter, 198-7). Many species show an endocrine response to behavior, such as increased LH and testosterone levels following mating (Cheng, 1983). Environmental factors also regulate the system (Wingfield, 1984). Finally, social order has powerful effects on the reproductive capability of the individual and may be seen in reduced testes weight and lower testosterone levels in subordinate males (Ottinger and Soares, 1982).

III. ORGANIZATION OF THE BRAIN-PITUITARY-GONAD AXIS

The BPG axis becomes functional in the mid to latter part of gestation in most homeotherms. Early differentiation is directed by expression of genetic sex via the histocompatibility antigen (HY antigen). There is a period of time early in development when the parts of the axis act independently. Hormones secreted by the embryonic gonads, and possibly the adrenal glands are responsible for later differentiation of the brain. This generally occurs perinatally, during the critical period. There is great species diversity in of the timing of and the endocrine events associated with sexual differentiation. However the outcome is organization of the system for later expression of male or female hormone patterns, morphology and behavior during maturation and in the adult.

A. Early Function of the BPG Axis

1. Mammals

At the outset, mammals and birds were thought to be exactly opposite in that the gonadal hormones of the heterogametic sex induces sexual differentiation. Although this oversimplifies the complexities of the processes involved, this is accurate for mammals. Testosterone, secreted mainly from the testes, acts to permanently organize target areas in the brain of the male mammal. The female develops essentially without hormonal exposure.

Both LH and FSH have been detected early in embryo development, prior to function of the BPG axis (for review, see Nemrod and Funkenstein, 1979). Receptors to both LH and FSH are found in the fetal testes after day 15 in the mouse (Warren et al., 1984). The appearance of these receptors coincide with increasing levels of gonadotropin and mirror the increasing steroidogenic responsiveness of the fetal gonad (Feldman and Bloch, 1978). Neuronal tracts containing GnRH are detected early in gestation in Rhesus monkey and rat embryos (Setalo et al., 1978; Goldsmith and Song, 1987). Similarly, LHRH containing neurons appear early in the rat (Setalo et al., 1978). The organizational impact of the gonadal hormones on the neuroendocrine system occurs perinatally for most species. Steroid effects on the brain have been studied at many levels. At the cellular level, *in vitro* application of gonadal steroid induced in sex differences in neuronal morphology (Raisman and Field, 1973). Similar morphological changes occur in the neuroendocrine system with steroid exposure later during maturation (Jones *et al*., 1985; Meisel and Pfaff, 1985). Studies *in vivo* verify the masculinizing effect of these steroids on behavior and endocrine response with perinatal exposure (Christensen and Gorski, 1978; Harlan *et al*., 1979; Jost, 1983). Changes in aromatase and reductase activity coincide with the timing of differentiation, suggesting that estradiol is the predominant active metabolite in differentiation (Tobet *et al*., 1985). There is evidence that the monoaminergic neurons may be a target of the steroid effect as evidenced by sex differences in monoamine levels and turnover (Harlan *et al*., 1979). These neuronal systems are found very early in development and therefore may be the basis of the perinatal neural change (Jacobson *et al*., 1985; Goffinet *et al*., 1986). Further Raum and Swerdloff (1981) demonstrated ß-adrenergic receptor stimulation prevented androgenization of the neonatal rat brain from exogenous testosterone, possibly due to inhibition of aromatization. Finally, disruption of the monoaminergic system in perinatal rats decreased plasma testosterone in adult male rats (Gupta *et al*., 1982; Hull *et al*., 1984).

2. Birds

By the second day of incubation, Rathke's pouch is present and by day 9, the infundibulum has formed and distinctive cell types are present (Tennyson *et al*., 1986). Morphologically, the hypothalamo-hypophyseal system develops early (Blahser and Henricks, 1982). Estradiol and testosterone are measurable in both male and female chicken and quail embryos and follow distinct profiles with higher levels of estradiol in the female, and elevated testosterone in males in the last half of incubation (Woods *et al*., 1977; Woods and Brazzill, 1981; Ottinger and Bakst, 1981; Abdel Nabi and Ottinger, 1986). Based on these hormone profiles and *in vitro* data (Haffen *et al*., 1975), sexual differentiation in the avian embryo depends on an appropriate ratio of testosterone to estradiol rather than a presence or absence of gonadal steroids. This is further supported by the effect of exogenous androgen or estrogen which demasculinizes the male and results in reduced fertility of the female (for review, see Adkins-Regan, 1981; Balthazart and Schumacher, 1984; Rissman *et al*., 1984). There is significant steroid content in the adrenal gland during development (Tanabe *et al*., 1979; Abdel Nabi and Ottinger, 1987) and it has been suggested that adrenal steroids are critical for differentiation. Finally, steroid induced sex differences in neuronal morphology have been found in specific areas of the brain which are important in reproduction (Arnold, 1980; Panzica et al., 1987). Monoaminergic neurons appear early in development are thought to orchestrate some of the metabolic and morphogenic processes in the embryo (for review, see Lauder and Krebs, 1984; Sarsa and Clement, 1987). Serotonergic systems were not found in the chick hypothalamus until relatively late in incubation (Wallace, 1985).

IV. ACTIVATION OF THE SYSTEM

The triggers that initiate sexual maturation are not understood. The traditional view has been that puberty requires physiological changes including 1. change in the hypothalamic "set point" for sensitivity to steroid feedback, 2. altered pituitary sensitivity for stimulation of LH and FSH production and 3. heightened gonadal response to promote steroidogenesis and gamete development. This means change in the response of the following: steroid, neurochemical and neuropeptide receptors, cellular enzymes systems, monoaminergic neurons, cellular machinery for synthesis and secretion of neuroendocrine products and sensory systems responsive to environmental factors.

Gonadal and pituitary hormones decline from the elevated levels associated with sexual differentiation and remain at basal levels until the initiation of sexual maturation. In man, testosterone levels fall by 7 months of age; FSH and LH begin to rise after 10 years of age (Ducharme et al., 1979; Giordana and Minuto, 1983), with a characteristic increase in nocturnal LH pulses prior to diurnal changes in LH (Judd, 1979). Testosterone increases in a similar manner in other mammals (Nazian, 1986). The mechanisms underlying these changes are perplexing, because the postnatal drop in testosterone is not associated with any change in the number of testicular LH receptors (Ducharme et al., 1979). However, later, with puberty, additional LH receptors are observed (Huhtaniemi et al., 1982). Part of the change in response over development is thought due to a transition between the fetal and adult Leydig cells.

Much evidence points to the CNS as the site for initiation of sexual maturation. With the change in pulsatile secretion of LH as a hallmark of puberty, it is likely that areas that time clock cyclic gonadotropin release are the site of initial change. This has been suggested with supporting data in hamsters by Smith and Stetson (1980). Precocious puberty can be arrested with an LHRH antagonist (Styne et al., 1985), giving further suggestion of the central role of this hormone. Finally, feedback mechanisms change as the testosterone levels rise (Matsumoto et al., 1985).

In birds, many of the same changes have been documented. Both LH and estradiol increase in the female chicken from 4 weeks of age to sexual maturity; LH and testosterone rise later in the male and peak at 34 weeks of age (Tanabe et al, 1981; Ottinger and Soares, unpublished data). Japanese quail maintained on a stimulatory photoperiod begin showing Leydig cell steroidogenesis at three weeks of age (Haffen et al., 1975; Ottinger and Bakst, 1981). Shortly thereafter, the testes undergo a period of rapid growth and spermatogenesis begins by five weeks of age (Ottinger and Brinkley, 1979a). Peripheral testosterone levels increase to adult levels preceded by rising LH levels (Hirano et al., 1978; Ottinger and Brinkley, 1979b). Courtship and then mating behavior initiate and increase in a correlated manner to the rising hormone levels (Ottinger and Brinkley, 1978). There is evidence for monoaminergic involvement (Duchala et al, 1985).

V. REPRODUCTIVE AGING: THE REVERSE OF MATURATION?

A. Mammals

The female mammal goes through sequential changes in cyclicity and endocrine status as her reproductive capability wanes. In the rat, there is a period of lengthening cycles, termed prolonged estrous (PE) followed by a state of constant estrous (CE). These changes have been well documented along with endocrine, morphological and behavioral parameters that accompany aging (for reviews, see Wise, 1983; Finch et al., 1984; Nelson and Felicio, 1985). There is change in chromatin, cytoplasmic factors, adenylate cyclase, and hormone receptors with age (Roth, 1979a; 1979b). There is debate as to the initial site of reproductive decline--is it breakdown of function at the level of the ovary, hypothalamus or pituitary gland? Some species, such as mongolian gerbils appear to have altered uterine function which interferes with pregnancy (Parkening et al., 1984). There is evidence for change in gonadal responsiveness and for increased incidence of pathology with age; however much of the data point to hypothalamic changes as critical. This is substantiated by recovery of cycling in rats by LHRH injections (for review, see Wise, 1983). Normal peripheral levels of LH and FSH were observed in old rats, however, with ovariectomy, LH and FSH levels do not become as elevated as in young animals (Wise and Ratner, 1980). Further, LHRH stimulated increase in LH is significantly lower in older females. Therefore, considerable attention has been paid to the neuroendocrine regulation of LHRH via estradiol receptors and monoaminergic systems.

There are age related changes in hypothalamic neuroendocrine systems in the mammalian female. Although there is no change in the distribution or hypothalamic content of LHRH, evidence points to a failure to secrete LHRH with age (Dorsa et al., 1984; Rubin et al., 1984; 1985), similar to anovulatory sterility (Leranth et al., 1986). Response to feedback from estradiol in middle aged rats was found to be slower than for young animals (Wise, 1984). This may explain for the lengthening of the cycle that occurs with age. There were measurable decreases in receptor concentration and in estradiol induced nuclear estradiol in the preoptic, medial basal and other areas of the hypothalamus (Wise and Parsons, 1984; Wise et al., 1984). Therefore, altered availability of estradiol receptors partially explain the decreased hypothalamic response.

The monoaminergic systems also become altered with age. Diurnal patterns of 5HT turnover in the suprachiasmatic area change in middle age implying altered timing of LHRH release (Wise et al., 1986). In addition, DA and 5HT receptors decline with age in some areas of the brain (Wong et al., 1984); there appears to be decreased biosynthesis of the DA receptor rather than membrane sequestration (Henry and Roth, 1986). The clearest regulatory relationship of a monoamine to LHRH production is that by NE, with Á1 receptors stimulating LHRH release and ß receptors are inhibitory to estradiol induced LH secretion (for review, see Weiland and Wise, 1987). In areas important for LHRH and behavior, the $ß_1$ receptors increase with little change in the $Á_1$ receptors, giving a net result of increased LHRH inhibition. In addition the aged female has less ability to up regulate receptors when NE decreases (Greenberg and Weiss, 1984). Finally, there is evidence for estradiol modulation of $Á_1$ receptors (Weiland and Wise, 1987)

The male mammal undergoes declining blood levels of the gonadal hormones coincident with decreased fertility (Pirke et al., 1978; Tsitouras et al., 1979; Kaler and Neaves, 1981; Warner et al., 1985). There is a loss in the episodic release of LH, but no change in LH binding in the testes associates with the lower production of testosterone (Bronson and Desjardins (1977; Steger et al., 1979). There are pathological changes in the testes which will explain some of the loss of gonadal function (Cohen et al., 1978). There is evidence of reduced NE and DA content of the hypothalamus and increased 5HT (Steger et al., 1985), implicating a declining NE system as a source of change. However, males that remained behaviorally active were found to retain higher endogenous testosterone than those that became inactive (Frankel, 1984). This also confirms that the reproductive decline is highly variable among individuals.

In birds, there are few data on the age related reproductive decline. A drop in fertility and hatchability with age has been well documented in poultry (Hays and Talmadge, 1949; Tomhave, 1956; Buvanendran, 1968; Ottinger and Soares, unpublished data). In the female there is a progressive decline in egg production, which is due to changing responsiveness of the hypothalamus (Williams and Sharp, 1978). Japanese quail have a similar drop in fertility with age (Woodard and Alplanalp, 1967; Ottinger et al., 1983). Sperm production decreases and testicular abnormalities, including tumor formation increases (Eroschenko et al., 1977; Gorham and Ottinger, 1986). The concentration of FSH receptors per gonad become reduced with age (Ottinger et al., 1987b). Plasma testosterone decreases well after a detectable drop in the frequency of mating behavior and in fertility. This is partially explained by the observed decrease in mating behavior as well as altered production of active testosterone metabolites (Ottinger et al., 1983; Balthazart et al., 1984; Ottinger and Balthazart, 1986). Changes in catecholamine content and turnover were found with age (Ottinger et al., 1985) Specific mechanisms underlying these changes must be further elucidated to understand their physiological basis.

In conclusion, there are many neuroendocrine parameters that contribute in a regulatory or modulatory way to reproduction. Often the neural basis of endocrine and behavioral regulation appears overlapping. The overall systems of mammals and birds are similar in a number of ways. However the hormonal control of sexual differentiation has rather different hormonal regulators. Additional research is critical elucidate the aging and neuroendocrine regulation of hormonal and behavioral components of reproduction.

ACKNOWLEDGEMENTS

This paper is Scientific Article #A-4765, Contribution #7768 of the Maryland Agricultural Experiment Station (Department of Poultry Science).

REFERENCES

Abdel Nabi, M.A., Ottinger, M.A., and Bakst, M.R. (1986). Poult. Sci. 65, 1.

Abdel Nabi, M.A., Pitts, S., and Ottinger, M.A. (1987). Poult. Sci. 66(1), 51.
Adkins-Regan, E. (1981). In "Neuroendocrinology of Reproduction: Physiology and Behavior" (N.T. Adler,ed),p.159-228. Plenum Press, New York.
Adkins, E.K., Boop, K K., Koutnik, D.L., Morris, J.B., and Pniewski, E.E. (1980). Phys. Behav. 24, 441-446.
Adkins-Regan,E. and Garcia,M. (1986). Phys. Behav.36,419-425.
Adkins-Regan, E., Pickett, P., and Koutnik, D. (1982). Horm. Behav. 16, 259-278.
Arnold, A. P. (1980). Amer. Scientist. 68, 165-173.
Arnold, A.P., Nottobohm, F., and Pfaff, D.W. (1976). J. Comp. Neurol. 165, 487-512.
Bahr, J.M., Wang, S.-C., Dial, O.K., and Calvo, F.O. (1983). Biol. Repr. 29, 326-334.
Balthazart, J. and Ottinger, M. A. (1984). J. of Endocrinol. 102, 77-81.
Balthazart, J. and Schumacher, M. (1984). Horm. Behav. 18, 287-297.
Balthazart, J., Marcelle, C., Sanna, P., and Schumacher, M. (1982). IRCS Med. Sci. 10, 267-268.
Balthazart, J., Prove, E., and Gilles, R. (1983). "Hormones and Behavior in Higher Vertebrates." (J. Balthazart, E, Prove, and R. Gilles, eds.), Springer Verlag, New York.
Balthazart, J., Turek, R., and Ottinger, M.A. (1984). Horm. Behav.18, 330-345.
Balthazart, J., Schumacher, M., and Ottinger, M. A. (1983). Gen. and Comp. Endocrinol. 51, 191-207.
Barclay, S.R., Johnson, A.L., and Cheng, M.-F. (1984). J. Ster. Biochem. 20(6B), 1528.
Barfield, R.J., Ronay, G., and Pfaff, D.W. (1978). Neuroendo. 26, 297-311.
Blahser, S. and Heinrichs, M. (1982). Cell Tissue Res. 223, 287-303.
Blaustein, J.D. and Letcher,B. (1987). Brain Res. 404, 51-57.
Boyd, S.K. and Moore, F.L. (1985). Horm. Beh. 19, 252-264.
Bronson, F.H. and Desjardins, C. (1977). Endocrinol. 101, 939-944.
Buvanendran, V. (1968). Poult. Sci. 47, 686-687.
Calas, A., (1985). Neurochem. Int. 7(6), 927-940.
Cheng, M.-F. (1983). In "Hormones and Behaviour in Higher Vertebrates." (J. Balthazart, E. Prove, and R. Gilles, eds.), p. 408-421. Springer-Verlag, Berlin.
Christensen, L.W., and Gorski, R.A. (1978). Brain Res. 146, 325-340.
Clark, J.T., Smith, E.R., and Davidson, J.M. (1985). Neuroendo. 41, 36-43.
Cohen, B.J., Anver, M.R., Ringler, D.H., and Adelman, R.C. (1978). Fed. Proc. 37, 2848-2850.

Cohen-Parsons, M. and Carter, C.S. (1987). Phys.Behav. 39, 309-314.
Connolly, P.B. and Callard, I.P. (1987). Biol. Repr. 36, 1238-1246.
Davidson, J. M. (1972) In "Hormones and Behavior" (S.Levine, ed),p. 64-104. Academic Press, New York.
Davies, D.T. and Follett, B.K. (1980). Gen. Comp. Endo. 40, 220-225.
Deviche, P. (1985). Pharm. Bioch. Behav. 22, 209-214.
Deviche, P., Delville, Y., and Balthazart, J. (1987). Brain Res. 21, 105-116.
DeVoogd, T. and Nottebohm, F. (1981). Science 214, 202-204.
Dorsa, D.M., Smith, E.R., and Davidson, J.M. (1984). Neurobiol. of Aging 5, 115-120.
Duchala, C. S., Ottinger, M. A., and Russek, E. (1984). Poult. Sci. 63, 1052-1060.
Ducharme, J.R., Catin-Savori, S., Tache, Y., Bourel, B., and Collu, R. (1979). J. Steroid Biochem. 11, 563-569.
Dudley, C. A., and Moss R. L. (1985). Pharmacol. Biochem. and Behav. 22, 967-972.
Dudley, C.A., Vale, W., Rivier, J., and Moss, R. L. (1983). Neuroendocr. 36, 486-488.
El Halawani, M.E., Burke, W.H., and Ogren, L.A. (1980). Gen. Comp. Endo. 41, 14-21.
Eroschenko, V.P., Wilson, W.O., and Siopes, T.D. (1977). J. Gerontol. 32(3), 279-285.
Erskine, M.S., MacLusky, N.J., and Baum, M.J. (1985). Biol. Repr. 33, 551-559.
Etgen, A. M. (1984). Horm. Beh. 18, 411-430.
Fehrer, S.C., Silsby, J.L., and El Halawani, M.E. (1987). Poult. Sci. 66(1), 98.
Feldman, S.C. and Block, E. (1978). Endo. 102, 999-1007.
Finch, C.E., Felicio, L.S., Mobbs, C.V., and Nelson, J.F. (1984). Endocr. Rev. 5, 467-497.
Frankel, A.I. (1984). Experim. Gerontol. 19, 345-348.
Fuxe, K. and Ljunggren, L. (1965). J. Comp. Neurol. 125, 355-382.
Gallo, R. V. (1980). Neuroendo. 31, 161-167.
Giordano, G. and Minuto, F. (1983). In "Recent Advances in Male Reproduction: Molecular Basis and Clinical Implications" (R. D'Agata, M.B. Lipsitt, P. Polosa, H.J. van der Molen, eds.), p. 287-298. Raven Press, New York.
Goffinet, A. M., Hemmendinger, L. M., and Caviness, V. S., Jr. (1986). Dev. Brain Res. 24, 187-191.
Goldsmith, P.C. and Song, T. (1987). J. of Comp. Neurol. 257,130-139.
Gorham, S.L. and Ottinger, M.A. (1986). Avian Dis. 30, 337-339.
Gray, G.D., Sodersten, P., Tallentire, D., and Davidson, J.

M. (1978). Neuroendo. 25, 174-179.
Greenberg, L.H. and Weiss, B. (1984). "Altered Endocrine Status During Aging" Alan Liss, Inc., New York, p. 57-70.
Gupta, C. and Jaffe, S.J. (1982). Science 216, 640-642.
Hafez, E.S.E. (1974). "Reproduction in Domestic Animals" Lea and Febiger, Philadelphia.
Haffen, K., Scheib, D., Guichard, A., and Cedard, L. (1975). Gen. Comp. Endo. 26, 70-78.
Harding, C.F., Walters, M., and Parsons, B. (1984). Brain Res. 306, 333-339.
Harlan, R.E., Gordon, J.H., and Gorski, R.A. (1979). Rev. of Neurosci. 4, 31-70.
Hattori, A., Ishii, S., and Wada, M. (1986). J. Endo. 108, 239-245.
Hays, F.A. and Talmadge, D.W. (1949). Agr. Res. 78, 285-290.
Henry, J.M. and Roth, G.S. (1986). J. of Gerontol. 2, 129-135.
Heritage, A.S., Stumpf, W.E., Sar, M., and Grant, L.D. (1980). Science. 207, 1377-1379.
Hirano, H., Nakamura, T., and Tanabe, Y. (1978). Jap. Poult. Sci. 15(5), 242-247.
Huang, E.S.-R., Kao, K.J., and Nalbandov, A.V. (1979). Biol. Repr. 20, 454-461.
Huhtaniemi, I.T., Nozu, K., Warren, D.W., Dufau, M.L., and Catt, K.J. (1982). Endocrinol. 111, 1711-1720.
Hull, E.M., Nishita, J.K., Bitran, D., and Dalterio, S. (1984). Science 224, 1011-1013.
Jacobson, C.D., Davis, F.C., and Gorski, R.A. (1985). Dev. Brain Res. 21, 7-18.
Johnson, A.L. and Advis, J.P. (1985). Biol. of Repr. 32, 813-819.
Johnson, A.L. and van Tienhoven, A. (1984). Biol Repr. 25, 153-161.
Jones, K.J., Pfaff, D.W., and McEwen, B.S. (1985). J. of Comp. Neurol. 239, 255-266.
Jost, A. (1983). Psychoneuroendo. 8(2), 183-193.
Joyce, J.N., Montero, E., van Hartesveldt, C. (1984). Pharm. Biochem. Behav. 21, 791-800.
Judd, H.L. (1979). In "Endocrine Rhythms" (D.T. Krieger, ed.), p. 299-324. Raven Press, New York.
Kaler, L.W. and Neaves, W.B. (1981). Endocrinol. 108, 712-719.
Kim, Y.S., Stumpf, W.E., Sar, M., and Martinez-Vargas, M.C. (1978). Amer. Zool. 18, 425-433.
King, S.A. and Millar, R.P. (1982). J. Biol. Chem. 257, 10729-10735.
Lauder, J. and Krebs, H. (1984). Adv. in Cell Neurob. 5, 3-51.
Leranth, C., Palkovits, M., MacLusky, N.J., Shanabrough, M., and Naftolin, F. (1986). Neuroendo. 43, 526-532.
Matsumoto, A.M., Karpas, A.E., Southworth, M.B., Dorsa, D.M.,and Bremner, W.J. (1986). Endocrinol. 119, 362-369.

Meisel, R.L. and Pfaff, D.W. (1985). Mol. Cell. Endo. 40, 159-166.

Miyamoto, K., Hasegawa, Y., Igarashi, M., Chinl, N., Sakakibara, S., Kangawa, K., and Matsuo, H. (1983). Life Sci. 32, 1341-47.

Myerson, B.J., Palis, A., and Sietnicks, A. (1979). In "Endocrine Control of Sexual Behavior" (C. Beyer, ed.) p.389-404. Raven Press, New York.

Nalbandov, A.V. (1976). "Reproductive Physiology of Mammals and Birds" W. H. Freeman and Co., San Francisco.

Naumenko, E.V. and Serova, L.I. (1976). Brain Res. 110, 537-545.

Nazian, S.J. (1986). J. Androl. 7, 49-54.

Nemrod, A. and Funkenstein, B. (1979). In "Development of responsiveness to steroid hormones" (A. Kaye and M. Kaye, eds.), p. 343-359. Pergamon Press, New York.

Nelson, J.F. and Felicio, L.S. (1985). Rev. Biol. Aging Res. 2, 251-314.

Oksche, A. and Hartwig, H.G. (1980). In "Avian Endocrinology" (A. Epple and M. Stetson, eds), Academic Press, New York

Ottinger, M.A. (1983). Poultry Sci. 62, 1690-1699.

Ottinger, M.A. (1983). In "Hormones and Behaviour in Higher Vertrebrates" (J. Balthazart, E. Prove, and R. Gilles, eds.), p. 351-367. Springer-Verlag, Berlin Heidelberg.

Ottinger, M.A., Adkins-Regan, E., Buntin, J., Cheng, M. F., DeVoogd, T., Harding, C., and Opel, H. (1984). J. of Experi. Zool. 232, 605-616.

Ottinger, M.A. and Bakst, M. R. (1981). Gen. and Comp. Endocrinol. 43, 170-177.

Ottinger, M.A. and Balthazart, J. (1986). Horm. and Beh. 20, 83-94.

Ottinger, M.A., Rawlings, C., Balthazart, J., and Rexroad, C.E., Jr. (1985). Poult. Sci. 64(1), 156.

Ottinger, M.A. and Brinkley, H.J., (1978). Horm. and Beh. 11, 175-182.

Ottinger, M.A. and Brinkley, H.J., (1979). Biol. of Repr. 20, 905-909.

Ottinger, M.A., Cortes, L., and Rawlings, C.S. (1987a). Am. Zool. 27(4), 154A.

Ottinger, M.A., Duchala, C.S., and Masson, M. (1983). Horm. and Beh. 17, 197-207.

Ottinger, M.A., Green, C., and Palmer, S.S. (1987b). Biol. Repr. 36(1), 158.

Ottinger, M.A. and Mahlke, K. (1984). Poultry Sci. 63, 1851-1854

Ottinger, M.A. and Soares, J.H.Jr. (1982). Proc. Conf Repr. Behav. Nashville, Tenn.

Panzica, G.C., Viglietti-Panzica, C., Calcagni, M., Anselmetti, G.C., Schumacher, M. and Balthazart, J.

(1986). Brain Res. 416, 59-68.
Parkening, T.A., Collins, T.J., and Smith, E.R. (1984). Experim. Gerontol. 19, 359-365.
Pirke, K.M., Geiss, M., and Sintermann, R. (1978). Acta Endocrinol. 89, 789-795.
Proudman, J.A. (1987). Poult. Sci. 66(1), 161.
Raeside, J.I., Robinson, D.J., and Naor, Z. (1984). Mol. Cell. Endo. 37, 191-196.
Raisman, G. and Field, P.M. (1973). Brain Res. 54, 1-29.
Raum, W.J., Glass, A.R., and Swerdloff, R.S. (1980). Endo. 106, 1253-1257.
Raum, W.J. and Swerdloff, R.S. (1981). Endo. 109, 273-278.
Rawlings, C.S. and Ottinger, M.A. (1986). Poult. Sci. 65, 170.
Rawlings, C.S. and Ottinger, M.A. (1987). Poult. Sci. 66, 163.
Renner, K.J., Allen, D.L., and Luine, V.N. (1986). Brain Res. Bull. 16, 469-475.
Rissman, E.F., Ascenzi, M., Johnson, P., and Adkins-Regan, E. (1984). J. Repr. Fert. 71, 411-417.
Roth, G.S. (1979a). Mech. of Ageing and Devel. 9, 497-514.
Roth, G.S. (1979b). Fed. Proc. 38, 1910-1914.
Rubin, B.S., Elkind-Hirsch, K., and Bridges, R.S. (1985). Neurobiol. of Aging 6, 309-315.
Sarsa, M. and Clement, S. (1987). J. Exp. Zool. 24, 181-190.
Savoy-Moore, R.T., Schwartz, N.B., Duncan, J.A., and Marshall, J. C. (1980). Sci. 209, 942-944.
Scanes, C.G., Stockell Hartree, A., and Cunningham, F.J. (1984). In"Physiolgy Biochemistry Domestic Fowl" 5,40-84
Selmanoff, M.K., Wise, P.M., and Barraclough, C.A. (1980). Brain Res. 192, 421-425.
Setalo, G., Antalicz, M., Saarossy, K., Arimura, A., Schally, A. V., and Flerka, B. (1978). Acta. Biol. Sci. Hung. 29, 285-290.
Sharp, P.J. (1975). Br. Poult. Sci. 16, 79-82.
Simpkins, J.W., Kalra, P.S., and Kalra, S.P. (1980). Endo. 107, 573-577.
Singh, P. and Muldoon, T.G. (1986). Endocrin. 118, 2355-2361.
Smith, S.G. III and Stetson, M.H. (1980). Endocrinol. 107, 1334-1337.
Snyder, S. (1984). Sci. 224, 22-31.
Steger, R.W., Peluso, J.J., Bruni, J.F., Hafez, E.S.E., and Meites, J. (1979). Endokrinologie 73, 1.
Steger, R.W., DePaolo, L.V., and Shepherd, A.M. (1985). Neurobiol. of Aging 6, 113-116.
Sterling, R.J. and Sharp, P.J. (1982). Cell Tiss. Res. 222, 283-298.
Stockmeier, C.A., Martino, A.M., and Kellar, K.J. (1985). Sci. 230, 323-325.
Stumpf, W.E. and Sar, M. (1978). Amer. Zool. 18, 435-445.

Styne, D.M., Harris, D.A., Egli, C.A., Conte, F.A., Kaplan, S.L., Rivier, J., Vale, W., and Grumbach, M.M. (1985). J. Clin. Endo. Metab. 61, 142-151.

Tanabe, Y., Nakamura, T. Fujioka, K., and Doi, O. (1979). Gen. Comp. Endo. 39, 26-33.

Tanabe, Y., Nakamura, T., Tanase, H., and Doi, O. (1981). Endo. Jap. 28(5), 605-613.

Tennyson, V.M., Nilaver, G., Hare-Yu, A., Valiguette, G., and Zimmerman, E.A. (1986). Cell Tissue Res. 243, 15-31.

Tobet, S.A., Shim, J.H., Osiecki, S.T., Baum, M.J., and Canick, J.A. (1985). Endo. 116, 1869-1877.

Tomhave, A.E. (1956). Poult. Sci. 35, 236-237.

Tsitouras, P.D., Kowatch, M.A., and Harman, S.M. (1979). Endocrinol. 105, 1400-1405.

Vijayan, E. and McCann, S. M. (1978). Neuroendo. 25, 221-235.

Wallace, J.A. (1985). J. of Comp. Neurol. 236, 443-453.

Warner, B.A., Dufau, M.L., and Santen, R.J. (1985). J. Clin. Endocrinol. Metab. 60, 263-268.

Warren, D.W., Huhtaniemi, I.T., Tapanainen, J., Dufau, M.L., and Catt, K.J. (1984). Endo. 114, 470-476.

Weick, R.F. (1978). Neuroendo. 26, 108-117.

Weiland, N.G. and Wise, P.M. (1986). Brain Res. 398, 305-312.

Weiland, N.G. and Wise, P.M. (1987). Endocrin. 121, 1751-58.

Wise, P.M. (1983). Rev. of Biol. Res. in Aging 1, 195-222.

Wise, P.M. (1984). Endocrinol. 115, 801-809.

Wise, P.M., Cohen, I.R., and Weiland, N.G. (1986). In "Integrative Neuroendocrinology: Molecular, Cellular and Clinical Aspects. (eds. McCann and Weiner). S. Karger, Basel, p. 80-91.

Wise, P.M., McEwen, Parsons, B., and Rainbow, T.C. (1984). Brain Res. 321, 119-126.

Wise, P.M. and Parsons, B. (1984). Endocrinol. 115, 810-816.

Wise, P.M. and Ratner, A. (1980). J. Gerontol. 35, 506-511.

Williams, H. and Nottebohm, F. (1985). Sci. 229, 279-282.

Williams, J.B. and Sharp, P.J. (1978). J. Reprod. Fert. 53, 141-146.

Wingfield, J.C. (1984). Gen. and Comp. Endocr. 56, 406-416.

Wong, D.F., Wagner, H.N., Dannals, R.F., Links, J.M., Frost, J.J., Ravert, H.T., Wilson, A.A., Rosenbaum, A.E., Gjedde, A., Douglas, K.H., Petronis, J.D., Folstein, M.F., Toung, J.K.T., Burns, H.D., and Kuhar, M.J. (1984). Science, 226, 1393-1396

Woodard, A.E. and Alplanalp, H. (1976). Poult. Sci. 46, 383-388.

Woods, J.E. and Brazzill, D.M. (1981). Gen. Comp. Endo. 44, 37-43.

Woods, J.E., Podczaski, E.S., Erton, L.H., Rutherford, J.E., and McCarter, C.F. (1977). Gen. Comp. Endo. 32, 390-394.

9

Neuroendocrine Regulation of the Mammalian Fetus[1]

Peter W. Nathanielsz, M.D., Ph.D.

Laboratory for Pregnancy and Newborn Research
Department of Physiology
College of Veterinary Medicine
Cornell University
Ithaca, NY 14853

I. INTRODUCTION

The hypothalamo-pituitary-adrenal axis plays a critical role in the maturation of several fetal organ systems. Of special importance in the newborn's ability to survive in the extrauterine environment is the action of cortisol to mature surface active lipids in the fetal lung (1). Secretion of cortisol from the fetal adrenal is central to the processes that result in the initiation of parturition (2). Although much information has been obtained on fetal neuroendocrine function in many experimental species (3), the largest single body of carefully controlled experimental data relating to the maturation of fetal neuroendocrine function has been obtained from in vivo work with the chronically instrumented fetal sheep preparation (4-6). The advantage of this preparation is that from 90 days of gestation (normal term 150 days) the fetus can be instrumented with vascular catheters at various sites in the circulation (e.g. the carotid artery, jugular vein, umbilical vein, femoral artery and femoral vein). Thus a wide variety of blood samples of different composition are

---

[1]This work was supported by a grant from the National Institutes of Health, HD-21350.

available. Catheters can also be placed in the amniotic and allantoic cavities to obtain samples of these fluids. Samples appropriate to the specific scientific question being asked can be obtained.

In addition to measurement of hormones and their metabolites in blood, classical isotopic clearance techniques can be performed to assess metabolic clearance rates, half lives, blood production rates for different hormones and biochemical pathways of synthesis and degradation. Isotope studies have been combined with in vitro tissue dispersion and enzyme investigations to demonstrate how fetal hormones modify fetal and placental endocrinology and metabolism. Thus, cortisol has been shown to modify the enzymatic pathways that convert progesterone to estrogens in the ovine placenta at term resulting in prostaglandin production and parturition (7).

These isotopic and in vitro methodologies can be combined with ablation of specific and essential endocrine organs, such as the fetal adrenal or fetal pituitary and the metabolic and endocrine deficits observed (8-10). Following these interventions, blood can be obtained from the fetus and mother for several days until the time of normal or delayed delivery if necessary. This paper will focus on current ideas of the development of the hypothalamo-pituitary-adrenal axis.

A. The Potential Role of the Hypothalamus in Regulation of ACTH Secretion in the Fetus

Section of the pituitary stalk of the fetus in late gestation results in prolongation of pregnancy. ACTH can still be measured in the plasma of stalk sectioned fetuses (11). In the adult it has been clearly demonstrated that various factors present in the hypothalamus can stimulate the secretion of ACTH from the pituitary gland in vitro. Corticotropin releasing factor (CRF) has been isolated from sheep hypothalami, characterized and synthesized. It is a 41-amino acid peptide which will stimulate the release of ACTH from pituitaries in vivo and in vitro. CRF is distributed in many hypothalamic areas but a major location is the parvocellular division of the paraventricular nucleus (PVN) (12,13). Studies in which these hypothalamic nuclei were electrically stimulated have demonstrated the importance of the PVN in regulating ACTH release (14,15). AVP will also stimulate ACTH release in the adult animal. AVP has been demonstrated to be present in the PVN and the staining for AVP increases following adrenalectomy (16). AVP has been shown to synergize with CRF in stimulating ACTH release from rat corticotrophs in vitro (17). There is a good correlation between fetal plasma ACTH and AVP concentrations at 100-120 days of gestational age (18).

AVP has been shown to stimulate ACTH release in the fetal sheep and to synergize with CRF (19). We have demonstrated that AVP is present within the hypothalamus in the fetal sheep and that there are vasopressinergic neurons which extend to the anterior pituitary (Fig. 1). Intracerebroventricular administration of AVP to the fetal sheep at 125-134 days of gestational age will stimulate fetal plasma ACTH and cortisol production (20). Therefore, the intriguing possibility exists that there may be a direct neural regulation of ACTH secretion in the fetus.

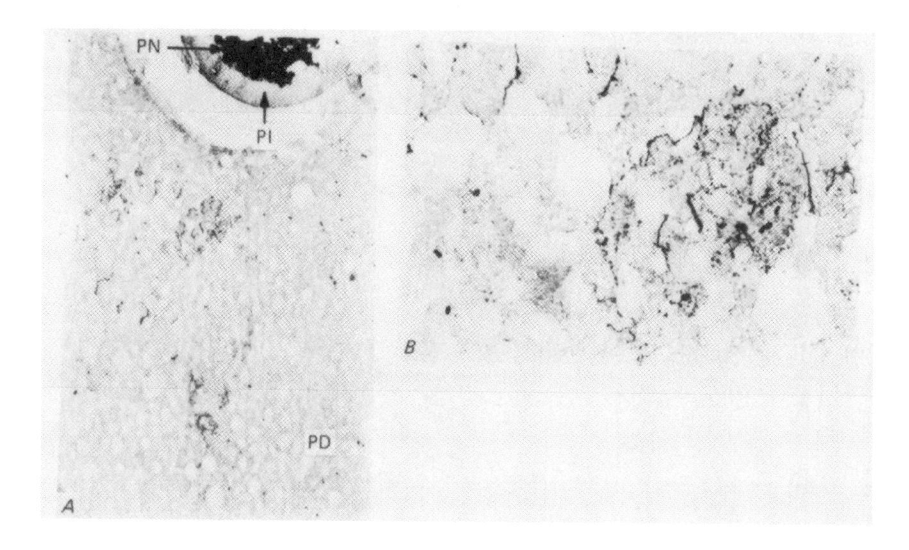

Fig. 1. A, AVP-NP in pars nervosa (PN), pars intermedia (PI) and pars distalis (PD) of a 134 days gestational age fetal sheep. B, immunoreactive axons in PD. (Reproduced with permission from Figueroa, J.P., et al., 1986, J. Physiol. 381,107P).

B. Timing of the Increase in the Level of Fetal Adrenal Cortisol Production That Produces the Critical Events of Fetal Maturation and Labor and Delivery

The classical paper of Bassett and Thorburn (21) demonstrated there was a rise in fetal plasma cortisol at the end of pregnancy. This rise appeared to be confined to the last few days of gestation. We conducted investigations designed to determine exactly when the increment of cortisol that finally leads to maturation of important fetal systems and the initiation of parturition begins. A detailed study using a carefully validated assay for cortisol in fetal sheep plasma demonstrated that the increase in the hypothalamo-pituitary adrenal axis appeared to occur as early as 20 days before delivery (22) (Fig. 2).

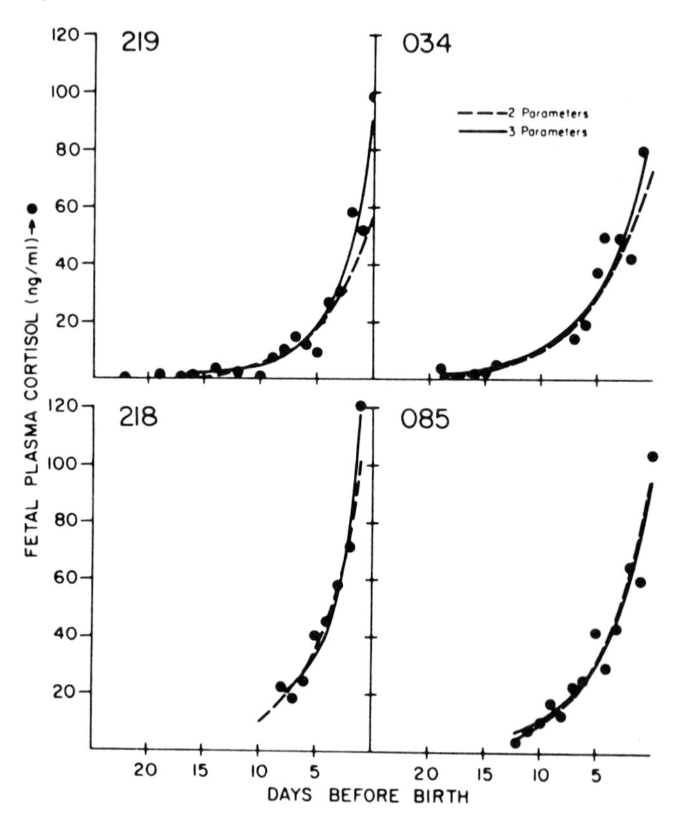

Fig. 2. Individual fetal plasma cortisol concentrations in four fetuses over the last 22 days of gestation ( ). Two best fit exponential type curves have been fitted through the data from each fetus according to a two-parameter model (---) and according to a three-parameter model (___). (Reproduced with permission from 22).

Utilizing a different experimental paradigm Wintour and her colleagues (23) showed that adrenalectomy at a wide range of gestational ages from 90 - 120 days was followed by a rise in fetal plasma ACTH at about 125 days gestation. Thus it would appear that critical processes in the regulation of the fetal hypothalamo-hypophyseal-adrenal axis occur around 20 days before delivery.

Although there are several reports of increasing fetal plasma cortisol concentrations in late gestation, there is controversy regarding the question whether fetal plasma ACTH increases over the last 20-30 days of gestation. There is a definite rise in fetal plasma ACTH in the very last days of gestation (24), however no longitudinal study has clearly demonstrated an increase in fetal plasma ACTH concentration in individual animals before the cortisol rise at 125-130 days of gestation. The studies available generally contain cross-sectional data with grouped means and thus differ from the curve fitting techniques for time-trends for fetal plasma cortisol concentrations obtained from data for individual animals (22). A rise in fetal plasma ACTH may be difficult to demonstrate in view of the known pulsatility of ACTH secretion. However, one study demonstrated a rise in fetal plasma ACTH over the period 25 to 8 days prior to delivery (25). In addition, there is a well established increase in fetal adrenal cortical sensitivity to ACTH as gestation progresses (26). Thus small fetal plasma ACTH increments adequate to produce the rise in fetal plasma cortisol may be difficult to detect. In addition to changes in fetal adrenal sensitivity to ACTH, there may be changes in pituitary response to CRF and AVP as gestation progresses. Some studies show a marked increase in fetal pituitary ACTH response to a 1 μg bolus of CRF (25) whilst the response in other studies has been a decrease in older gestational age fetuses when compared with younger fetuses (27).

C. Histological Changes in the Pituitary at Critical Stage of Maturation

ACTH is a 39 aminoacid peptide and does not cross the placenta. ACTH has been demonstrated in fetal sheep plasma as early as 59 days of gestation (28). ACTH immunoreactive cells have been demonstrated in the pituitary as early as 40 days of gestation (29). The morphology of cells in the pituitary that stain for ACTH changes in a dramatic fashion around 130 days of gestation. Prior to 90 days gestation, the predominant cell type is large and generally columnar in shape and has been referred to as the fetal type corticotroph (29). This type of cell is absent from the anterior pituitary of adult sheep. In late term fetuses, the predominant cell type is a stellate corticotroph. These changes in cell morphology may be related to alterations in biosynthetic function required to produce the different molecular weight forms of immunoreactive ACTH demonstrated in the fetal pituitary (30) and in fetal sheep plasma (31).

D. Considerations of Cortisol Feedback on the Fetal Hypothalamus and Pituitary: Level of Sensitivity of the System

The increasing drive by ACTH to cortisol production in late fetal life in the presence of increasing fetal plasma cortisol concentrations suggests a decrease in negative feedback of cortisol, either at the hypothalamic or pituitary level in late gestation. Stimuli to the hypothalamo-pituitary adrenal cortical axis have been classified as "steroid insuppressible" or "steroid suppressible". Stimuli such as laparotomy with intestinal traction are relatively steroid insuppressible. It may be that the late gestation drive to activation of the fetal adrenal is steroid insuppressible. However, an alternative explanation is that there is a resetting of the feedback mechanism in late gestation. Cortisol has been demonstrated to inhibit the ACTH release in response to hypotension at 117 to 131 days gestation (32). It may be hypothesized that a decrease in cortisol feedback is a major feature in the increased activity of the system. Further experiments are required to test this suggestion.

The delayed increase in fetal plasma ACTH concentration following bilateral adrenalectomy (23) also suggests changes in feedback during gestation. We have demonstrated that when pharmacological adrenalectomy is performed with metapyrone, fetal plasma compound S, 11-deoxycortisol does not increase until 125 days gestational age (33), also suggesting that the effects of lowered fetal plasma cortisol concentration are not reflected in increased fetal ACTH secretion until some critical maturation has occurred.

E. Experimental Studies of the Role of the Paraventricular Nucleus (PVN) in the Fetal Sheep

Recent availability of the stereotaxic atlas for the fetal sheep brain has made possible the production of specific and carefully localized lesions in discrete areas of the fetal hypothalamus with subsequent investigation of the deficit produced (Fig. 3). The original stereotaxic atlas for the fetal sheep brain was produced by Gluckman and Parsons (34). Cannulae can also be fabricated that allow access to the fetal lateral ventricles thereby permitting infusions into the cerebrospinal fluid of hormones, secretagogues and antagonists that alter fetal neuroendocrine function in specific ways.

The effect of bilateral fetal PVN lesions on baseline fetal plasma ACTH concentrations, the response of fetal ACTH to hypotension and hypoxemia, as well as the pituitary response to CRF have been studied in fetal sheep in which the PVN has been destroyed at 108 to 110 days of gestational age. Electrolytic lesions were produced using monopolar electrodes with an anodal current of 5.6 mA passed through the electrode for 30 seconds.

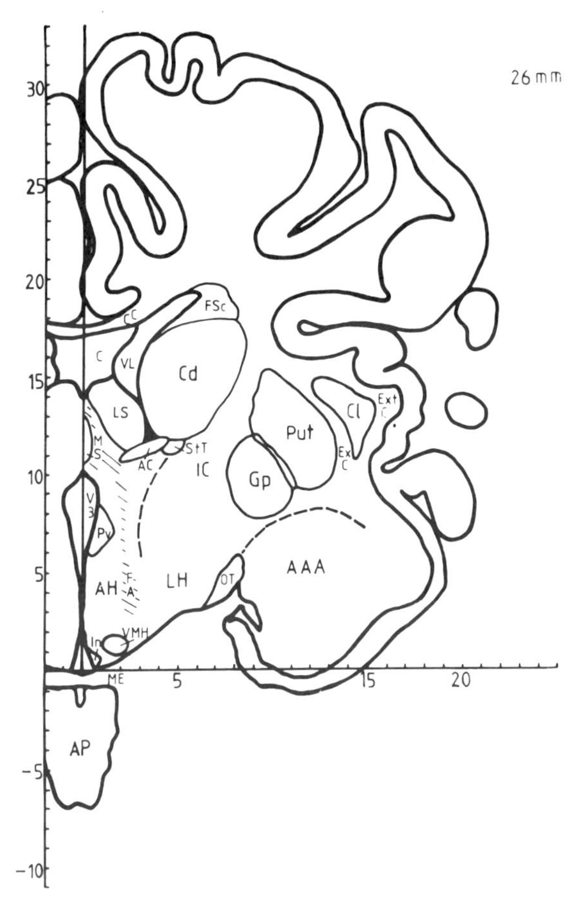

Fig. 3. Tracing of thionin stained coronal section of section of a representative fetal sheep brain at 120 days gestation. Section taken at 26 mm anterior to stereotaxic zero. The horizontal and vertical coordinates are in mm. (Reproduced with permission from 34.)

In fetuses in which bilateral PVN lesions have been placed, the fetal plasma ACTH increment that normally follows a reduction of fetal blood pressure of 50% was abolished (Fig. 4). In addition, the ACTH increment that normally occurs during experimentally induced fetal hypoxemia was abolished by bilateral lesions (Fig. 5). It is of interest that the bilateral PVN lesions diminish the response of the fetal pituitary to a bolus injection of synthetic CRF (Fig.6).

This observation suggests that the fetal PVN plays a trophic role in maintaining pituitary corticotrope function.

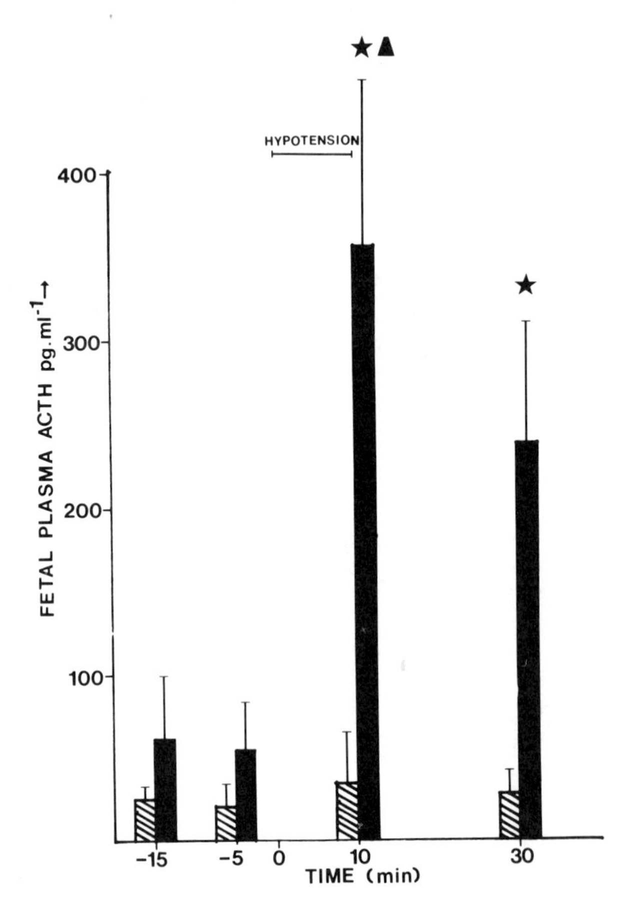

Fig. 4. Fetal plasma ACTH response (mean ± S.D.) of PVN lesioned fetuses ▧ (n = 4) and sham lesioned control ■ (n = 3) to 50% hypotension induced by nitroprusside infused intravascularly to the fetus from 0 to 10 min at 112-117 days gestation. Surgery was performed at 108 - 110 days gestation. *$p < 0.01$ when the lesioned and sham lesioned groups are compared at the same time. ▲ Significantly greater than mean resting concentration ($p < 0.05$). (Reproduced with permission from 48.)

F. Sensitivity of the Fetal Hypothalamo-Pituitary-Adrenal Axis to Mild Degrees of Hypoxemia

We have demonstrated that minor decreases in fetal arterial PO_2 occur during maintained episodes of low amplitude myometrial activity, contractures, which occur

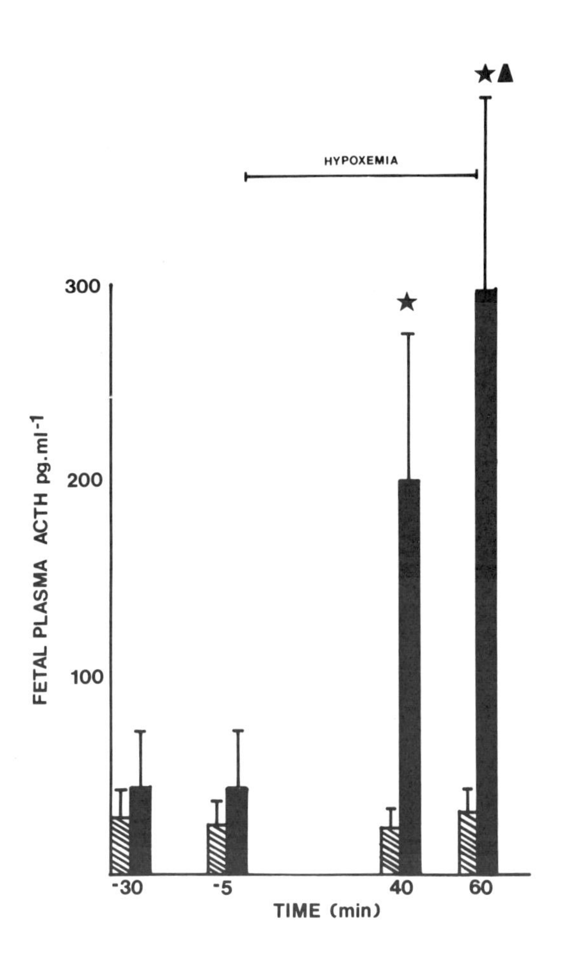

Fig. 5. Fetal plasma ACTH response (mean ± S.D.) of PVN lesioned fetuses ▧ (n = 4) and sham lesioned controls ■ (n = 3) to hypoxemia (ΔPO_2 approx. 7 mm Hg) at 112-117 days gestation. Surgery was performed at 108-110 days gestation. *$P < 0.05$ when the lesioned and sham lesioned groups are compared at the same time. ▲ Significantly greater than mean resting concentration ($p < 0.05$). (Reproduced with permission from 48.)

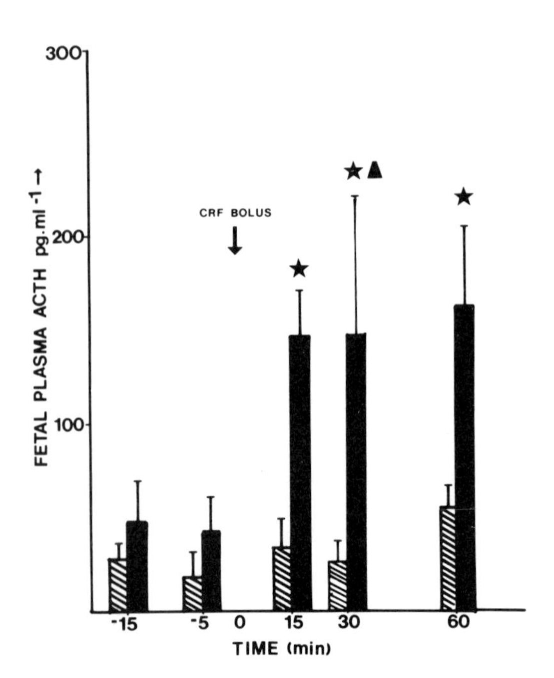

Fig. 6. Fetal plasma ACTH response (mean ± S.D.) of PVN lesioned fetuses ▧ (n = 4) and sham lesioned controls ■ (n = 3) to a single injection of 1 μg CRF to the fetal jugular vein at 112 - 117 days gestation. Surgery was performed at 108 - 110 days gestation. *$P < 0.05$ when the lesioned and sham lesioned groups are compared at the same time. ▲ Significantly greater than mean resting concentration ($p < 0.05$). (Reproduced with permission from 48.)

throughout the last part of pregnancy (35). Contractures have been shown to exist during pregnancy in all mammalian species studied. In the sheep they have been shown to alter the fetal movements that can be observed with the ultrasound. Contractures also alter the position of the pregnant uterus in the abdomen (36). During contractures PO_2 may fall 2 to 8 Torr (35). In a recent study we have demonstrated that during experimentally produced fetal hypoxemia in which the fetal PO_2 has been decreased by 5 Torr for 24 hours, the fetal hypothalamo-hypophyseal-adrenal system was activated with a concomitant significant increase in fetal plasma cortisol (Fig. 7) (37). Thus, small degrees of fetal hypoxemia may influence fetal neuroendocrinology.

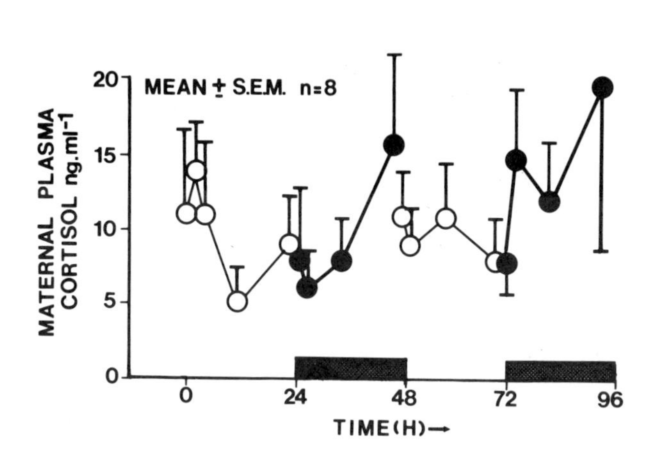

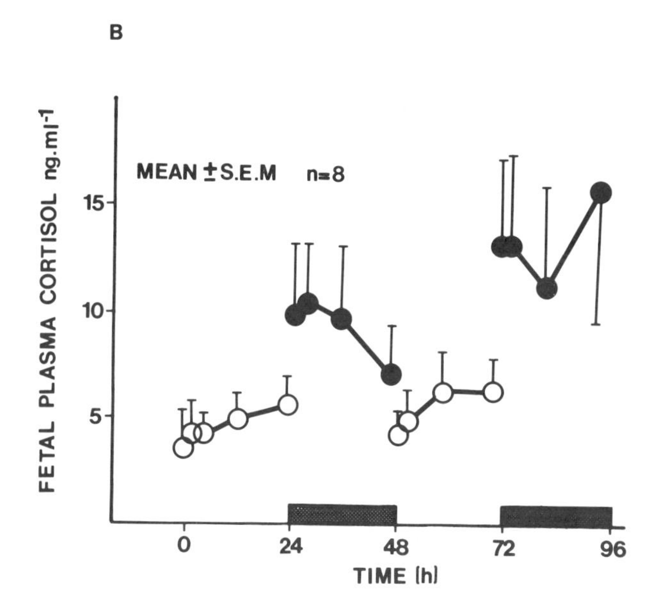

Fig. 7. Maternal arterial plasma cortisol concentration (A) and fetal carotid arterial plasma cortisol concentration (B) in eight experiments at 122 to 139 days' gestation. Open circles (○) indicate samples obtained during infusion of air and closed circles (●) during infusion to produce hypoxemia (■).

G. Effect of Contractures on the Fetal Hypothalamo-Hypophyseal-Adrenal Axis

In our initial description of contractures we postulated that they might influence fetal neuroendocrine function (38). These influences may differ at different times of gestation due to variation in the amount of amniotic fluid and the degree of contact the fetus has with the myometrium. During a contracture fetal chest wall dimensions are considerably altered (Fig. 8). Thus, sensory stimulation may be one pathway whereby fetal neuroendocrine function is altered. The other separate or interacting potential pathway is the fall in fetal PaO_2 (Fig. 9). Challis and coworkers have produced evidence that contractures that follow a bolus injection of oxytocin in the pregnant ewe result in a fall in fetal PaO_2 and are accompanied by a rise in fetal plasma ACTH. We have some unpublished evidence that late in gestation spontaneous contractures are accompanied by an increase in fetal plasma ACTH; contractures are also accompanied by an increase in fetal plasma catecholamines (Aarnoudse - personal communication).

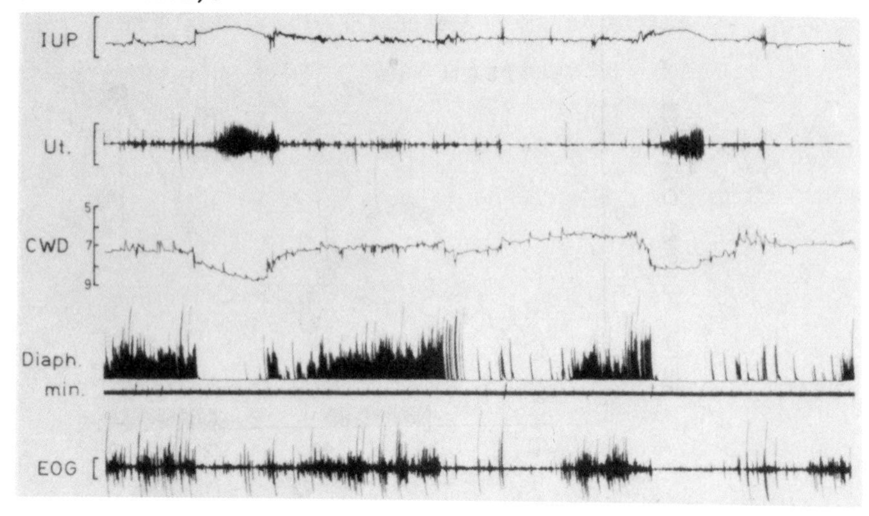

Fig. 8. Relationship of IUP, uterine EMG (Ut.) and CWD. Diaphragmatic EMG (Diaph.) and EOG at 137 days' gestation. Calibration bars as in previous figures. CWD = distance, in centimeters, between bilaterally placed transducers in the midaxillary line at the level of the seventh intercostal space. (Reproduced with permission from 46.)

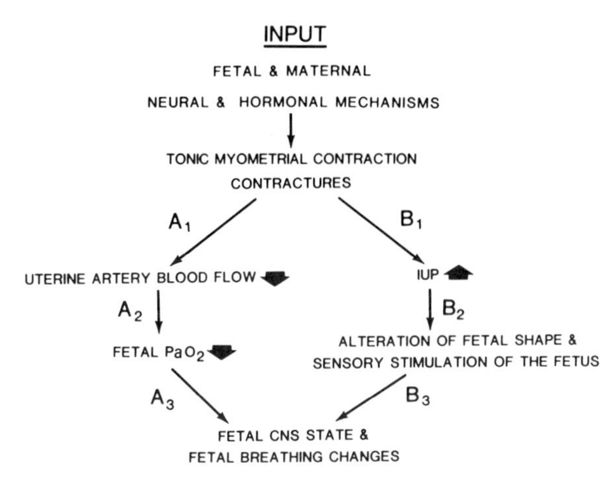

Fig. 9. Two hypotheses of the mechanism of production of changes in fetal central nervous system (CNS) state and breathing changes induced by contractures. (Reproduced with permission from 47.)

H. Observations in the Non-Human Primate

In the pregnant non-human primate there is evidence for increased growth of the fetal adrenal gland late in gestation and an increment in fetal plasma DHEA sulphate prior to delivery (39). Unfortunately, chronic instrumentation of the pregnant non-human primate fetus is much more difficult than the pregnant sheep. However, the available studies show both similarities and differences in fetal neuroendocrine function (40, 41). One of the most interesting features is the presence of distinct circadian rhythms in fetal neuroendocrine function, driven by the maternal circadian changes in plasma cortisol (41). In addition, there are marked circadian rhythms in myometrial activity in the pregnant rhesus monkey (42).

I. Other Neuroendocrine Axes in Fetal Development

Although this review has focused on the hypothalamo-pituitary-adrenal axis a considerable amount of work has been performed in other systems. The regulation of growth hormone secretion in the fetus has been studied extensively. In ovine fetuses, plasma growth hormone concentrations may be 20 fold higher than after birth. Shortly after birth, growth hormone values have fallen to low levels comparable to adults and these decreases can also be produced during premature labor induced by cortisol infusion to the fetus as 128 days gestation (43). Since growth hormone does not exert much influence upon fetal growth, its regulation may not appear to be of major importance in fetal life. Growth hormone receptors are absent from the fetal liver (44). There are several different maturational changes within the hypothalamo-pituitary axis during fetal life that regulate growth hormone secretion (45).

II. SUMMARY AND CONCLUSIONS

The availability of immunohistochemical techniques, specific peptides demonstrating agonist and antagonist activity within the axis and direct lesions of the various nuclei in the hypothalamus suggest that further information on this important neuroendocrine system will be forthcoming in the near future.

ACKNOWLEDGEMENTS

I would like to thank Karen Moore for her help with the production of this manuscript. I would also like to thank my colleagues especially Dr. Thomas McDonald, Dr. Jorge Figueroa and Dr. Angela Massmann for their help and involvement in the studies reported here.

REFERENCES

1. Liggins, G.C. and Kitterman, J.A. (1981) In: The fetus and independent life. CIBA Foundation Symposium 86. Pp. 308-357.
2. Nathanielsz, P.W. (1978) Ann. Rev. Physiol. 40, 411-455.
3. Ellendorff, F., Gluckman, P.D. and Parvizi, N., eds. (1984) Research in Perinatal Medicine (II), Fetal Neuroendocrinology. Pub: Perinataology Press, Ithaca, New York.
4. Gluckman, P.D. (1984) J. Develop. Physiol. 6(3), 301-
5. Nathanielsz, P.W. and Fisher, D.A. (1979) Ani. Reprod. Sci. 2, 57-62.
6. Liggins, G.C., Fairclough, R.J., Grieves, S.A., Kendall, J.Z. and Knox, B.S. (1973) Rec. Prog. Horm. Res. 29, 111-159.
7. Steele, P.A., Flint, A.P.F. and Turnbull, A.C. (1972) J. Endocrinol. 69, 239-246.

8. Liggins, G.C., Kennedy, P.C. and Holm, L.W. (1967) Am. J. Obstet. Gynecol. 98, 1080-1085.
9. Drost, M. and Holm, L.W. (1968) J. Endocrinol. 40, 293-296.
10. Barnes, R.J., Comline, R.S. and Silver, M. (1978) J. Physiol. 275, 567-579.
11. Nathanielsz, P.W., Abel, M.H., Bass, F.G., Krane, E.J., Thomas, A.L. and Liggins, G.C. (1978) Quart. J. Exp. Physiol. 63, 211-219.
12. Bloom, F.E., Battenbergy, E.L.P., Rivier, J. and Vale, W. (1982) Reg. Peptides 4, 43-48.
13. Swanson, L.W., Sawchenko, P.E., Rivier, J. and Vale, W.W. (1983) Neuroendocrinol. 36, 165-186.
14. Gann, D.S. (1978) Rec. Prog. Horm. Res. 34, 357-400.
15. Dornhorst, A., Carlson, D.E., Seif, S.M., Robinson, A.G., Zimmerman, E.A. and Gann, D.S. (1981) Endocrinol. 108, 1420-1424.
16. Zimmerman, E.A., Carmel, P.W., Husain, M.K., Ferin, M., Tannenbaum, M., Frantz, A.G. and Robinson, A.G. (1973) Science 182, 925-927.
17. Watanabe, T. and Orth, D.N. (1987) J. Endocrinol. 38, 83-84.
18. Wintour, E.M. (1984) J. Develop. Physiol. 6, 291-299.
19. Norman, L.J. and Challis, J.R.G. (1987) Endocrinol. 120, 1052-1058.
20. McDonald, T.J., Figueroa, J.P., Reimers, T.J. and Nathanielsz, P.W. (1988) Society for Gynecologic Investigation, March 17-20, 1988, abstract 470.
21. Bassett, J.M., Thorburn, G.D. (1969) J. Endocrinol. 44, 285-286.
22. Magyar, D.M., Fridshal, D.., Elsner, C.W., Glatz, T., Eliot, J., Klein, A.H., Lowe, K.C., Buster, J.E. and Nathanielsz, P.W. (1980) Endocrinol. 107, 155-159.
23. Wintour, E.M., Coghlan, J.P., Hardy, K.J., Hennessy, D.P., Lingwood, B.E. and Scoggins, B.A. (1980) Acta Endocrinol. 95, 546-552.
24. Rees, L.H., Jack, P.M.B., Thomas, A.L. and Nathanielsz, P.W. (1975) Nature 253, 274-275.
25 MacIsaac, R.J., Bell, R.J., McDougall, J.G., Tregear, G.W., Wang, X. and Wintour, E.M. (1985) J. Develop. Physiol. 7, 329-338.
26. Glickman, J.A. and Challis, J.R.G. (1980) Endocrinol. 106, 1371-1376.
27. Brooks, A.N., Challis, J.R.G. and Norman, L.J. (1987) Endocrinol. 120, 2383.
28. Alexander, D.P., Britton, H.G., Forsling, M.L., Nixon, D.A. and Ratcliffe, J.G. (1973a) In: The Endocrinology of Pregnancy and Parturition - Experimental Studies in the Sheep (ed. Pierrepoint, C.G.) pp. 112-125. Alpha Publishing, Cardiff.

29. Perry, R.A., Mulvogue, H.M., McMillen, I.C. and Robinson, P.M. (1985) J. Develop. Physiol. 7(6), 397-404.
30. Silman, R.E., Holland, D., Chard, T., Lowry, P.J., Hope, J., Rees, L.H., Thomas, A. and Nathanielsz, P. (1979) J. Endocrinol. 81, 19-34.
31. Jones, C.T. and Roebuck, M.M. (1980) J. Steroid Biochem. 12, 77-82.
32. Wood, C.E. (1986) Am. J. Physiol. 250: R795.
33. Carson, G.D., Buster, J.E. and Nathanielsz, P.W. (1981) In: Society for Gynecological Investigation, 1981. Abstract 217.
34. Gluckman, P.D. and Parsons, Y. (1984) In: Animal Models in Fetal Medicine (III). Pp. 69-107. Nathanielsz, P.W., ed. Pub: Perinatology Press, Ithaca, New York.
35. Jansen, C.A.M., Krane, E.J., Thomas, A.L., Beck, N.F.G., Lowe, K.C., Joyce, P., Parr, M. and Nathanielsz, P.W. (1979) Am. J. Obstet. Gynecol. 134, 776-783.
36. Taverne, M.A.M. and Scheerboom, J.E.M. (1985) Res. Vet. Sci. 146, 557-567.
37. Towell, M.E., Figueroa, J.P., Markowtiz, S., Elias, B. and Nathanielsz, P.W. (1987) Am. J. Obstet. Gynecol. (Get journal # and page #'s from PWN reprint files).
38. Nathanielsz, P.W., Ratter, S., Thomas, A.L., Rees, L. and Jack, P.M.B. (1976) In: The Fetus and Birth. CIBA Foundation Symposium No. 47, Elsevier, Pp. 73-91.
39. Seron-Ferre, M., Taylor, N.F., Rotten, D., Koritnik, D. and Jaffe, R.B. (1983) J. Clin. Endocrinol. Metab. 57, 1173.
40. Walsh, S.W. (1975) In: Perinatal Endocrinology. Albrecht, E. and Pepe, G.J., eds. Pp. 219-241. Pub: Perinatology Press, Ithaca, New York.
41. Novy, M.J. and Walsh, S.W. (1983) Am. J. Obstet. Gynecol. 145, 920-931.
42. Taylor, N.F., Martin, M.C., Nathanielsz, P.W. and Seron-Ferre, M. (1983) Am. J. Obstet. Gynecol. 146, 557-567.
43. Lowe, K.C., Gluckman, P.D., Jansen, C.A.M. and Nathanielsz, P.W. (1986) Am. J. Obstet. Gynecol. 154, 420-423.
44. Gluckman, P.D., Buller, J.H. and Elliott, T.B. (1983) Endocrinol. 112, 1607-1612.
45. Gluckman, P.D. (1984) J. Develop. Physiol. 6, 301.
46. Nathanielsz, P.W., Bailey, A., Poore, E.R., Thorburn, G.D. and Harding, R. (1980) Am. J. Obstet. Gynecol. 138, 653-659.

47. Nathanielsz, P.W., Jansen, C.A.M., Yu, H.K., Cabalum, T. (1984) In: Fetal Physiology and Medicine: The Basis of Perinatology (Beard, R.W. and Nathanielsz, P.W., eds), Perinatology Press, Pp. 629-653.
48. McDonald, T.J., Rose, J.C., Figueroa, J.P., Gluckman, P.D. and Nathanielsz, P.W. (1988) J. Develop. Physiol. In press.
49. Lye, S.J., Wlodek, M.E. and Challis, J.R.G. (1985) J. Endocrinol. 106, R9-R11.

10

PRENATAL AND POSTNATAL FUNCTIONS OF THE BIOLOGICAL CLOCK IN REPRODUCTIVE DEVELOPMENT

M. H. Stetson, T. H. Horton and R. S. Donhan

Physiology and Anatomy Program
School of Life and Health Services
University of Delaware
Newark, Delaware 19716

I. INTRODUCTION: Photoperiod and Circadian Involvement in Mammalian Reproductive Rhythms

Periodic processes convey functional advantages to biological systems at any level of organization, from the temporal coordination of biochemical processes within a cell to the coordination of physiological and behavioral processes with daily and annual environmental rhythms (Rapp, 1987). Reproduction is but one example of a complex activity in which period of complete quiescence alternate with periods of intense commitment. Our goal is to examine the development of neuroendocrine rhythms and photoperiodic processes which contribute to the regulation of periodic reproduction in mammals. Herein we examine 1) how maternal factors modulate the effects of photoperiod on reproductive development in juvenile Siberian hamsters (*Phodopus sungorus*), and 2) the postnatal development of circadian rhythms of gonadotropin secretion in immature Syrian hamsters (*Mesocricetus auratus*).

A. Photoperiodic and Seasonal Effects on Reproductive Maturation

Seasonal variation in reproductive success has led to

selection of individuals that can most reliably coordinate reproduction with environmental conditions favorable to the survival of the young. In north temperate latitudes seasonal variation in day length is the most reliable predictor of seasonal changes in environmental quality and is capable of regulating reproductive activity of many species. The exact role of photoperiod may be related to the ecology of a species (Negus and Berger, 1972; Horton, 1984a). Many mammals attain sexual maturity in a matter of weeks and have the capability to reproduce in the year of their birth; for these species seasonal reproduction may be regulated by altering juvenile development as well as adult reproductive function. Of the species discussed herein, photoperiod alters pubertal development and adult reproductive function of the Siberian hamsters, *Phodopus sungorus*, but only adult reproductive activity in the Syrian hamster, *Mesocricetus auratus* (Brackmann and Hoffmann, 1977; Hoffman, 1978; Darrow *et al.*, 1980; Rollag *et al.*, 1982; Yellon and Goldman, 1984; Sisk and Turek, 1987).

As the mechanism for photoperiodic time keeping is intimately involved with the circadian system of mammals, we consider the developing photoperiodic and circadian systems together. After a brief discussion of the neuroendocrine components comprising the photoperiodic time keeping mechanism, we consider a mechanism by which photoperiodic information is used to regulate pubertal development. Following this we discuss the role of the maturing circadian system in coordinating estrus cycles in pubertal rodents.

B. Development of the Circadian System

The earliest studies of maternal effects on the circadian system demonstrated that development of rhythmicity, including period length and whether the animal is diurnally or nocturnally active, is genetically programmed (Davis, 1981). However, the environment impresses itself at an early age via entrainment of fetal rhythms by the maternal system (Deguchi, 1975, 1977; Reppert and Schwartz, 1983, 1986a,b; Davis and Gorski, 1983).

The components required for a functional circadian system include an afferent sensory pathway, a time-keeping mechanism (oscillator) and output pathways to drive the observed behavioral and physiological rhythms. The suprachiasmatic nuclei (SCN) of the hypothalamus are recognized to be the site of the circadian clocks in mammals (Stephan and Zucker, 1972; see reviews by Rusak and Zucker, 1979; Stetson and Watson-Whitmyre, 1984; Moore and Card, 1985; Rosenwasser and Adler, 1986). Afferent retinal information reaches the SCN by either a direct projection, the retinohypothalamic tract (RHT), or an indirect projection, the geniculohypothalamic tract (Moore and Lenn, 1972; Moore and Card, 1975). The efferent pathways from the SCN project primarily to adjacent regions of the hypothalamus. An efferent projection of the SCN to the paraventricular nuclei has been recently shown to be a key component of the photoperiod control mechanism of the hamster (Pickard and Turek, 1983; Lehman *et al.*, 1984).

The development of the circadian system has been most thoroughly studied in the laboratory rat. The SCN are anatomically distinct by gestational day 18 (Ifft, 1972) and, day-night rhythms of glucose metabolism are detectable by the twentieth day of gestation (Reppert and Schwartz, 1983) but maturation of the synaptic connections of the SCN with its afferent and efferent pathways occurs postnatally. The RHT reaches the SCN by postnatal days 3 or 4 (Hendrickson *et al.*, 1972; Stanfield and Cowan, 1976) but, in the hamster at least, retinal projections to the SCN are not complete until about the end of the second week (Frost *et al.*, 1979). The development of efferent pathways has been inferred by the emergence of overt rhythms in hormones, metabolism or behavior. The daily rhythm of pineal melatonin production is apparent in the second week of life in Syrian and Siberian hamsters and the laboratory rat (Fuchs and Moore, 1980; Tamarkin *et al.*, 1980b; Rollag and Stetson, 1981) and corresponds with the development of adrenergic innervation of the pineal gland (Yuwiler *et al.*, 1977; van Veen *et al.*, 1978). Similarly, the appearance of rhythmic electrical activity of the SCN appears postnatally (Shibata *et al.*, 1983). Other metabolic functions, hormone levels, and neurotransmitter receptor numbers become apparent at different ages (Reppert and Uhl, 1987; see reviews by Davis, 1981 and Wirz-Justice, 1987, see also Section III).

These results suggest that the central oscillator is functioning prior to entrainment by environmental information and also prior to the appearance of rhythmic output. Blockage of Na^+-dependent action potentials by infusion of tetrodotoxin (TTX) into the SCN of adult male rats held in constant darkness produces arrhythmic activity and blocks the ability of light to phase-shift the ciracadian system (Schwartz *et al.*, 1987). Following cessation of TTX infusion, rhythmic activity resumes at the time preducted had the circadian systems been free running. The authors conclude that while daily rhythms of electrical activity are necessary for the synchronization of overt circadian rhythms, they may not be an integral part of the central pacemaker.

II. Prenatal Circadian and Photoperiodic effects on postnatal Maturation

A. Interaction between Circadian System and Melatonin

The rhythm of pineal melatonin is under direct control of the circadian system, free-running with a period slightly different from 24 hours in hamsters held on constant darkness (Klein and Weller, 1970; Tamarkin et al., 1979; Darrow and Goldman, 1986). When entrained by light-dark cycles, production and release of the hormone normally occur during the dark period (Klein, 1978; Yellon et al., 1982). The pineal functions as a neuroendocrine transducer, receiving photoperiodic information from the SCN, via the paraventricular nuclei, in the form of sympathetic neural discharge, which results in the production and release of melatonin. The precise means by which the melatonin signal regulates reproductive activity remains a mystery; two theories have been advanced, each supported by a volume of data in one or more mammalian species. One theory proposes that duration of the melatonin production signals daylength to the organism and is based on the observation that the duration of melatonin production is proportional to the duration of the dark phase; i.e. the longer the night, the longer the duration of melatonin production (see Goldman and Darrow, 1983; Stetson and Watson-Whitmyre, 1986). Experimental evidence supporting this theory derives from work in which melatonin is infused into pinealectomized animals; infusions of longer durations produce short day (long night) effects while shorter duration infusions produce long day (short night) effects (Carter and Goldman, 1983a,b; Goldman et al., 1984; Bittman and Karsch, 1984; Bittman et al., 1983).

The second theory proposes that the important factor determining the reproductive response to day length is the coincidence in time of melatonin release and sensitivity of melatonin target tissue to the hormone. Two types of experimental evidence support this theory. Firstly, a daily rhythm in the binding of labelled melatonin to membrane fractions of the brains of at least two mammalian species has been reported (Vacas and Cardinali, 1979; Laudon and Zisapel, 1986). These data correlate well with the effects of melatonin injections in three species of hamster; single injections of melatonin late in the afternoon or late in the night (times when melatonin binding to brain membrane fractions is high) caused gonadal regression while melatonin administered at any other time was without effect (Watson-Whitmyre and Stetson, 1983; Stetson and Tay, 1983;

Stetson *et al.*, 1986a; Hong and Stetson, 1987). Secondly, administration of melatonin to pinealectomized hamsters by infusion or multiple daily injections (at any time of day) or by single daily injection (only at a restricted period of the day) are *all* effective in inducing gonadal regression (Goldman *et al.*, 1984; Watson-Whitmyre, 1984) or delay of gonadal maturation (Carter and Goldman, 1983b). Since a single injection of melatonin at restricted periods is as effective as multiple injections several hours apart, these data do not support the duration hypothesis. The data suggest that single injections are effective if coincident with a daily rhythm of sensitivity to melatonin and multiple injections (or long duration infusions) at any time are effective because they overlap with periods of tissue sensitivity.

B. Maternal Transfer of Photoperiodic Information

An inherent problem in the use of day length as a cue is the fact that all but the shortest and longest photoperiods occur twice each year. Polyestrous species may produce litters over a span of several months, encountering the same absolute day lengths twice during the breeding season. Young born at the beginning of the breeding season are often observed to develop rapidly under the same photoperiods that inhibit development of young born at the end of the breeding season. How can a day length act to stimulate development in young born at one time, but inhibit development of young born at another stage of the breeding season?

One difference is that the direction of change in day length differs between the two times of year. That animals are capable of utilizing this kind of information was first demonstrated experimentally in the microtine rodent, *Microtus montanus* and subsequently in the Siberian hamster, *Phodopus sungorus* (Horton, 1984b; Stetson *et al.*, 1986b). Vole pups gestated in long days (16 hours light:8 hours dark, 16L:8D) have retarded somatic growth and reproductive maturation when reared in a photoperiod of intermediate length (14L:10D), while gestation in short days (8L:16D) yields accelerated growth and maturation of pups exposed to 14L:10D postnatally (Horton, 1984b, 1985). Thus, postnatal maturation reflects the prenatal photoperiod; hypothetically such information might be transmitted by the mother either during gestation or during lactation. To determine when the mother transfers information about the gestational photoperiod to her young, pregnant females were exposed to long or short days; neonates were switched between dams on the day of birth so that all were raised by foster mothers, some having foster mothers that had been exposed to long days and others having foster mothers that had been exposed to short days. All young and their foster mothers were moved to an intermediate photoperiod (14L:10D) on the day of birth. Because all litters were raised in the same postnatal photoperiod, differences in the growth rates between litters must be due to information received from the mother; the cross-fostering paradigm allowed us to determine whether that information was received during gestation, lactation or both periods. Pups that were gestated on short days grew more rapidly than those gestated on long days, indicating that information was transferred from mother to fetus during gestation. While it was hypothesized that information about the prenatal photoperiod might be transferred during lactation, as a result of the female experiencing a change in photoperiod at parturitation, the data demonstrated that this did not occur (Horton, 1985).

The experiments presented in the preceding paragraph demonstrated that effects of _gestational_ photoperiod result from factors acting prenatally, not during lactation as a result of the mother responding to a change in photoperiod. Since all mothers and litters were exposed to 14L:10D during lactation, these experiments did not preclude the possibility that information about the lactational photoperiod was transferred to the pups; if there were a continual reprogramming of the neonatal neuroendocrine responses during development, the lactational photoperiod could potentially modify effects of the gestational photoperiod. Further experiments in the Siberian hamster have demonstrated that no photoperiodic information is received during lactation (postnatal days 1-15) (Stetson, unpublished). When pups were temporarily held in a variety of photoperiods between postnatal days 1 and 15 there was no modification of the growth pattern that would have been predicted based on the pups' prenatal photoperiod. This suggests that pups receive information from their mothers _in utero_ and that the information is not modified until pups become capable of responding directly to photoperiod, which occurs at about day 15 (van Veen _et al._, 1978; Tamarkin _et al._, 1980b; Yellon _et al._, 1985).

Other experiments have exploited the fact that continuous light (LL) is neither stimulatory nor inhibitory, but, since it provides no periodic information, is neutral with respect to stimulation of the reproductive system. Thus, development reflects the photoperiod that animals were exposed to previously; animals housed in LL are, in a sense, "coasting" on previous information. When pups are reared in LL after birth their testicular weight at 28 days of age directly reflects the gestational photoperiod; if mothers saw 16L:8D, pups develop larger testes in LL than if the mother saw 12L:12D. This implies that prenatal photoperiodic information is used to program reproductive development and does not serve solely to provide a reference day length for comparison with the postnatal photoperiod. Therefore we are left with a model in which the maternal system programs the reproductive development of the fetus *in utero* than, later in life, at the time the pup's own photoperiod-sensing mechanism begins to respond to the environmental light-dark cycle, photoperiod can modify developmental programming of the pup. But the pup's interpretation of the extant photoperiod as stimulatory or inhibitory is dependent on the gestational photoperiod information received from its mother.

C. Mechanism for Maternal Transfer of Photoperiodic Information

Since the pineal effectively transduces photoperiodic information through secretion of melatonin, and since melatonin can cross the placenta (Reppert and Klein, 1978), it was the first substance implicated as a possible effector of the transfer of photoperiodic information between mother and fetus. Subsequent work suggested that the number of hours each night that melatonin is produced by the maternal pineal, and presumably crossing the placenta, conveys information to the fetus about the length of night (Elliott and Goldman, 1986, Weaver and Reppert, 1986).

We have injected pregnant siberian hamsters with melatonin and found that an injection paradigm similar to those that induce gonadal regression in adult males also inhibits testicular development of pups born to these mothers. Twenty-four groups of females housed in a long photoperiod were injected at times corresponding to each hour of the day throughout pregnancy. At birth the dams and litters were transferred to constant light until litters were sacrificed at 28 days of age. Only pups born to females injected late in the afternoon, prior to lights out, or at a single time point late at night, just prior to when the lights came on, showed inhibited testicular development characteristic of pups gestated in a short photoperiod (see SECTION II, A; Horton *et al.*, 1987a). Similar results from injections to adult male Siberian hamsters have been interpreted to reflect a daily rhythm in sensitivity of target tissues to melatonin (Stetson *et al.*, 1986a).

If the daily duration of melatonin, acting directly on the fetus, is the mechanism by which day length information is transferred, one would then predict that if fetuses were exposed to constant high levels of melatonin *in utero* the reproductive response would be independent of photoperiod. However, in adult male Siberian hamsters the effect of constant-release melatonin capsules is dependent on the photoperiod to which the males are exposed; males receiving melatonin capsules in long photoperiod undergo testicular regression, while testicular regression is prevented when implanted males are transferred to a short photoperiod (Hoffmann, 1974). We implanted female Siberian hamsters with melatonin (melatonin-clamped) or beeswax capsules; each group was subdivided with part of the females retained on long days, the others were transferred to short days. All females were paired with a male. At birth, dams and litters were moved to constant light. If the effects of melatonin during gestation were independent of other actions of photoperiod the testicular development of the pups born to melatonin-clamped females should have been the same, regardless of the gestational photoperiod. This was not the case. When reared in LL, young born to melatonin-clamped females housed in a long photoperiod had inhibited testicular development, while pups born to melatonin-clamped females in a short photoperiod had large testes. The results show that the effects of the melatonin implants during gestation, on peripubertal gonadal development, depend on gestational photoperiod (Horton *et al.*, 1978b). The results of these experiments support the proposal that melatonin is a component in the mechanism for transfer of photoperiodic information from the mother to the fetus. They also suggest that it is only part of a more complex system, perhaps including rhythms in tissue receptivity.

III. Circadian Oscillators and Postnatal Maturation

Thus far, we have shown that: 1) photoperiodic information is translated by a neuroendocrine oscillator system, and 2), this system functions in the mother during gestation to synchronize maturation of her offspring to the seasonal environment. Next we consider oscillator function and gonadotropin secretion in immature and pubertal Syrian hamsters.

A. Estrous Cycles: Dependence on a Circadian Oscillator

Everett and Sawyer (1949, 1950; Everett, 1972) were the first to report advance or delay of ovulation following an appropriately-timed injection of progesterone or barbiturate into the rat. The period of advance or delay was always about 24 hours, thus suggesting a rhythmic component that had a functional periodicity of a day. They further suggested that the diurnal rhythm might serve as a mechanism for synchronization of the estrous cycle with the environmental photocycle. Alleva *et al.*, (1971) demonstrated that, in Syrian hamsters, the free-running estrous cycle has a periodicity of about 96 hours, thus suggesting that an endogenous pacemaker regulates the duration of the estrous cycle. Since cycle extensions, whether spontaneous or induced, were always about 1 day, Alleva *et al.*, (1971) proposed that the pacemaker mechanism incorporated a circadian oscillator whose periodicity was entrained by the environmental photocycle. Fitzgerald and Zucker (1976) and Stetson *et al.*, (1977) showed that the period of the free-running estrous cycle is 4 times the period of the simultaneously recorded rhythm of locomotor activity. In addition, preovulatory surges of luteinizing hormone (LH) and follicle-stimulating hormone (FSH) in free-running hamsters have a fixed relationship to the endogenous circadian clock as reflected in the rhythm of locomotor activity (Stetson and Gibson, 1977; Stetson and Anderson, 1980).

As for other behavioral and physiological circadian rhythms, evidence suggests that the locus of the circadian pacemaker(s) is the SCN (Rusak and Zucker, 1979; Stetson and Watson-Whitmyre, 1984; Rosenwasser and Adler, 1986). Thus, lesions of the SCN that disturb the daily rhythm of locomotor activity also result in an anestrous condition in both hamsters and rats (Stetson and Watson-Whitmyre, 1976; Brown-Grant and Raisman, 1977).

B. Postnatal Development of Hormonal Rhythms

Levels of circulating LH and FSH in adult Syrian hamsters directly reflect the activity of the circadian neuro-oscillator system upon exposure to short photoperiods (<12.5 hours light/day): the animals become anovulatory, but each day there is a mid-afternoon surge of both gonadotropins (Bridges and Goldman, 1975; Seegal and Goldman, 1975; Bittman and Goldman, 1979). A daily rhythm of gonadotropin release occurs in other anovulatory conditions, both natural (e.g., lactation: DiPinto and Stetson, 1981) and experimental (ovariectomy: Stetson *et al.*, 1978; melatonin injections: Stetson and Hamilton, 1981). Thus, it appears that the circadian oscillator system that regulates the timing of preovulatory LH and FSH release during the estrous cycle may also function during short-day induced anestrous and during nursing. More recently, it has become clear that the functioning of this system significantly precedes puberty in the hamster.

The first reported endocrine expression of the neuroendocrine circadian oscillator is the appearance, in the female Syrian hamster, of a daily, mid-afternoon surge of LH and FSH (Smith and Stetson, 1980; Donham et al., 1984). The surges begin between 15 and 17 days of age and persist until they are replaced by a 4-day pattern of gonadotropin release at puberty, which occurs at about 35 days of age in our colony. Although of lesser amplitude, the phase of the daily surges with respect to the daily photocycle is identical to that of the LH and FSH preovulatory surges in the adult. Barbiturate injections block the daily surges in immature animals but they reappear 24 hours later; in the adult, barbiturate injections also block the preovulatory LH and FSH surges, again, they reappear the next day. Sterilization by treatment of the neonate with androgens also abolishes the daily surges of LH and FSH (Donham and Stetson, 1985).

As a result of these observations, it seems likely that the appearance of daily LH and FSH surges on 15-17 days of age signals the maturation of mechanisms that continue to function in the adult and are responsible for the timing of the preovulatory gonadotropin surges every 4th day. In all cases, the timing of the daily gonadotropin surges is similar to the timing of preovulatory surges of the estrous cycle.

Postnatal maturation of the adrenal glucocorticoid rhythms develops gradually in rodents, and usually somewhat later than the appearance of the gonadotropin rhythms (Allen and Kendall, 1967; Takahashi et al., 1979; Donham et al., 1988). In addition to the somewhat different ontogenetic patterns of the pituitary-ovarian axis and the pituitary-adrenal axis in the immature hamster, surgical intervention into either system appears to not affect rhythmicity of the other. Each of these rhythmicities probably represents an independent expression of the neural circadian oscillator.

C. Effect of Rhythmic Release of LH and FSH in Immature Hamsters.

Following the appearance of the daily afternoon surges at about 16 days of age, ovarian progesterone secretion increases and becomes rhythmic; the rhythm is dependent on the rhythm of LH and FSH (Donham et al., 1984). Maximum progesterone levels approach those of the non-pregnant adult by about 30 days of age, probably as a result of production by an abundant interstitium (Greenwald and Peppler, 1968; Shaha and Greenwald, 1983). Within a week after appearance of the gonadotropin rhythms, estradiol secretion also increases and becomes rhythmic (Donham et al., 1987; Donham and Stetson, unpublished) and, like the progesterone rhythm, there is a daily afternoon surge of estradiol. These results suggest that one important effect of the daily surges of gonadotropin is to stimulate and temporally organize the activity of the prepubertal ovaries. Follicular development is accelerated during the 2nd through the 4th week after birth with antral follicles appearing at about the age of first ovulation (Greenwald and Peppler, 1968).

It is difficult to determine the significance of the specific patterning of LH and FSH release as opposed to a tonically augmented release. Urbanski and Ojeda (1985a) recently report that the peripubertal period of development in the female rat is also characterized by a daily rhythm of LH in the blood. Like the hamster, a surge occurs each afternoon. *In vitro* exposure of ovaries taken from the peripubertal rat to stimulated "misinsurges" suggests that the diurnal rhythm of LH secretion is as effective in inducing estradio and progesterone release as is continuous exposure (Urbanski and Ojeda, 1985b). Since considerably less LH is required to effect similar steroid release, the authors suggest that this the rhythmic mode of release is functionally more efficient than is continuous release. In any case, the normal pattern of gonadotropin release in the immature hamster is rhythmic and the result is that ovarian function is also temporally organized as well as stimulated. A further suggestion of the importance of the rhythm of LH and FSH release in hamsters is shown by experiments in which the age of onset of rhythmic release is advanced by daily injections of GnRH beginning at 8 days of age, which is about a week earlier than normal rhythmic release would be expected (Donham *et al*., 1986). As a result of this experimental induction of a daily gonadotropin surges, the age of first ovulation was advanced. In a converse experiment, when the appearance of daily afternoon surges was delayed for about a week by daily phenobarbital injections, the age of puberty was delayed by a corresponding amount. Recently, similar results were reported after experimental advancement of the age of onset of mid-afternoon LH surges in the immature rat (Urbanski and Ojeda, 1987).

D. Neuroendocrine Mechanisms Regulating Daily Endocrine Rhythms in Rodents.

What event precipitates the rather abrup appearance of daily surges of LH and FSH on days 15-17? The components required, in addition to the oscillator and its afferent connections, as discussed in Section I, B, are, 1) a neuroendocrine pathway which links the output of the neural clock to the LHRH neurons, and 2) the pituitary gonadotropes. The appearance of the daily LH and FSH surges may signal maturation of one or more of these components. If so, the critical maturation may not be the oscillator itself (Section I, B), nor the putitary gonadotropes, since the pituitary responds to LHRH challenge by the 8th day of life (1Smith et al., 1982). In the rat, there is postnatal transformation of LHRH cell subtypes (Wray and Hoffman, 1986) and synaptic formation in the hypothalamus occurs at least until the time of puberty (Matsumoto and Arai, 1976). These results may suggest that, apart from the oscillator itself and the putitary, most of the elements required for the endocrine expression of the neural clock develop postnatally.

Removal of the ovaries before 15-17 days of age prevents appearance of the daily LH and FSH surges (Donham et al., 1985); this effect is prevented by simultaneous implantation of tonic-release capsules of estradiol (Donham et al., 1987). Progesterone is ineffective. It appears unlikely that ovarian steroids are, by themselves., initiating the daily rhythm since steroid levels of the serum neither dramatically increase prior to the appearance of the daily rhythm of LH and FSH, nor are they rhythmic. Alternatively, estradiol may be required for the expression of a neural oscillator that matures in the absence of the ovaries. Consistent with this view, we have observed that effects of ovariectomy on days 10-134 of age (i.e., prior to the normal appearance of daily surges) are reversed by implantation of estradiol capsules on day 21 of age (after surges would normally be expected to appear; Donham and Stetson, unpublished).

E. Transition of Daily Rhythms into 4-Day Estrous Cycles

Sequential sampling of individuals as they approach pubertal age (indicated by the first appearance of a characteristic vaginal discharge that, in the adult, is always observed on the morning of estrus) provides an indication of the shifting hormonal patterns during the peripubertal period (Donham and Stetson, unpublished). As early as 9 days prior to the first vaginal estrus, there is a large amplitude afternoon surge of LH, significantly larger than the surge of the previous day or the one following. Subsequently, another amplified LH surge appears 4 days later, i.e., at 5 days prior to first vaginal estrus and, finally, the normal preovulatory surge occurs on the day before first estrus. During this period, there is a continuance of the daily surges so that, for about a week before puberty, the two rhythms are superimposed. Serum progesterone primarily reflects the daily rhythm while, estradiol, like LH, shows both a daily and a 4-day pattern.

The phase of the daily surges was identical to that of the larger surges occurring at 4-day intervals. However, given the limited sampling schedule it is certainly possible that shifts in phase occur during this period, especially given the ability of estradiol to shift phase of daily surges in the adult (Moline *et al.*, 1986). Clearly, however, the amplitude of the surges is being affected in some manner. The data suggest that, as the daily surges of LH are being gradually dampened, the daily surges of estradiol are increasing. It seems likely that at 9, 5 and 1 days before first vaginal discharge, the ratio of estradiol to progesterone increases and elicits an amplified surge of LH.

Jorgenson and Schwartz (1987) argue that, in the adult hamster, since barbiturate blockade for 3 days in short-day anestrous females does not evoke measurable ovarian development and since daily injections of ovine LH and FSH for 24 days did not reliably interrupt estrous cycles in long-day females, that rhythmic release of gonadotropins simply reflects the low estrogen levels typical of anestrous females. Reduction of circulating estrogen to non-measurable levels by ovariectomy plus adrenalectomy in both adults (Bittman and Goldman, 1979) and prepubertal females (Donham *et al.*, in press) does not eliminate the daily surges, suggesting that they reflect the ability of the neural oscillator system to evoke rhythmic gonadotropin release. Higher levels of estradiol (20-60 pg/ml of serum) are apparently incompatible with expression of the 24 hour rhythm and thus as levels of estrogen increase in the peripubertal female and in the anestrus-estrus transition the estrogen/progesterone ratio increases and results in the transition to 4-day estrous cycles. The impetus for estrogens to increase and elicit this chain of events is as yet undefined.

IV. References

Allen, C., and Kendall, J.W. (1967). Maturation of the circadian rhythm of plasma corticosterone in the rat. *Endocrinology* *80*, 926-930.

Alleva, J.J., Walewski, M.V., and Alleva, F.R. (1971). A biological clock controlling the estrous cycle of the hamster. *Endocrinology* *88*, 1368-1379.

Bittman, E.L., and Goldman, B.D. (1979). Serum levels of gonadotrophins in hamsters exposed to short photoperiods: effects of adrenalectomy and ovariectomy. *J. Endocrinol.* *83*, 113-118.

Bittman, E.L., Karsch, F.J. (1984). Nightly duration of pineal melatonin secretion determines the reproductive response to inhibitory daylength in the ewe. Biol. Reprod. 30, 585-593.

Bittman, E.L., Dempsey, R.J., Karsch, F.J. (1983). Pineal melatonin secretion drives the reproductive response to daylength in the ewe. Endocrinology 113, 2276-2283.

Brackmann, M., and Hoffmann, K. (1977). Pinealectomy and photoperiod influence testicular development in the Djungarian hamster. Naturwissenschaften 64, 341.

Bridges, R.S., and Goldman, B.D. (1975). Diurnal rhythms in gonadotropins and progesterone in lactating and photoperiod induced acyclic hamsters. Biol. Reprod. 13, 617-622.

Brown-Grant, K., and Raisman, G. (1977). Abnormalities in reproductive function associated with the destruction of the suprachiasmatic nuclei in female rats. Proc. R. Soc. London Ser.B. 198, 279-296.

Carter, D.S., and Goldman, B.D. (1983a). Antigonadal effects of timed melatonin infusion in pinealectomized male Djungarian hamsters (Phodopus sungorus sungorus): Duration is the critical parameter. Endocrinology 113, 1261-1267.

Carter, D.S., and Goldman, B.D. (1983b). Progonadal role of the pineal in the Djungarian hamster (Phodopus sungorus sungorus): Mediation by melatonin. Endocrinology 113, 1268-1273.

Darrow, J.M., and Goldman, B.D. (1986). Circadian regulation of pineal melatonin and reproduction in the Djungarian hamster. J. Biol. Rhythms 1, 39-54.

Darrow, J.M., Davis, F.C., Elliott, J.A., Stetson, M.H., Turek, F.W., and Menaker, M. (1980). Influence of photoperiod on reproductive development in the golden hamster. Biol. Reprod. 22, 443-450.

Davis, F.C. (1981). Ontogeny of Circadian Rhythms. In "Handbook of Behavioral Neurobiology, Volume 4, Biological Rhythms" (J. Aschoff, ed.) pp, 257-270. Plenum Press, New York.

Davis, F.C., and Gorski, J. (1983). Entrainment of circadian rhythms in utero: Role of the maternal suprachiasmatic nucleus. Soc. Neurosci. Abst. 8, 625.

Deguchi, T. (1975). Ontogenesis of a biological clock for serotonin:acetylcoenzyme A N-acetyltransferase in pineal gland of rat. Proc. Ntl. Acad. Sci. U.S.A. 72, 2814-2818.

Deguchi, T. (1977). Circadian rhythms of enzyme and running activity under ultradian lighting schedule. Am. J. Physiol. 232, E375-E381.

DiPinto, M.N., and Stetson, M.H. (1981). Clock-timed gonadotropin release in lactating but not pregnant hamsters. In "Photoperiodism and Reproduction: (R. Ortavant, J. Pelletier, and J.P. Ravault, eds..), pp. 83-97. Les Colloques de l'INRA.

Donham, R.S., and Stetson, M.H. (1985). Neonatal androgen abolishes clock-timed gonadotrophin release in prepubertal and adult female hamsters. J. Reprod. Fert. 73, 215-221.

Donham, R.S., DiPinto, M.N., and Stetson, M.H. (1984). Twenty-four hour rhythms of gonadotropin release induces cyclic progesterone secretion by the ovary of prepubertal and adult hamsters. Endocrinology 114, 821-826.

Donham, R.S., DiPinto, M.N., and Stetson, M.H. (1985). Effects of ovariectomy on clock-timed daily gonadotropin rhythms in prepubertal golden hamsters. Biol. Reprod. 32, 284-289.

Donham, R.S., Creyaufmiller, N., Lyons, T.J., and Stetson, M.H. (1986). Temporal relationship between the onset of daily gonadotrophin surges and of puberty in female golden hamsters. J. Endocrinol. 108, 219-224.

Donham, R.S., Posern, F.V., and Stetson, M.H. (1987). Daily rhythms of serum luteinizing hormone in the immature hamster are estradiol-dependent. Biol. Reprod. 36, 864-870.

Donham, R.S., Rollag, M.D., and Stetson, M.H. (1988). Daily rhythms of pituitary-ovarian function in the immature hamster are independent of adrenal and pineal infuence. J. Reprod. Fert. in press.

Elliott, J.A., and Goldman, B.D. (1986). Pineal gland of pregnant Djungarian hamsters mediates reception of photoperiodic information by the developing fetus. Biol. Reprod. 34, (Suppl. 1, 221.

Everett, J.W. (1972). The third annual Carl G. Hartman lecture. Brain, pituitary gland, and the ovarian cycle. Biol. Reprod. 6, 3-12.

Everett, J.W., and Sawyer, C.H. (1949). A neural timing factor in the mechanism by which progesterone advances ovulation in the cyclic rat. Endocrinology 45, 198-218.

Everett, J.W., and Sawyer, C.H. (1950). A 24-hour periodicity in the "LH release apparatus" of female rats, disclosed by barbiturate sedation. Endocrinology 47, 198-218.

Fitzgerald, K.M., and Zucker, I. (1976). Circadian organization of the estrous cycle of the golden hamster. Proc. Natl. Acad. Sci. U.S.A. 73, 2923-2927.

Frost, D.O., So, K.-F., and Schneider, G.E. (1979). Postnatal development of retinal projections in Syrian hamsters: a study using autoradiographic and anterograde degeneration techniques. Neuroscience 4, 1649-1677.

Fuchs, J.L., and Moore, R.Y. (1980). Development of circadian rhythmicity and light responsiveness in the rat suprachiasmatic nucleus: A study using the 2-deoxy-[1-^{14}C] glucose method. Proc. Natl. Acad. Sci. U.S.A. 77, 1204-1208.

Goldman, B.D., and Darrow, J.M. (1983). The pineal gland and mammalian photoperiodism. Neuroendocrinology 37, 386-396.

Goldman, B.D., Darrow, J.M., and Yogev, L. (1984). Effects of timed melatonin infusions on reproductive development in the Djungarian hamster (Phodopus sungorus). Endocrinology 114, 2074-2083.

Greenwald, G.S., and Peppler, R.D. (1968). Prepubertal and pubertal changes in the hamster ovary. Anat. Rec. 161, 447-458.

Hendrickson, A.E., Wagner, N., and Cowan, W.M. (1972). An autoradiographic and electron microscope study of retinohypothalamic connections. Z. Zellforsh 135, 1-26.

Hoffmann, K. (1974). Testicular involution in short photoperiods inhibited by melatonin. Naturwissenschaften 61, 364-365.

Hoffmann, K. (1978). Effects of short photoperiods on puberty, growth and moult in the Djungarian hamster. (Phodopus sungorus). J. Reprod. Fertil. 54, 29-35.

Hong, S.M., and Stetson, M.H. (1987). Detailed diurnal rhythm of sensitivity to melatonin injections in Turkish hamsters, Mesocricetus brandti. J. Pineal Res. 4, 69-78.

Horton, T.H. (1984a). Variability in response to photoperiod by the vole, Microtus montanus. Ph.D. dissertation, The University of Utah, Salt Lake City, Utah.

Horton, T.H. (1984b). Growth and reproductive development of male Microtus montanus is affected by the prenatal photoperiod. Biol. Reprod. 31, 499-504.

Horton, T.H. (1984c). Growth and maturation in Microtus montanus: Effects of photoperiods before and after weaning. Can. J. Zool. 67, 1741-1746.

Horton, T.H. (1985). Cross-fostering of voles demonstrates in utero effect of photoperiod. Biol. Reprod. 33, 934-939.

Horton, T.H., Ray, S.L., Fry, L., and Stetson, M.H. (1987a). Daily pattern of sensitivity to prenatal melatonin injections in Djungarian hamsters, Phodopus sungorus. Biol. Reprod. 36, (Suppl. 1), 175.

Horton, T.H., Ray, S.L., and Stetson, M.H. (1987b). Gestational photoperiod alters effect of melatonin capsules. Amer. Zool. in press.

Ifft, J.D. (1982). An autoradiographic study of the time of final division of neurons in rat hypothalamic nuclei. J. Comp. Neur. 144, 193-204.

Jorgenson, K.L., and Schwartz, N.B. (1987). Dynamic pituitary and ovarian changes occurring during the anestrus to estrus transition in the golden hamster. Endocrinology 120, 34-42.

Klein, D.C. (1978). Pineal gland as a model of neuroendocrine control mechanisms. In "The Hypothalamus" (S. Reichlin, R.J. Baldessarini, and J.B. Martin, eds.), Association for Research in Nervous and Mental Disease 56, 303-326. Raven Press, New York.

Klein, D.C., and Weller, J.L. (1970). Indole metabolism in the pineal gland: A circadian rhythm in N-acetyltransferase. Science 169, 1093-1085.

Lehman, M.N., Silver, R., Gladstone, W.R., Kahn, R.M., Gibson, M., and Bittman, E.L. (1984). Circadian rhythmicity restored by neural transplant. Immunocytochemical characterization of the graft and its integration with the host brain. J. Neurosci. 7, 1626-1638.

Laudon, M., and Zisapel, N. (1986). Characterization of central melatonin receptors using 125I-melatonin. FEBS Lett. 197, 9-12.

Matsumoto, A., and Arai, Y. (1976). Developmental changes in synaptic formation in the hypothalamic arcuate nucleus of female rats. Cell Tiss. Res. 169, 143-156.

Moline, M.L., Albers, H.E., and Moore-Ede, M.C. (1986). Estrogen modifies the circadian timing and amplitude of the luteinizing hormone surge in female hamsters exposed to short photoperiods. Biol. Reprod. 35, 516-523.

Moore, R.Y., and Lenn, N.J. (1972). A retinohypothalamic projection in the rat. J. Comp. Neurol. 146, 1-14.

Moore, R.Y., and Card, J.P. (1985). Visual pathways and the entrainment of circadian rhythms. In "The Medical and Biological Effects of Light" (R.J. Wurtman, M.J. Baum, and J.T. Potts, Jr., eds.), Ann. N.Y. Acad. Sci. 453, 123-133.

Negus, N.C., and Berger, P.J. (1972). Environmental factors and reproductive processes in mammalian populations. In "Biology of Reproductions, Basic and Clinical Studies" (J.T. Velardo and B.A. Kaspro, eds.), pp. 89-98, Third Pan American Congress of Anatomy. New Orleans, Louisiana.

Negus, N.C., Berger, P.J., and Brown, B.W. (1986). Microtine population dynamics in a predictable environment. Can. J. Zool. 64, 785-792.

Pickard, G.E., and Turek, F.W. (1983). The hypothalamic paraventricular nucleus mediates the photoperiodic control of reproduction but not the effects of light on the circadian rhythm of activity. Neurosci. Lett. 43, 67-72.

Rapp, P.E. (1987). Why are so many biological systems periodic? Prog. Neurobiol. 29, 261-273.

Reppert, S.M., and Klein, D.C. (1978). Transport of maternal $[^3H]$melatonin to suckling rats and the fate of $[^3]$melatonin rat. Endocrinology 102, 582-588.

Reppert, S.M., and Schwartz, W.J. (1983). Maternal coordination of the fetal biological clock in utero. Science 220, 969-971.

Reppert, S.M., and Schwartz, W.J. (1986a). Maternal endocrine extirpations do not abolish maternal coordination of the fetal circadian clock. Endocrinology 119, 1763-1767.

Reppert, S.M., and Schwartz, W.J. (1986b). Maternal suprachiasmatic nuclei are necessary for maternal coordination of the developing circadian system. J. Neurosci. 6, 2724-2729.

Reppert, S.M., and Uhl, G.R. (1987). Vasopressin messenger ribonucleic acid in supraoptic and suprachiasmatic nuclei: Appearance and circadian regulation during development. Endocrinology 120, 2483-2487.

Rollag, M.D., and Stetson, M.H. (1981). Ontogeny of the pineal melatonin rhythm in golden hamsters. Biol. Reprod. 24, 311-314.

Rollag, M.D., DiPinto, M.N., and Stetson, M.H. (1982). Ontogeny of the gonadal response of golden hamsters to short photoperiod, blinding and melatonin. Biol. Reprod. 27, 898-902.

Rosenwasser, A.M., and Adler, N.T. (1986). Structure and function in circadian timing systems: evidence for multiple coupled circadian oscillators. Neurosci. Biobehav. Rev. 10, 432-448.

Rusak, B., and Zucker, I. (1979). Neural regulation of circadian rhythms. Physiol. Rev. 59, 449-526.

Schwartz, W.J., Gross, R.A., and Morton, M.T. (1987). The suprachiasmatic nuclei contain a tetrodotoxin-resistant circadian pacemaker. Proc. Natl. Acad. Sci. U.S.A. 84, 1694-1698.

Seegal, R.F., and Goldman, B.D. (1975). Effects of photoperiod on cyclicity and serum gonadotropins in the Syrian hamster. Biol. Reprod. 12, 223-231.

Shaha, C., and Greenwald, G.S. (1983). Development of steroidogenic activity in the ovary of the prepubertal hamster. 1. Response to in vivo or in vitro exposure to gonadotropins. Biol. Reprod. 28, 1231-1241.

Shibata, S., Liou, S.Y., and Ueki, S. (1983). Development of the circadian rhythm of neuronal activity in suprachiasmatic nucleus of rat hypothalamic slices. Neurosci. Lett. 43, 231-234.

Sisk, C.L., and Turek, F.W. (1987). Reproductive responsiveness to short photoperiod develops postnatally in male golden hamsters. J. Androl. 8, 91-96.

Smith III, S.G., and Stetson, M.H. (1980). Maturation of the clock-timed gonadotropin release mechanism in hamsters: a key event in the pubertal process? Endocrinology 107, 1334-1337.

Smith III, S.G., Matt, K.S., Prestowitz, W.F., and Stetson, M.H. (1982). Regulation of tonic gonadotropin release in prepubertal female hamsters. Endocrinology 110, 1262-1267.

Stanfield, B., and Cowan, W.M. (1976). Evidence for a change in the retinohypothalamic projection in the rat following early removal of one eye. Brain Research 104, 129-136.

Stephan, F.K., and Zucker, I. (1972). Circadian rhythms in drinking behavior and locomotor activity of rats are eliminated by hypothalamic lesions. Proc. Natl. Acad. Sci. U.S.A. 69, 1583-1586.

Stetson, M.H., and Anderson, P.J. (1980). Circadian pacemaker times gonadotropin release in free-running female hamsters. Am. J. Physiol. 238, R23-R27.

Stetson, M.H., and Hamilton, B. (1981). The anovulatory hamster: a comparison of the effects of short photoperiod and daily melatonin injections on the induction and termination of ovarian cyclicity. J. Exp. Zool. 215, 173-178.

Stetson, M.H., and Gibson, J.T. (1977). The estrous cycle in golden hamsters: a circadian pacemaker times preovulatory gonadotropin release. J. Exp. Zool. 201, 289-294.

Stetson, M.H., and Tay, D.E. (1983). Time-course of sensitivity of golden hamsters to melatonin injections throughout the day. Biol. Reprod. 29, 432-438.

Stetson, M.H., and Watson-Whitmyre, M. (1976). Nucleus suprachiasmaticus: the biological clock in the hamster? Science 191, 197-199.

Stetson, M.H., and Watson-Whitmyre, M. (1984). Physiology of the pineal and its hormone melatonin in annual reproduction in rodents. In "The Pineal Gland" (R.J. Reiter, ed.), pp. 109-153. Raven Press, New York.

Stetson, M.H., and Watson-Whitmyre, M. (1986). Effects of exogenous and endogenous melatonin on gonadal function in hamsters. J. Neural Transm. (Suppl.) 21, 55-80.

Stetson, M.H., Watson-Whitmyre, M., and Matt, K.S. (1977). Circadian organization in the regulation of reproduction: timing of the 4-day estrous cycle of the hamster. J. interdiscipl. Cycle Res. 8, 350-352.

Stetson, M.H., Watson-Whitmyre, M., and Matt, K.S. (1978). Cyclic gonadotropin release in the presence and absence of estrogenic feedback in ovariectomized golden hamsters. Biol. Reprod. 19, 40-50.

Stetson, M.H., Watson-Whitmyre, M., DiPinto, M.N., and Smith III, S.G. (1981). Daily luteinizing hormone release in ovariectomized hamsters: effect of barbiturate blockade. Biol. Repro. 24, 139-144.

Stetson, M.H., Sarafidis, E., and Rollag, M.D. (1986a). Sensitivity of adult male Djungarian hamsters (Phodopus sungorus) to melatonin injections throughout the day: Effects on the reproductive system and the pineal. Biol. Reprod. 35, 618-632.

Stetson, M.H., Elliott, J.A., and Goldman, B.D. (1986b). Maternal transfer of photoperiodic information influences the photoperiodic response of prepubertal Djungarian hamsters (Phodopus sungorus sungorus). Biol. Repro. 34, 664-670.

Takahashi, K., Hanada, K., Kobayashi, K., Hayafuji, C., Otani, S., and Takahashi, Y. (1979). Development of the circadian adrenocortical rhythm in rats: studied by determination of 24- or 48-hour patterns of blood corticosterone levels in individual pups. Endocrinology 104, 954-961.

Tamarkin, L., Reppert, S.M., and Klein, D.C. (1979). Regulation of pineal melatonin in the Syrian hamster. Endocrinology 104, 385-389.

Tamarkin, L., Reppert, S.M., Klein, D.C., Pratt, B., and Goldman, B.D. (1980a). Studies on the daily pattern of pineal melatonin in the Syrian hamster. Endocrinology 107, 1525-1529.

Tamarkin, L., Reppert, S.M., Orloff, D.J., Klein, D.C., Yellon, S.M., and Goldman, B.D. (1980b). Ontogeny of the pineal melatonin rhythm in the Syrian (Mesocricetus auratus) and Siberian (Phodopus sungorus) hamsters and in the rat. Endocrinology 107, 1061-1064.

Urbanski, H.F., and Ojeda, S.R. (1985a). In vitro simulation of prepubertal changes in pulsatile luteinizing hormone release enhances progesterone and 17β-estradiol secretion from immature rat ovaries. Endocrinology 117, 638-643.

Urbanski, H.F., and Ojeda, S.R. (1985b). The juvenile-peripubertal transition period in the female rat: establishment of a diurnal pattern of pulsatile luteinizing hormone secretion. Endocrinology 117, 644-649.

Urbanski, H.F., and Ojeda, S.R. (1987). Activation of uteinizing hormone-releasing hormone release advances the onset of female puberty. Neuroendocrinology 46, 273-276.

Vacas, M.I., and Cardinali, D.P. (1979). Diurnal changes in melatonin binding site of hamster and rat brains: Correlation with neuroendocrine responsiveness to melatonin. Neurosci. Lett. 15, 259-263.

van Veen, Th., Brackmann, M., and Moghimzadeh, E. (1978). Post-natal development of the pineal organ in the hamsters Phodopus sungorus and Mesocricetus auratus. Cell Tiss. Res. 189, 241-250.

Watson-Whitmyre, M. (1985). Photoperiodism in the golden hamster: Dependence on rhythmic sensitivity to melatonin. Ph.D. Dissertation, University of Delaware.

Watson-Whitmyre, M., and Stetson, M.H. (1983). Simulation of peak pineal melatonin release restores sensitivity to evening melatonin injections in pinealectomized hamsters. Endocrinology 112, 763-765.

Weaver, D.R., and Reppert, S.M. (1986). Maternal melatonin communicates daylength to the fetus in Djungarian hamsters. Endocrinology 119, 2861-2863. (Rapid Communication).

Wirz-Justice, A. (1987). Circadian rhythms in mammalian neurotransmitter receptors. Progr. Neurobiol. 29, 219-259.

Wray, S., and Hoffman, G. (1986). Postnatal morphological changes in rat LHRH neurons correlated with sexual maturation. Neuroendocrinology 43, 93-97.

Yellon, S.M., and Goldman, B.D. (1984). Photoperiod control of reproductive development in the male Djungarian hamster (Phodopus sungorus). Endocrinology 114, 664-670.

Yellon, S.M., Tamarkin, L., Pratt, B.L., and Goldman, B.D. (1982). Pineal melatonin in the Djungarian hamster: Photoperiodic regulation of a circadian rhythm. Endocrinology 111, 488-492.

Yellon, S.M., Tamarkin, L., and Goldman, B.D. (1985). Maturation of the pineal melatonin rhythm in long- and short-day reared Djungarian hamsters. Experientia 41, 651-652.

Yuwiler, A., Klein, D.C., Buda, M., and Weller, J.L. (1977). Adrenergic control of pineal N-acetyltransferase activity: Developmental aspects. Am. J. Physiol. 233, E141-E146.

11

Comparative Aspects of Female Puberty[1]

Sergio R. Ojeda
Henryk F. Urbanski

Division of Neuroscience
Oregon Regional Primate Research Center
Beaverton, OR

I. *INTRODUCTION*

During recent years a renewed interest in studying the neuroendocrine control of sexual development has arisen. Several investigators using different animal models and various contemporary techniques have begun to explore in great depth the mysteries of the central nervous system (CNS) in an attempt to disclose how the brain may influence reproductive development. This is indeed a formidable and exciting task, particularly in our times when new and powerful methodologies are being added to our scientific arsenal.

A simple, and perhaps correct view is that the basic mechanisms which underlie the acquisition of reproductive capacity are the same in all eutherian mammals. Some may not be expressed as clearly, and for as long a duration, in one species as compared to another but they may still occur as an integral part of the cascade of developmental events that lead to the attainment of puberty. This seemingly naive assumption provides the basis for the view that some animal models can be of a greater value than others when trying to unravel the mechanisms underlying different physiological events. For example, the juvenile hiatus of gonadotropin secretion can be best studied in sub-human

[1]Supported by grants from the NIH (HD-09988, Project IV) and the NSF (BNS 831807).

primates; the pubertal activation of estradiol positive feedback, even though it occurs in all species, can be examined more readily and in greater detail in the laboratory rat than in larger animals. The developing lamb offers an excellent model to study the effect of photoperiod and nutrition on sexual maturation.

The present article will not attempt to provide a comprehensive account of all that is known about the process of puberty in different species. Instead, it will discuss the progress made in this field based on data collected from the human and from three select animal models: the rat, the sheep, and the rhesus macaque. The resemblances and differences between these models and humans will be stressed and interesting aspects concerning the sexual development of other species will also be mentioned. For a more detailed account the interested reader is referred to several recent reviews (Bronson and Rissman, 1986; Plant, 1987; Foster, 1987; Ojeda and Urbanski, 1987).

II. THE IMMATURE STAGE

During the interval between birth and the attainment of sexual maturity, the neuroendocrine reproductive system undergoes a series of developmental changes which unfold at a pace much slower than during the actual transition to sexual maturity. More importantly, however, it is during this time that the basic regulatory mechanisms governing neuroendocrine reproductive competence become firmly established.

A. Length of Prepubertal Development and Life Expectancy

The temporal pattern of sexual development varies considerably between species, and within the same species it may show considerable individual variation (Ramirez, 1973). Although there is little doubt that the primary factors controlling the length of prepubertal development are genetic (Land, 1978), a variety of environmental, nutritional and social factors are powerful supplementary determinants of the timing of puberty (Bronson and Rissman, 1986). In general, a correlation appears to exist between the life expectancy of a particular species and the rate of sexual development. Thus, female rats which in the wild have a life expectancy of no longer than 1-2 years, reach puberty within 5 to 6 weeks after birth. Ewes have a life span of several years with spring-born lambs reaching puberty between 8-9 months of age. Female rhesus macaques (Macaca mulatta) which may live for several decades reach maturity between 3-4 years of age. The human female, on the other hand, lives much longer (an average life expectancy of 70 years), and does not normally become sexually mature before the age of 10. The marked differences in the timing of puberty of several mammalian species are demonstrated in Table I.

TABLE I. Differences in the timing of puberty of several mammalian species

Species	Gestation, days	Puberty [a]
Primates		
man	270	12-13 yr
chimpanzee	238	8 yr
macaque	168	3- 4 yr
Farm animals		
horse	336	12-15
cattle	280	8-12
goat	151	4- 8
sheep	150	5- 7
swine	112	4- 7
Small laboratory animals		
hamster	16	1
mouse	20	1, 2
rat	21	1, 3
rabbit	30	5- 6
ferret	42	1, 3
dog	60	8-24
guinea pig	63	2- 3
cat	63	7-12

[a]Months unless otherwise indicated. Modified from: Ramirez, 1973 (with permission).

B. The Hypothalamic-Pituitary Unit

Measurement of circulating gonadotropin levels has been extensively used to estimate the activity of the luteinizing hormone-releasing hormone (LHRH) secretory system during both sexual maturation and adulthood. In the four major species considered in this article there exists a period during postnatal development in which gonadotropin secretion appears to proceed independently from the presence of the ovary. In the laboratory rat, such a period is very short and limited to the first few days after birth (Goldman et al., 1971). A similar situation exists in the female lamb (Foster et al., 1975) in which luteinizing hormone (LH) secretion fails to increase in response to ovariectomy for the first 4 weeks after birth (first ovulation: 26-50 weeks of age). A different picture emerges in the case of the human and rhesus monkey in which a transient postnatal period of LH response to gonadectomy is followed by a prolonged phase of insensitivity which extends throughout most of the infantile-juvenile phases of development (Winter and Faiman, 1972; Plant et al., 1974). A similar prolonged period of unresponsiveness has been described in the guinea pig (first ovulation: 60 days of age) which fails, for up to two weeks, to respond with an increase in plasma LH when the ovaries are removed on postnatal day 10, but responds readily if ovariectomy is performed on day 30 (Nass et al., 1984). These data, though interesting because they suggest that the guinea pig could be a viable animal model in which to examine the juvenile hiatus in gonadotropin secretion seen in humans, have been recently questioned (Fraser and Plant, 1987). These latter authors found that if guinea pigs were kept with their mothers beyond day 10 instead of being weaned as Nass et al. (1984) did, the animals responded to ovariectomy with a vigorous increase in LH secretion. The issue is, however, still unsettled because maintaining the animals with their mothers may have allowed for maternal milk LHRH to stimulate LH secretion from the offspring's pituitary (for review see Smith and Ojeda, 1986).

Elucidation of the mechanisms underlying the juvenile reduction in gonadotropin secretion in humans and sub-human primates is of considerable importance, because of the profound repercussions it will have on current efforts to understand the pubertal process. Of potential importance in this regard is the recent finding (Plant, 1986a) that neonatal ovariectomy of rhesus monkeys results in a sustained hypersecretion of follicle-stimulating hormone (FSH), a pattern of response that is in marked contrast to the biphasic profile normally observed in males (Plant, 1986), i.e., an initial, marked infantile increase in FSH followed by a decrease to basal levels throughout juvenile development (Fig. 1). Interestingly, the LH response to ovariectomy, though truncated in females, shows a pattern that is basically similar in both sexes. These findings and the observation that the frequency of pulsatile LH release is slower in females than in males have led to the conclusion that the so-called hypothalamic GnRH pulse generator is not fully operative in immature female monkeys (Plant, 1986a).

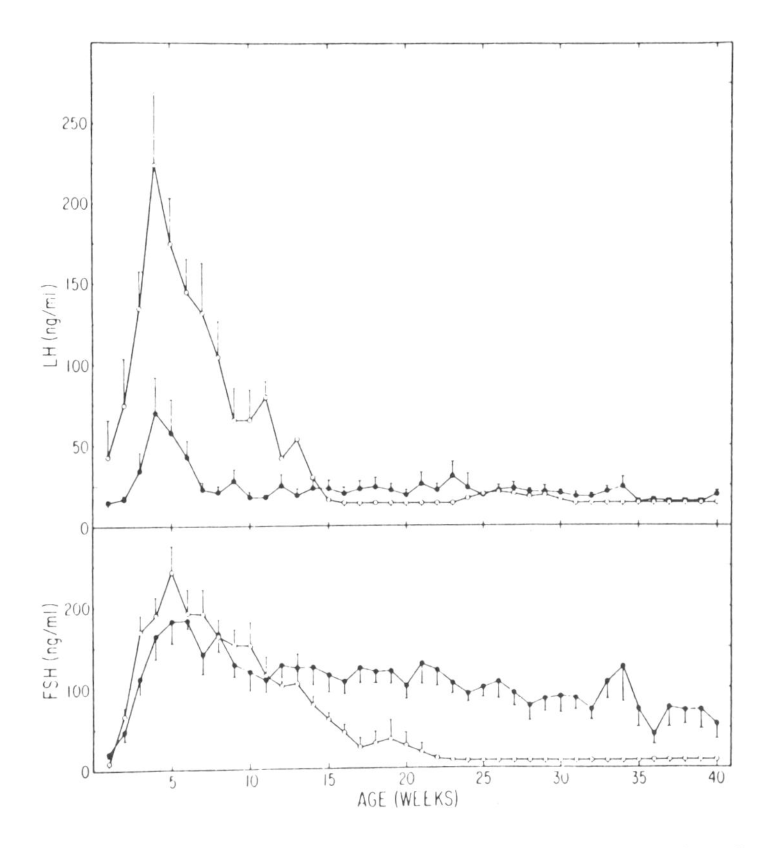

Fig. 1. A comparison of the time courses of circulating concentrations (mean ± SEM) of LH (top panel) and FSH (bottom panel) in agonadal male (n=4; O-O) and female (n=6; ●-●) rhesus monkeys during the first 40 weeks of postnatal life. Gonadectomy in both sexes was performed between 1-2 weeks of age. From: Plant (1986). Endocrinology 119: 539 (with permission).

If this is the case, it would also suggest that the frequency of LHRH discharges in juvenile ovariectomized animals is too slow for LH release to be sustained, but sufficiently fast for FSH secretion to be maintained at a high rate.

For many years the view has been maintained that the reduced rate of gonadotropin secretion seen in juvenile primates is due to the predominance of inhibitory tone(s) on LHRH secretion, a phenomenon known as the "central restraint" (for a review see Reiter and Grumbach, 1982). While it is entirely possible that maturation of excitatory and inhibitory inputs may occur at different postnatal stages there is no experimental evidence that supports the concept of a central restraint operating during juvenile development other than the findings that hypothalamic lesions, of experimental or pathological nature, in rats and humans, respectively, can advance the onset of puberty (Donovan and Van der Werff ten Bosch, 1956; Gellert and Ganong, 1960; Marks and Elders, 1979). Experimental lesions of the monkey hypothalamus (Norman and Spies, 1981; Terasawa et al., 1984a), however, have failed to advance puberty as dramatically as seen in human females with idiopathic precocious puberty. Indeed, the lesion of certain hypothalamic regions rather than removing an inhibitory tone may set in motion repair or compensatory mechanisms which may activate, rather than depress, the secretory activity of LHRH neurons. Clearly, much more research is needed to resolve this issue, but it is revealing that pharmacological removal of a potential inhibitory component such as the opioid system has failed to revert the quiescent mode of LHRH secretion in children (Fraioli et al., 1984; Sander et al., 1984). In contrast, pulsatile administration of the excitatory amino acid analog N-methyl-D-aspartic acid to juvenile male rhesus monkeys has been shown to elicit rhythmic discharges of LH release and to induce precocious puberty (Plant et al., 1987).

As indicated before, ovariectomy of juvenile rats or sheep results in a prompt increase in plasma gonadotropins indicating that the LHRH-LH secreting system is under strong gonadal inhibitory control, but that once this is removed LHRH secretion can increase without central restraint. It must be pointed out, however, that neonatal ovariectomy of rats produces a biphasic pattern of FSH release, levels first increasing for about 2 weeks and then decreasing during the juvenile period to intermediate values (Urbanski and Ojeda, unpublished). This pattern may reflect a changing frequency of pulsatile LHRH release.

C. *The Ovaries*

Although formation of ovulatory follicles fails to occur during prepubertal development, follicular growth and atresia are ongoing processes throughout prepubertal years. The human ovary undergoes cycles of follicular growth that appear to occur every 28-40 days (Winter et al., 1978). That, indeed, the primate ovary is steroidogenically active is demonstrated by the finding that concentrations of estradiol in the ovarian vein of juvenile rhesus monkeys is 3- to 4-fold greater than peripheral levels (Williams et al., 1982). Moreover, ovariectomy is followed by a decrease in circulating estradiol levels in these animals (Winter et al., 1977). The sheep ovary becomes responsive to gonadotropins between the 2nd to 4th week of postnatal life and can ovulate by 5 to 6 weeks of age when challenged with a large dose of gonadotropin (Worthington and Kennedy, 1979). Little is known, however, regarding the normal developmental changes in steroidogenic capacity and ovarian morphology that occurs during the prepubertal period of the female lamb.

In contrast to this sparcity of information, a wealth of information exists concerning the juvenile rat ovary (Schwartz, 1974; Richards, 1980; Ojeda et al., 1984). The responsiveness of the ovary to gonadotropins increases gradually during juvenile days (Advis and Ojeda, 1978) a phenomenon that appears to be related to an increase in LH (hCG) receptors. The actions of gonadotropins on the rat ovary are facilitated by prolactin and growth hormone which contribute to the maintenance of LH receptors (for a review see Ojeda et al., 1984). An intriguing aspect of ovarian development which has only recently been explored in more detail, concerns the possibility that in addition to the well characterized hormonal control, the central nervous system regulates the development of ovarian production via direct neural connections. Thus, evidence now exists that the immature ovary is innervated by noradrenergic nerve fibers (Burden, 1985) and by nerves containing vasoactive intestinal peptide (VIP), substance P (SP), neuropeptide Y (NPY) and calcitonin-gene related peptide (CGRP) (Ahmed et al., 1985; Dees et al., 1985; McDonald et al., 1987; Calka et al., 1987). While VIP reaches the ovary via the superior ovarian nerve NPY, SP and CGRP fibers are contained in the plexus nerve. Of these peptides only VIP was found to affect ovarian steroidogenesis. It stimulates the secretion of progesterone, estradiol, and androgens (Davoren and Hsueh, 1984; Ahmed et al., 1985), an effect that appears to be at least in part due to the capacity of VIP to induce the synthesis of two key enzyme complexes in the steroidogenic pathway, namely the cholesterol side chain cleavage enzyme and the aromatase enzyme (Trzeciak et al., 1986; George and Ojeda, 1987). Since VIP can induce aromatase enzyme activity before the development of primordial follicles and prior to the acquisition of ovarian responsiveness to FSH (George and Ojeda, 1987), it appears that VIP may represent an early neuroendocrine signal controlling the initial phases of ovarian development.

III. THE ONSET OF PUBERTY

A. The Initiating Events

Although the first cellular manifestations of puberty are likely to occur well before the first changes in plasma hormone levels no specific markers for such early events have been identified that may permit an accurate prediction of the onset of puberty. It is clear, however, that in most species the earliest manifestation of the advent of sexual maturity is a change in the mode of LH release.

1. The First Overt Hormonal Changes

In all four species considered herein, the onset of puberty is characterized by the appearance of changes in pulsatile LH release. In humans, rhesus monkeys, and rats, the predominant change observed is an increase in LH pulse amplitude, although an acceleration of pulse frequency is to be suspected because basal LH levels increase concomitantly (Boyar et al., 1972; Terasawa et al., 1984; Urbanski and Ojeda, 1985). In female sheep an increase in LH pulse frequency is more readily apparent (Foster et al., 1985) (Fig. 2), perhaps because of the prior (juvenile) slow frequency of the LHRH pulse generator.

In both the human and rhesus monkey, the augmentation of LH release occurs during sleep (human) or at night (monkey). In rats the change is observed in the afternoon (Fig. 3) and in sheep it does not appear to be related to the time of day. Although changes in pituitary responsiveness to LHRH may contribute to modulating these changes in LH release, there is little doubt that they are ultimately determined by changes in the secretion of LHRH from the hypothalamus. Indeed, recent evidence has established conclusively that the onset of female puberty in the rhesus monkey is characterized by a gradual increase in LHRH pulse amplitude as measured in push-pull cannula perfusates of the medial basal hypothalamus (Watanabe et al., 1987).

2. The Activation of LHRH-Secreting Neurons

The mechanisms underlying the activation of LHRH secretion at puberty are poorly understood. For many years the hypothesis was accepted that an age-related decrease in sensitivity to gonadal steroid negative feedback was responsible for the pubertal rise in circulating gonadotropins. This, however, can no longer be accepted as a tenable explanation as experiments performed in both rats and primates, and clinical data obtained in humans, indicate that a gonadal-independent change in "central drive" is most

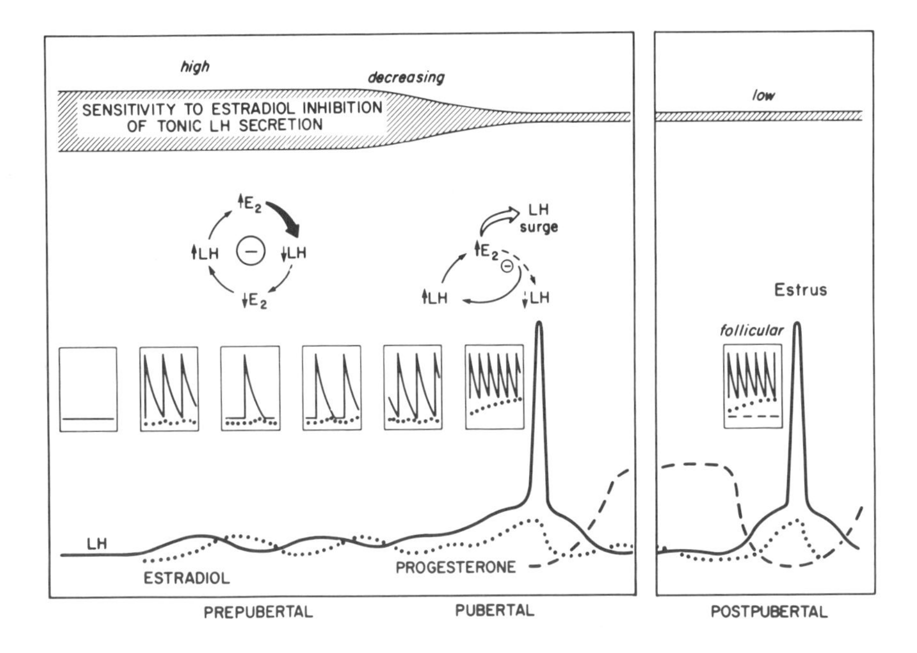

Fig. 2. Schematic hypothesis for the initiation of the gonadotropin surge in the female lamb. Changing baseline of LH and estradiol reflects their discontinuous secretion. Insets depict detailed patterns of LH and estradiol over a 6-hour period. From: Foster (1987). In "The Physiology of Reproduction" (E. Knobil and J. D. Neill, eds.). p. 1737, Raven Press, New York (with permission).

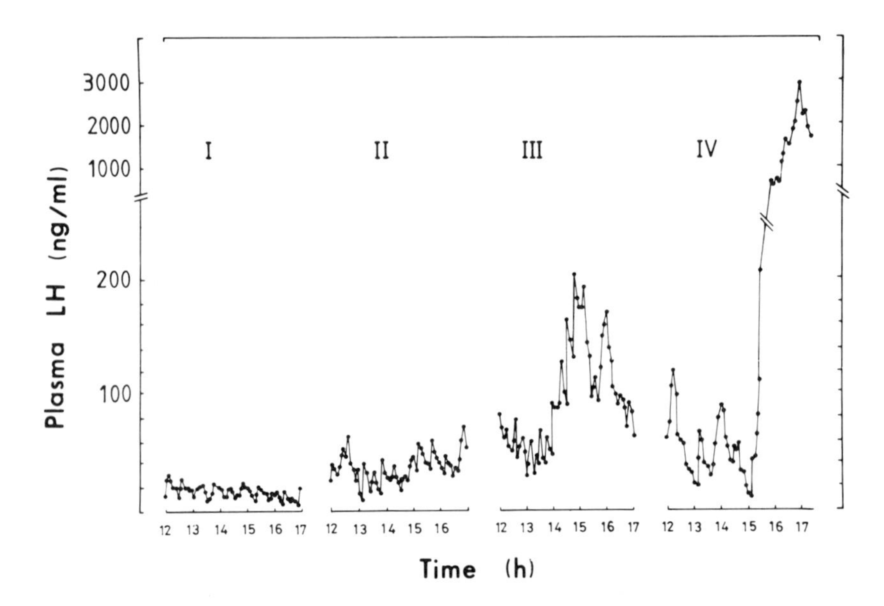

Fig. 3. Postulated sequence of changes in the mode of LH release during the onset of puberty in the female rat. Roman numerals indicate the phases in which different afternoon patterns of LH release were observed in conscious free-moving peripubertal animals, bled every 5 minutes using an automated bleeding technique. Each profile is derived from a different animal. I, Low amplitude pulses similar to those seen in the morning; II, increased basal LH release and LH pulse amplitude; III, minisurge of LH secretion; IV, proper, proestrous surge of LH. From: Ojeda et al. (1986). Rec. Prog. Horm. Res. 42, 434 (with permission).

likely responsible for the initiation of puberty in these species (Fig. 4) (for a review see Plant, 1987; Ojeda et al., 1984). In contrast, attempts to demonstrate a gonadal-independent activation of gonadotropin release in the pubertal sheep have failed (Foster and Yellon, 1987), suggesting that a decreased sensitivity to steroid negative feedback (Fig. 2) may indeed be the principal determinant for the initiation of puberty in this species (Foster, 1987).

Studies in monkeys have shown that the content of LHRH in the hypothalamus is not different between juvenile and peripubertal animals (for references see Plant, 1987), so that reduced stores of this decapeptide cannot be responsible for the failure of LHRH to be released in a pubertal

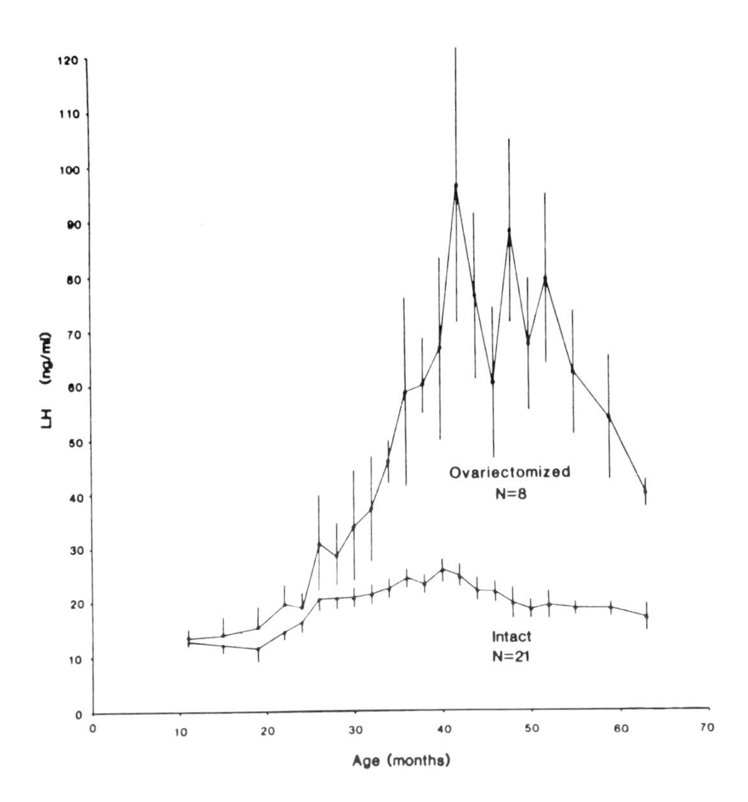

Fig. 4. Developmental changes in the basal level of LH (mean ± SEM) in intact and neonatally ovariectomized female rhesus monkeys. From: Terasawa et al. (1984). Endocrinology 115, 223 (with permission).

manner earlier in development. Studies in the rat, on the other hand, have revealed that the capacity of LHRH neurons to release LHRH in response to intracellular probes such as prostaglandin E_2 (PGE_2), stimulators of cyclic-AMP formation or activators of protein kinase C is already developed before puberty (for a review see Ojeda and Urbanski, 1988). These observations and the finding that, in both rhesus monkeys and female rats, precocious puberty can be induced by pulsatile administration of the excitatory amino acid analog N-methyl-D-aspartic acid (Urbanski and Ojeda, 1987; Plant et al., 1987), permit the conclusion that immaturity of the intracellular machinery responsible for LHRH release is not the limiting factor for puberty to occur. Rather, the initiation of puberty depends on the development of components located beyond these intracellular pathways.

3. The Change in Neural Inputs to the LHRH Secreting Cells

Such a change may involve several components including the appearance of specific receptors on the LHRH cell membrane, the establishment of the proper synaptic circuitry connecting the LHRH neurons to relevant neurotransmitter systems, or a change in excitatory/inhibitory inputs normally involved in regulation of the secretion activity of the LHRH neurons.

While very little is known about this entire issue there is now convincing evidence that LHRH neurons receive synaptic contacts from other LHRH neurons (Leranth et al., 1985a; Pelletier, 1987; Thind and Goldsmith, in press), GABAergic neurons (Leranth et al., 1985), and opioid neurons (Thind and Goldsmith, in press). Whether functional completion of these synaptic contacts is causally related to the time of puberty is unknown. Of interest in this regard, however, is a recent finding that in rats, and rhesus monkeys, the morphological characteristics of LHRH neurons change during postnatal development (Steiner et al., 1983; Wray and Hoffman, 1986) in a gonadal-independent manner (Wray and Gainer, 1987). In the rat there exists two types of LHRH neurons: "irregular" that have spine-like processes on their surface, and "smooth" which lack these processes. While the number of smooth cells decreases during prepubertal development, there is a proportional increase in the number of irregular cells (Wray and Hoffman, 1986). This has been interpreted as indicating a greater number of synaptic contacts being established during sexual development, a concept that would fit well with the possibility that completion of the synaptic circuitry to LHRH neurons is a prerequisite for the pubertal increase in LHRH release to occur. Of considerable interest in this regard is the recent demonstration that the number of irregular LHRH neurons contacted by catecholaminergic nerve fibers increases during sexual development while the catecholaminergic contacts with smooth cells remain unaltered (Wray and Hoffman, 1986). This suggests that a subpopulation of LHRH neurons becomes more densely connected to specific neurotransmitter circuitries in the hypothalamus and therefore is subjected to the attending changes in incoming information associated with a greater synaptic density.

B. *The Ovarian Response*

1. Gonadotropin Control

There is no doubt that ovarian development is under the firm control of circulating gonadotropins. Recent experiments in sheep, monkeys, and rats have revealed that experimental simulation in juvenile individuals of the plasma pattern of LH levels observed in peripubertal animals is an effective stimulus for the activation of ovarian secretory activity. When 20-week-old juvenile lambs were injected hourly for 48 h with LH, simulating the high frequency LH pulses observed during the follicular phase of the estrous cycle, plasma estradiol levels increased to follicular phase levels and a pre-ovulatory surge of gonadotropins occurred followed by the formation of an active corpus luteum (Foster et al., 1984). High amplitude, low frequency pulses of exogenous LH were ineffective. Administration of LHRH to rhesus monkeys in a pulsatile pattern at hourly intervals was also effective in activating the ovary and inducing precocious puberty (Wildt et al., 1980). A similar result was obtained when the release of LHRH was directly stimulated by the pulsatile administration of N-methyl-D-aspartic acid at intervals that mimicked the pattern of LH release in adult animals (Plant, 1987). In rats, the simulation of the afternoon peripubertal pattern of LH pulses, in an *in vitro* perfusion system, resulted in a significant enhancement of both estradiol and progesterone secretion from immature ovaries (Urbanski and Ojeda, 1985a). More remarkably, reproducing the pattern of LH release *in vivo*, via stimulation of LHRH release with N-methyl-D-aspartic acid markedly advanced the onset of puberty after only 4 days of treatment (Urbanski and Ojeda, 1987).

It is obvious, therefore, that if the juvenile ovary is physiologically stimulated it will produce a pattern of steroid secretion adequate to trigger a pre-ovulatory surge of gonadotropins. It is also clear that once the pubertal process is initiated by the neural mechanisms directly regulating the activity of LHRH neurons, the most important endocrine component that determines the timing of puberty is the ovary.

Only when the ovary has become able to produce estradiol in sufficient amounts to exert a pre-ovulatory surge of gonadotropin and can respond to this surge with ovulation, will puberty occur. In fact, the completion of puberty in monkeys and sheep does not occur at the time of the first pre-ovulatory surge of gonadotropins (Foster, 1987; Plant, 1987), in part because the ovary is still insufficiently mature to form an active, functionally competent corpus luteum.

2. Direct Neural Control

As indicated before, the mammalian ovary is innervated by extrinsic nerves of noradrenergic and peptidergic nature. While most of these studies have been performed in the rat there is immunocytochemical and morphological evidence that the ovary of other species, including the human, is also innervated (for a review see Burden, 1985). Little is known regarding alterations in activity of ovarian nerves during the time when the pattern of LH release begins to change. That an increase in impulse traffic through these nerve fibers may occur is suggested by the observation that the ovarian content of norepinephrine as well as tyrosine hydroxylase activity, the rate limiting step in catecholamine synthesis, increase as puberty approaches (Ben-Jonathan, 1984). More recently, it has been observed that the ovarian content of VIP, which remains fairly stable throughout neonatal, infantile and juvenile development increases significantly around day 30 (Ahmed et al., 1986), i.e., coinciding with the initiation of the diurnal pattern of LH secretion. This change, however, cannot be attributed to gonadotropins because neither LH nor FSH can induce it in hypophysectomized rats. Intriguingly, a unilateral lesion in the anterior hypothalamic area of hypophysectomized immature rats resulted in an increase of VIP content in the ovary ipsilateral to the lesion, suggesting that VIP levels in the ovary are regulated by a direct CNS-ovarian pathway (Ahmed et al., unpublished).

IV. THE TIMING OF PUBERTY

A. Influence of Nutritional Factors

From a survival point of view it makes sense for a female mammal not to breed until she is sufficiently mature somatically to be able to maintain a pregnancy and give birth to viable offspring. It is therefore not surprising that undernutrition will severely retard sexual maturation. For example, whereas ewe lambs with *ad libitum* feeding show signs of first ovulation at about 30 weeks of age, their feed-restricted counterparts are still anovulatory after 48 weeks. This suppressive effect of severe undernutrition appears to be exerted through a decrease in LH pulse frequency which in turn fails to support the high level of estradiol secretion that is necessary for the production of a normal pre-ovulatory LH surge (Foster et al., 1985). In the female rat the effects of food restriction on gonadotropin secretion and the onset of puberty are equally impressive. When prepubertal females were maintained at 45% of their expected 50-day body weight pulsatile LH release was completely suppressed and the animals failed to reach sexual maturity. Interestingly, complete sexual development could be rapidly resumed in these animals either by providing them with unlimited access to food or by administering LHRH in a pulsatile manner (Bronson, 1986). It remains to be determined exactly how undernutrition affects LHRH release and the function of the hypothalamic pulse generator during puberty. However, based on studies performed using crab-eating macaques (Macaca fascicularis), one possibility is that humoral metabolic signals may play a significant role (Steiner et al., 1983).

B. *Environmental Regulation*

Most vertebrate species breed only during certain times of the year, especially in non-tropical latitudes where environmental conditions show marked seasonal changes. These breeding cycles usually display a particular phase relationship with a reliable seasonal cue, such as changing day-length, so that pregnancy and subsequent parturition occur at the most optimum time for survival of the species. It is not surprising, therefore, that in many animals the onset of puberty is also influenced by seasonal environmental signals, a subject that has been studied most extensively in the ewe (Dyrmundsson, 1973; Foster et al., 1985).

In general, sheep are described as being short-day breeders because they become sexually active in the autumn, when day-length is decreasing; the lambs are subsequently born 5 months later, in the spring. As indicated before, when spring-born lambs are raised outdoors they become sexually mature by about 30 weeks of age. However, autumn-born lambs do not, despite having achieved somatic maturity. Instead, first ovulation in these animals is delayed until the following autumn (i.e., the time of the adult breeding season). More importantly, when autumn-born lambs are reared under artificial photoperiods, which mimic the natural light cycles perceived by lambs born in the spring, puberty is not delayed. This finding supports the view that it is the photoperiod itself, rather than some other environmental variable, that exerts the suppressive effect on the developing reproductive system. Also, as with under-nutrition, the suppressive effect of photoperiod on the timing of puberty is mediated through the gonadotropin-releasing system.

Female rhesus monkeys, like most breeds of sheep, breed only during the autumn and winter, when reared in their natural environment. Likewise, the timing of puberty in the rhesus monkey is also under environmental control (Wilson et al., 1984, 1986). Normally the ovaries do not appear to reach full maturity until several months after menarche (first menstruation). Thus menarche is usually followed by a period of "adolescent sterility," lasting for 3-15 months, which is typically characterized by anovulatory cycles with extended periods of amenorrhea. It is during this adolescent period that sexual maturation is most susceptible to seasonal environmental influences which may suppress ovulation despite a sufficiently mature gonadotropin-releasing system. It has been observed that when female rhesus monkeys are reared outdoors they will ovulate for the first time either in the autumn-winter immediately following menarche or in the following autumn, but not in the intervening spring-summer period. In contrast, when adolescent monkeys are housed indoors first ovulation will occur at any time between 31-50 months of age. Therefore, development of the reproductive system of the rhesus monkey, like the lamb, appears to be environmentally suppressed during the spring and summer. As outlined above, in the lamb this environmental suppression is exerted primarily by the photoperiod, but it has yet to be determined whether the same is also true for the rhesus monkey, or whether in fact some other seasonal environmental cues are involved.

In contrast, the rat is typically considered to be a non-seasonal breeder even in the feral condition. Furthermore, reproductive development in this species is not influenced by the seasonal changes in day-length. Surprisingly, the same is also true for the prepubertal golden hamster even though the reproductive system of the adult is profoundly affected by the photoperiod (Darrow et al., 1980). However, in another long-day breeding rodent, the Djungarian hamster, sexual maturation can be delayed by about 10 weeks by rearing the animals under short, as opposed to long, days (Hoffman, 1978). A similar situation exists in the ferret where animals reared under long or short days will become sexually mature by 20-22 weeks and 30-50 weeks, respectively (Ryan and Robinson, 1987). The mechanism by which photoperiodic information is transduced into a particular pattern of gonadotropin secretion has not been completely elucidated. However, there is good evidence to suggest that in mammals the neural circuitry involves the eyes, the retinohypothalamic tract, the suprachiasmatic nuclei, hypothalamospinal fibers, the peripheral sympathetic nervous system, the pineal gland, and more than likely melatonin (a pineal hormone) (Reiter, 1980).

V. *THE FIRST PRE-OVULATORY SURGE OF GONADOTROPINS*

In both sheep and rats the capacity of the LHRH-LH releasing system to respond to estradiol with a pre-ovulatory surge of secretion develops long before puberty (for review see Ojeda et al., 1986; Foster, 1987). However, the surge does not occur because the immature ovary has not reached a developmental state at which it can secrete sufficiently elevated levels of estradiol.

In contrast, humans and rhesus monkeys undergo menarche several months before their LHRH-LH releasing system becomes responsive to estradiol positive feedback. Since by the time of menarche the levels of estradiol produced by the ovary are sufficiently elevated to support growth of the uterine endometrium, the conclusion may be reached that the failure of an LH surge to occur in early puberty is exclusively due to a central inability to respond to estradiol. This is supported by the observation that circulating estradiol levels in human females wax and wane throughout the pre- and post-menarcheal period (Faiman and Winter, 1974) probably reflecting consecutive waves of follicular development in the absence of ovulation. That the central component of estradiol positive feedback could indeed be activated earlier if the ovary had the capacity to produce adequate levels of estradiol is demonstrated by the finding that exogenous administration of estradiol to juvenile ovariectomized monkeys evokes a pre-ovulatory surge of LH by the time of the expected menarche (Terasawa, 1985).

A. *The Acquisition of Ovarian Pre-Ovulatory Competence*

Most of the information regarding the peripubertal development of ovarian function derives from the rat and has been reviewed earlier (Ojeda and Urbanski, 1987).

Several maturational changes are involved in hastening the acquisition of pre-ovulatory competence by the rat ovary. While FSH receptor content is already maximal by the end of juvenile development, the number of LH receptors in granulosa cells increases dramatically during the days preceding the first pre-ovulatory surge of gonadotropins (Smith-White and Ojeda, 1981). Concomitant with this increase, a decline in LHRH receptor content occurs (Smith-White and Ojeda, 1983), suggesting a reduction of an inhibitory tone. The relevance of these changes for the pubertal activation of ovarian function is suggested by the fact that the steroidal responsiveness of the ovary to gonadotropins increases dramatically at this time (Advis et al., 1979), reflecting the development of follicles destined to ovulate at the first estrus.

The neurogenic component of the ovary also undergoes changes. The content of β-adrenergic receptors increases before the LH surge (Aguado et al., 1982). Paralleling the increase in receptor content, the release of progesterone in response to β_2-adrenergic stimulation also becomes more prominent.

In addition, the steroidogenic response to VIP undergoes profound changes at the time of puberty (Ahmed et al., 1985). The estradiol response to VIP, already distinct in juvenile rats, increases noticeably before the LH surge. The progesterone response to the peptide increases only moderately at this time, and then strikingly after ovulation. Radioimmunoassayable SP content in the ovary also increases before the LH surge (Ojeda et al., 1985). Although the role that SP may play in the ovary is not known it is tempting to speculate that it may be involved in the regulation of blood flow.

The net outcome of these developmental changes is an increased production of estradiol from the ovary. The strength and duration of this increase is decisive for the central component of estradiol positive feedback to be activated in all species so far examined.

B. *The Activation of Estradiol Positive Feedback*

It has been demonstrated that the treatment of immature rats with a dose of estradiol (via Silastic capsules) that produces pre-ovulatory serum levels can induce an LH surge as early as day 22 of postnatal life, i.e., at the beginning of the juvenile period (Andrews et al., 1981). Greater estradiol levels are needed to induce LH release in younger rats, but no response can be obtained before postnatal day 16. A similar early development of the central-component of estradiol positive feedback is observed in sheep which are able to respond to exceedingly small doses of estradiol as early as 19 weeks of age, i.e., 10-12 weeks before spontaneous puberty (Foster, 1984).

The inability of estradiol to elicit an LH surge before menarche in humans and rhesus monkeys has been attributed to a strong inhibitory control exerted by estradiol at this time of development (Foster et al., 1983). This idea, however, has been contested by Terasawa (1985) who demonstrated that neonatal removal of the ovaries followed by administration of estradiol during the juvenile period results in an LH surge at around the time of expected menarche, i.e., much earlier than in intact animals. The expression of this LH surge correlates well with the basal levels of LH; animals with the highest basal LH levels respond more effectively to the stimulatory effect of estradiol. Since the decrease in sensitivity to estradiol negative feedback takes place after menarche, and because an age-related increase rather than decrease in estradiol negative feedback effectiveness was observed in these experiments, the conclusion was reached that a gonadal-independent maturation of the LHRH system rather than a "resetting of the gonadostat" underlies the development of estradiol positive feedback.

The sites where estradiol acts to induce the first preovulatory surge of gonadotropins are well documented in the rat. However, in monkeys (and by inference humans) a controversy exists as to whether or not estradiol acts on the hypothalamus to stimulate LHRH release or only on the pituitary to increase the gonadotropin responsiveness to an invariable pattern of LHRH secretory episodes. While Knobil and his associates provided evidence for this latter concept in a series of elegant experiments (review by Pohl and Knobil, 1982), other investigators have demonstrated that estradiol increases LHRH release under both *in vivo* and *in vitro* conditions (Levine et al., 1985, 1985a) Recent data obtained in the sheep are also supportive of the view that, like the rat, estradiol in this species acts on both the hypothalamus to enhance LHRH release and on the pituitary to enhance the LH response to LHRH (Clarke et al., 1987). The sequence of events leading to the first pre-ovulatory surge of gonadotropin in the female rat and human (primate) female, is represented in Figs. 5 and 6, respectively.

During the last few years efforts to elucidate the cellular and molecular mechanisms underlying the effect of estradiol on the hypothalamus have been intensified. A detailed discussion of this subject is beyond the scope of this article, but it is important to mention some of the most relevant aspects. The interested reader is referred to two recent reviews of the matter (Ojeda et al., 1986; Ojeda and Urbanski, 1988).

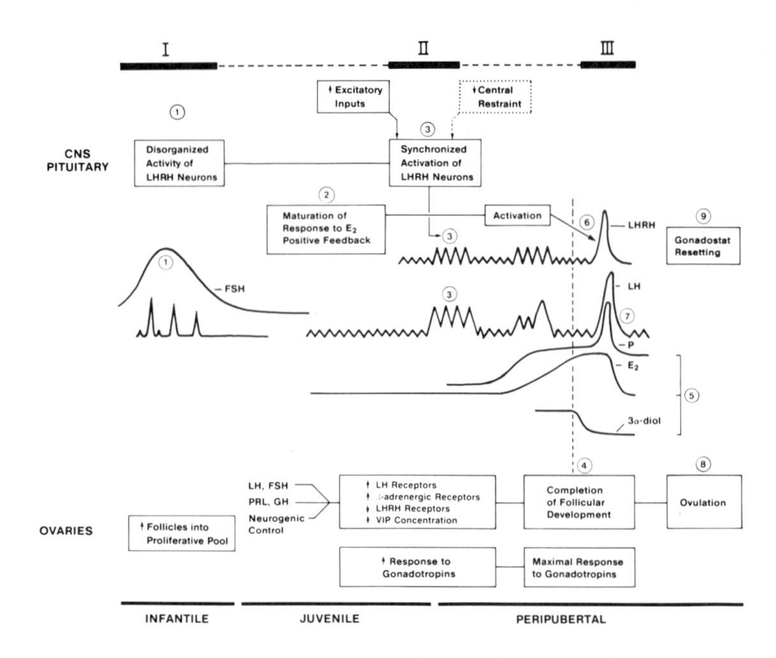

Fig. 5. Proposed sequence of developmental events leading to the first pre-ovulatory LH surge in the female rat. The numbers indicate the sequence in which the events may occur. The dotted line represents 1200 hours on the day of first proestrus. The box outlined by interrupted lines indicates that a loss in central restraint may not be a predominant factor for the synchronized activation of LHRH-secreting neurons. From: Ojeda and Urbanski (1987). In "The Physiology of Reproduction (E. Knobil and J. D. Neill, eds.), p. 1697. Raven Press, New York (with permission).

Immunohistochemical evidence exists that LHRH neurons do not contain nuclear estrogen receptors (Shivers et al., 1984). Therefore, most of the effects of estradiol on LHRH secretion may be exerted on neuronal or glial populations associated with LHRH neurons. There is, however, evidence that estradiol can facilitate LHRH release by acting at a non-genomic site(s) (Drouva et al., 1984). Whether this (presumably) membrane effect is exerted directly on the LHRH neuron itself is unknown.

The facilitatory effect of estradiol on LHRH release appears to involve an increased trans-synaptic flow of neurotransmitters known to be excitatory for LHRH release, such as norepinephrine (Paul et al., 1979), increased synthesis of the neuropeptide, non-genomic membrane effects, and activation of processing enzymes that yield the mature

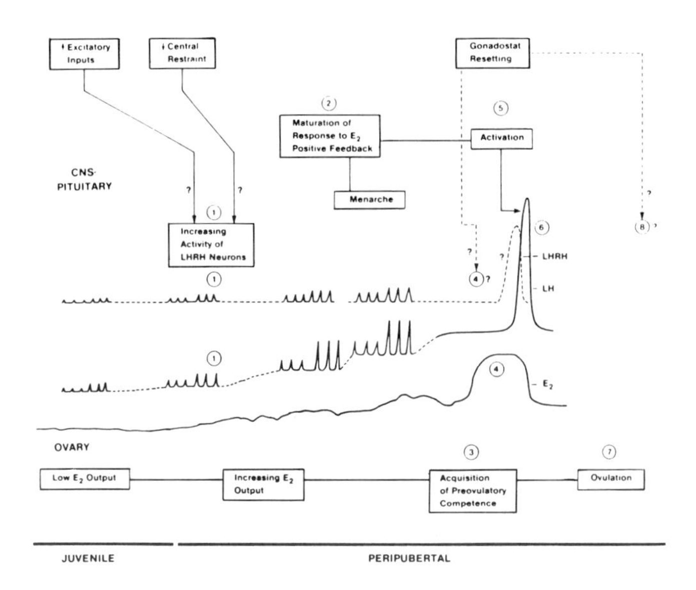

Fig. 6. Proposed sequence of events leading to the first pre-ovulatory LH surge in primates. The numbers indicate the sequence in which the events may occur. Present evidence does not permit a firm conclusion as to the timing of gonadostat resetting. It may occur shortly before the first ovulation or after it. A pre-ovulatory LHRH surge may (Norman et al., 1982), or may not (Knobil, 1980), be necessary for the pre-ovulatory gonadotropin surge to occur. Although the magnitude of LHRH pulses may increase during development an increase in baseline is not necessary. From: Ojeda et al. (1984). In "Neuroendocrine Perspectives," Vol. 3 (E. E. Muller, and R. M. MacLeod, eds.). p. 264. Elsevier, Amsterdam and New York (with permission).

peptide (for references see Drouva et al., 1986). The intracellular mechanisms leading to LHRH release appear to depend on the activation of two independent, but complementary, pathways; a prostaglandin E_2 (PGE_2)-cyclic AMP (cAMP)-mediated route and a protein kinase C-dependent pathway (Ojeda and Urbanski, 1987). Estradiol increases the LHRH response to both PGE_2 and cAMP and to the neurotransmitter norepineprine which activates the PGE_2-cAMP signal transduction mechanism. Whether estradiol also facilitates the effect of protein kinase C activators is not yet known, but is very likely, because the facilitatory actions of the steroid appears to affect the overall responsiveness of the LHRH neurons to stimulatory inputs.

The mechanisms by which a major discharge of LHRH occurs at puberty are poorly understood, but it would not be unreasonable to speculate that they involve a direct synchronization between LHRH neurons through their anatomical connections (Leranth et al., 1985; Pelletier, 1987) and simultaneous activation of the two above mentioned intracellular signal-transduction pathways.

VI. THE SEARCH FOR THE "PRIMUS MOVENS"

Elucidation of the mechanism(s) responsible for the activation of LHRH release at puberty remains a fundamental challenge to developmental neuroendocrinologists. Even in sheep that appear to enter puberty in the absence of a gonadal-independent change in "central drive," LHRH secretion increases at puberty apparently as a consequence of a still poorly understood change in "sensitivity" to estradiol negative feedback. One may suspect that the basis of this change is very similar to those underlying the change in "central drive" observed in species such as the human, rhesus monkey and rats.

Just as important as the issue of the pubertal activation of LHRH release, is that of the juvenile hiatus of gonadotropin secretion that characterizes sexual development of primates. Why is the LHRH-secreting system shut off after the initial, neonatal period of enhanced activity? Why is it reactivated at puberty? In a more general sense if the peculiarities of each species are not considered then one is left with three fundamental questions: a) Is the LHRH neuron itself the most decisive limiting factor for puberty to occur? b) If not, what is the nature of the inputs to these neurosecretory cells that determine the increase in LHRH secretory activity? c) If the brain substance(s) responsible for activating the LHRH network are identified, what factor(s) determine the enhancement of their operativity at the end of juvenile development and not before?

A partial answer to the first question is already at hand. The LHRH neurons of both rats and monkeys can be prematurely activated to cause puberty if challenged with an excitatory neurotransmitter/neuromodulatory substance. Whether the anatomical and biochemical substrates underlying the action of other neurotransmitters on LHRH secretion are also operative during juvenile development is not known.

Since immaturity of the LHRH neurons does not appear to be the limiting factor for puberty to occur, such limitation must reside on the neuronal circuitries impinging upon the

LHRH network. It now seems certain that removal of an inhibitory opioid tone is not the cause of the pubertal activation of LHRH release. That activation of an excitatory tone may play a role is suggested by the above mentioned observation that an excitatory amino acid administered in a pulsatile fashion can induce precocious puberty. Though of potential importance, this is an initial finding that does not preclude the involvement of other inhibitory or excitatory systems in the process.

Even if excitatory amino acids are the physiological activators of LHRH release at puberty the question of what determines their operativity remains to be answered. At this point the long-recognized possibility that the states of somatic growth are monitored by a brain "somatometer" (Steiner et al., 1983; Plant, 1987) which is responsible for the activation of excitatory circuits associated with LHRH cells should be given proper consideration. Future work along these lines should prove fruitful for the understanding of female mammalian puberty.

REFERENCES

Advis, J.P., and Ojeda, S.R. (1978). Endocrinology 103, 924.

Advis, J.P., Andrews, W.W., and Ojeda, S.R. (1979). Endocrinology 104, 653.

Aguado, L.I., Petrovic, S.L., and Ojeda, S.R. (1982). Endocrinology 110, 1124.

Ahmed, C.E., Dees, W.L., and Ojeda, S.R. (1985). Endocrinology 118, 1682.

Ahmed, C.E., Trzeciak, W.H., and Ojeda, S.R. (1986). Prog. 68th Ann. Mtg. Endo. Soc. p. 145 (abs).

Andrews, W.W., Mizejewski, G.J., and Ojeda, S.R. (1981). Endocrinology 109, 1404.

Ben-Jonathan, N., Arbogast, L.A., Rhoades, T.A., and Bahr, J.M. (1984). Endocrinology 115, 1426.

Boyar, R., Perlow, M., Hellman, L., Kapen, S., and Weitzman, E. (1972). J. Clin. Endocrinol. Metab. 35, 73.

Bronson, F.H. (1986). Endocrinology 118, 2483.

Bronson, F.H., and Rissman, E.F. (1986). Biol. Reprod. 61, 157.

Burden, H.W. (1985). In "Catecholamines as Hormone Regulators" (N. Ben-Jonathan, J.M. Bahr and R.I. Weiner, eds.), p. 261. Raven Press, New York.

Calka, J., McDonald, J.K., Jordan, L., Costa, M.E., and Ojeda, S.R. (1987). Neurosci. Abs. p. 1577 (abs).

Clarke, I. J., Thomas, G. B., Yao, B., and Cummins, J. T. (1987). Neuroendocrinology 46, 82.
Darrow, J. M., Davis, F. C., Elliott, J. A., Stetson, M. N, Turek, F. W., and Menaker, M. (1980). Biol. Reprod. 28, 373.
Davoren, J. B., and Hsueh, A. J. W. (1984). Biol. Reprod. 33, 37.
Dees, W. L., Kozlowski, G. P., Dey, R., and Ojeda, S. R. (1985). Biol. Reprod. 33, 471.
Dierschke, D. J., Karsch, F. J., Weick, R. F., Weiss, G., Hotchkiss, J., and Knobil, E. (1974). In "Control of the Onset of Puberty" (M. M. Grumbach, G. D. Grave, and F. E. Mayer, eds.), p. 104. John Wiley and Sons, New York.
Donovan, B. T., and Van der Werff ten Bosch, J. J. (1956). Nature 178, 745.
Drouva, S. V., Laplante, E., Gautron, J-P., and Kordon, C. (1984). Neuroendocrinology 38, 152.
Drouva, S. V., Gautron, J-P., Pattou, E., Laplante, E., and Kordon, C. (1986). Neuroendocrinology 43, 32.
Dyrmundsson, O. R. (1973). Anim. Breed. Abs. 41, 273.
Faiman, C., and Winter, J. S. D. (1974). In "The Control of the Onset of Puberty" (M. M. Grumbach, G. D. Grave, and F. E. Mayer, eds.), p. 32. John Wiley and Sons, New York.
Foster, D. L., (1984). Endocrinology 115, 1186.
Foster, D. L. (1987). In "The Physiology of Reproduction" (E. Knobil and J. D. Neill, eds.), p. 1737. Raven Press, New York.
Foster, D. L., Jaffe, R. B., and Niswender, G. D. (1975). Endocrinology 96, 15.
Foster, D. L., Rapisarda, J. J., Bergman, K. S., Lemons, J. A., Steiner, R. A., and Wolf, R. C. (1983). In "Neuroendocrine Aspects of Reproduction" (R. L. Norman, ed.), p. 103. Academic Press, New York.
Foster, D. L., Ryan, K. D., and Papkoff, H. (1984). Endocrinology 115, 1179.
Foster, D. L., Yellon, S. M., and Olster, D. H. (1985). J. Reprod. Fert. 75, 327.
Fraioli, F., Cappa, M., Fabbri, A., Gnessi, L., Moretti, C., Borrelli, P., and Isidon, A. (1984). Clin. Endocrinol. 20, 299.
Fraser, M. O., and Plant, T. M. (1987) Neurosci. Abs. 13, 1527.
Gellert, R. J., and Ganong, W. F. (1960). Acta Endocrinol (Copenh) 33, 569.
George, F. W., and Ojeda, S. R. (1987). Proc. Natl. Acad. Sci. (USA) 84, 5803.

Goldman, B. D., Grazia, Y. R., Kamberi, I. A., and Porter, J. C. (1971). Endocrinology 88, 771.
Hoffmann, K. (1978). J. Reprod. Fert. 54, 29.
Knobil, E. (1980). Rec. Prog. Horm. Res. 36, 53.
Land, R. B. (1978). J. Reprod. Fert. 52, 427.
Leranth, C., MacLusky, N. J., Sakamoto, H., Shanabrough, M., and Naftolin, F. (1985). Neuroendocrinology 40, 536.
Leranth, C., Segura, L. M. G., Palkovits, M., MacLusky, N. J., Shanabrough, M., and Naftolin, F. (1985a). Brain Res. 345, 332.
Levine, J. E., Bethea, C. L., and Spies, H. G. (1985). Endocrinology 116, 431.
Levine, J. E., Norman, R. L., Gliessman, P. M., Oyama, T. T., Bangsberg, D. R., and Spies, H. G. (1985a). Endocrinology 117, 711.
Marks, S. R., and Elders, M. J. (1979). J. Natl. Med. Assoc. 71 153.
McDonald, J. K., Dees, W. L., Ahmed, C. E., Noe, B. D., and Ojeda, S. R. (1987). Endocrinology 121, 1703.
Nass, T. E., Terasawa, E., Dierschke, D. J., and Goy, R. W. (1984). Endocrinology 115, 220.
Norman, R. L., and Spies. H. G. (1981). Endocrinology 108, 1723.
Norman, R. L., Gliessman, P., Lindstrom, S. A., Hill, I., and Spies, H. G. (1982). Endocrinology 111, 1874.
Ojeda, S. R., and Urbanski, H. F. (1987). In "The Physiology of Reproduction" (E. Knobil and J. D. Neill, eds.), p. 1697. Raven Press, New York.
Ojeda, S. R., and Urbanski, H. F. (1988). In "Neuroendocrine Control of the Hypothalamic-Pituitary System" (H. Imura, ed.), in press. Japan Sci. Soc. Press, Tokyo.
Ojeda, S. R., Smith-White, S. S., Urbanski, H. F., and Aguado, L. I. (1984). In "Endocrine Perspectives" (E. E. Muller and R. M. MacLeod, eds.), p. 225. Elsevier, Amsterdam.
Ojeda, S. R., Costa, M. E., Katz, K. H., and Hersh, L. B. (1985). Biol. Reprod. 33, 286.
Ojeda, S. R., Urbanski, H. F., and Ahmed, C. E. (1986). Rec. Prog. Horm. Res. 42, 385.
Paul, S. M., Axelrod, J., Saavendra, J. M., and Skolmick, P. (1979). Brain Res. 178, 499.
Pelletier, G. (1987). Neuroendocrinology 46, 457.
Plant, T. M. (1986). Endocr. Rev. 7, 75.
Plant, T. M. (1986a). Endocrinology 119, 539.
Plant, T. M. (1987). In "The Physiology of Reproduction" (E. Knobil and J. D. Neill, eds.), p. 1761. Raven Press, New York. Plant, T. M. (1987). Endocrinology 119, 539.
Plant, T. M., Gay, V. L., Marshall, G. R., and Arslam, M. (1987). Proc. 69th Mtg. Endo. Soc., p. 37 (abs).

Pohl, C. R., and Knobil, E. (1982). Ann. Rev. Physiol. 44, 583.

Ramirez, V. D. (1973). In "Handbook of Physiology" (R. O. Greep and E. B. Astwood, eds.), p. 1. American Physiological Society, Washington, D. C.

Reiter, E. O., and Grumbach, M. M. (1982). Ann. Rev. Physiol. 44, 595.

Reiter, R. J. (1980). Endocr. Rev. 1, 109.

Richards, J. S. (1980). Physiol. Rev. 60. 51.

Ryan, K. D., and Robinson, S. L. (1987). Biol. Reprod. 36, 333.

Sander, S. E., Case, G. D., Hopwood, N. J., Kelch, R. P., and Marshall, J. C. (1984). Ped. Res. 18, 322.

Schwartz, N. B. (1974). Biol. Reprod. 10, 236.

Shivers, B. D., Harlam, R., Morrell, J., and Pfaff, D. W. (1984). Nature 304, 345.

Smith-White, S., and Ojeda, S. R. (1981). Endocrinology 109, 152.

Smith-White, S., and Ojeda, S. R. (1983). Neuroendocrinology 36, 449.

Smith-White, S., and Ojeda, S. R. (1986). Endocr. Exp. 20, 147.

Steiner, R. A., Cameron, J. L., McNeill, T. H., Clifton, D. K., and Bremmer, W. J. (1983). In "Neuroendocrine Aspects of Reproduction" (R. L. Norman, ed.), p. 183. Academic Press, New York.

Terasawa, E. (1985). Endocrinology 117, 2490.

Terasawa, E., Bridson, W. E., Nass, T. E., Noonan, J. J., and Dierschke, D. J. (1984). Endocrinology 115, 2233.

Terasawa, E., Noonan, J. J., Nass, T. E., and Loose, M. D. (1984a). Endocrinology 115, 2241.

Thind, K. K., and Goldsmith, P. C. (1988). Neuroendocrinology, in press.

Trzeciack, W. H., Ahmed, C. E., Simpson, E., and Ojeda, S. R. (1986). Proc. Natl. Acad. Sci. 83, 7490.

Urbanski, H. F., and Ojeda, S. R. (1985). Endocrinology 117, 644.

Urbanski, H. F., and Ojeda, S. R. (1985a). Endocrinology 117, 638.

Urbanski, H. F., and Ojeda, S. R. (1987). Neuroendocrinology 46, 273.

Watanabe, G., Schultz, N. J., and Terasawa, E. (1987). Proc. 69th Mtg. Endo. Soc., p. 37 (abs).

Wildt, L., Marshall, G., and Knobil, E. (1980). Science 207, 1373.

Williams, R. F., Turner, C. K., and Hodgen, G. D. (1982). J. Clin. Endocrinol. Metab. 55, 660.

Wilson, M. E., Gordon, T. P., Blank, M. S., and Collins, D. C. (1984). J. Reprod. Fertil. 70, 625.
Wilson, M. E., Gordon, T. P., and Collins, D. C. (1986). Endocrinology 118, 293.
Winter, J. S. D., and Faiman, C. (1972). J. Clin. Endocrinol. Metab. 35, 561.
Winter, J. S. D., Ellsworth, L. R., Faiman, C., Reyes, F. I., and Hobson, W. C. (1977). Proc. 59th Ann. Mtg. Endo. Soc., p. 176 (abs).
Winter, J. S. D, Faiman, C., and Reyes, F. I. (1978). Clin. Obstet. Gynecol. 21, 67.
Worthington, C. A., and Kennedy, J. P. (1979). Aust. J. Biol. Sci. 32, 91.
Wray, S., and Hoffman, G. (1986). Neuroendocrinology 43, 93.
Wray, S., and Gainer, H. (1987). Neuroendocrinology 45, 413.

12

NEUROHUMORAL HYSTERESIS AS A MECHANISM FOR AGING IN HUMANS AND OTHER SPECIES: COMPARATIVE ASPECTS

Charles V. Mobbs
Rockefeller University
New York, NY 10021

The neuroendocrine control of some humoral substances, including certain hormones and glucose, progressively deteriorates with age in humans and other mammals; these impairments can lead to many other impairments in physiological function. Some neuroendocrine deterioration may be caused by the cumulative damage of regulatory neurons by the very substances that these neurons control. For example, estrogen, glucose, and glucorticoid levels are normally regulated by neurons in the ventromedial and arcuate nuclei of the hypothalamus (for glucocorticoids in rats, perhaps the hippocampus as well). These neuronal areas are impaired, and the regulation of these substances deteriorates, with age. Specific age-correlated neuroendocrine impairments can be accelerated by high levels of estrogen, glucose, and glucocorticoids, and some age-correlated impairments can be specifically attenuated by lowering these humoral substances. The phenomenon of progressive damage of a regulatory locus by the substance it regulates is called here "neurohumoral hysteresis", since hysteresis means "the influence of the previous history or treatment of a body on its subsequent response to a given force or changed condition." In the context of a cyclic (humoral) challenge, hysteresis constitutes both a memory and a clock, since each humoral cycle adds further damage to the neuroendocrine substrate, often leading to higher levels of the humoral substance, leading to further damage. Neurohumoral hysteresis may be a vestigial (pleiotropic) property of mature organisms related to the processes which set neuroendocrine negative feedback set-points during development, whose destructive effects are not selected against because they are expressed after procreation. This paper will compare some examples of neurohumoral hysteresis, and suggest that hyperglycemic hysteresis may be a mechanism for many age-correlated impairments, including diabetes and cardiovascular disease, in humans.

Development, Maturation, and Senescence of
Neuroendocrine Systems: A Comparative Approach

I. INTRODUCTION

The cascading effects of neuroendocrine impairments have long been noted as potential primary mechanisms in aging (Dilman, 1971; Finch, 1978). However, little is known of the cause of such impairments, so it is unclear if these neuroendocrine impairments are secondary to other age-correlated impairments. Recent studies have suggested a mechanism which could cause these neuroendocrine impairments. This mechanism, neurohumoral hysteresis, may contribute to the deterioration of neuroendocrine regulation of a number of physiological systems, including some in humans. The essential feature of neurohumoral hysteresis is a destructive positive feedback between a humoral substance and neurons which regulate, generally by negative feedback, that very substance.

A. CRITERIA FOR NEUROHUMORAL HYSTERESIS.

The minimal criteria for determining if neurohumoral hysteresis is a mechanism in the senescence of a physiological system are:

(1) The regulation of the relevant humoral substance must deteriorate with age; in general levels of the substance should increase with age, although due to gland "burn-out" hypersecretion may be followed by hyposecretion.

(2) Damage to the neurons which control the substance should induce age-like impairments in the regulation of the substance.

(3) Elevations of the substance should induce age-like impairments in the neuroendocrine loci controlling the substance; these impairments should be irreversible, i.e., should persist after normal levels of the substance are restored. Impairments should include neuronal damage and physiological sequalae of neuronal damage.

(4) Reduction of the substance during aging, e.g., by gland removal in youth, should attenuate age-correlated impairments of the neuroendocrine loci controlling the substance.

By these criteria age-correlated impairments in the regulation of estrogen, glucose, and glucocorticoids may be due in part to neurohumoral hysteresis. This paper will examine the evidence that neurohumoral hysteresis is a mechanism in aging, the pertinence of different forms of hysteresis across mammalian species, and the relationship of hysteresis to normal developmental processes.

II. ESTROGENIC HYSTERESIS

A. MEETS CRITERIA

Estrogenic hysteresis appears to contribute to reproductive senescence in both rats and mice by the following criteria:

(1) Reproductive senescence is marked by numerous neuroendocrine impairments, including impairment in negative feedback regulation of LH and damage to the arcuate nucleus. Estrogen is relatively elevated in the early phases of reproductive senescence, and the late phases are marked by hyposecretion of estrogen.

(2) Arcuate nucleus damage causes numerous age-like neuroendocrine reproductive impairments, including decreased negative feedback on LH.

(3) Elevated estrogen causes numerous irreversible age-like neuroendocrine reproductive impairments, including arcuate nucleus damage.

(4) Removal of estrogen by ovariectomy when young attenuates numerous age-correlated neuroendocrine reproductive impairments, including decreased negative feedback and arcuate nucleus damage.

B. RATS

Perhaps the earliest and most robust age-correlated physiological impairment in mammals is the loss of reproductive functions in females (Harman et al., 1984). The end-stage of reproductive senescence is often marked by decreased levels of the regulatory hormone estradiol, presumably due to terminal ovarian involution (Harman et al., 1984), but the early stages of reproductive senescence may be marked by relatively elevated levels of estradiol in rats (Lu et al., 1979) and mice (Nelson et al., 1981). It was long assumed, particularly in humans, that intrinsic ovarian senscence was the major, if not only, mechanism in the loss of fertility in females (Davidson et al., 1983).

However, Aschheim published a series of studies using ovarian transplants which suggested that reproductive senscence in female rats was due to neuroendocrine impairments, and that these age-correlated impairments could be attenuated if the rats were ovariectomized (effectively eliminating estrogen) when young (reviewed in Aschheim, 1983). Slightly earlier, in perhaps the first evidence of neurohumoral hysteresis as a mechanism of aging, a study by Kawashima (1960) reported that daily injections of very low levels of estrogen would cause premature loss of estrous cycles in female rats, although the levels used had no direct effect on estrous cycles. Furthermore, a single injection of a very high level of estradiol benzoate or estradiol valerate caused immediate and permanent loss of estrous cycles (Brown-Grant, 1975; Brawer et al., 1978). These results suggested that ovarian secretions might gradually damage the neuroendocrine substrate which regulates reproductive cycles, and thus that estrogenic hysteresis contributes to reproductive senescence in rats.

Regulation of luteinizing hormone (LH), an important component in the control of reproductive cycles, is impaired during reproductive senscence in rats, and in some respects these impairments are estrogen-dependent. The steroid-induced LH surge (reviewed in Wise, 1984), post-ovariectomy LH secretion (Huang et al., 1976), and the supression of LH by estradiol (negative feedback) (Gray et al., 1980) are all impaired with age. High levels of estrogen impair the secretion of LH in response to preoptic area stimulation (Brawer et al., 1983), and ovariectomy when young attenuates some impairments in LH secretion (Blake et al., 1983).

The identity of at least one neuroendocrine substrate involved in estrogenic hysteresis was suggested by studies of Brawer and colleagues, who reported that very high levels of estradiol could damage the arcuate nucleus of rats, leading to pituitary tumors and other neuroendocrine impairments (Brawer et al., 1975; 1978). Furthermore, the arcuate nucleus of female rats exhibits age-correlated damage, and this damage is attenuated by ovariectomy when young (Schipper et al. 1981). Arcuate nucleus damage is a plausible mechanism for the development of age-correlated reproductive impairments since arcuate nucleus lesions induce numerous age-like reproductive impairments including acyclicity, impaired post-ovariectomy LH rise, and impaired negative feedback of estrogen on LH secretion (Sridaran and Blake, 1978).

Several issues regarding estrogenic hysteresis in rats remain unclear. Since old and estrogen-treated rats are hyperprolactinemic, the extent to which this hyperprolactinemia contributes to reproductive impairments requires further investigation. Since ovariectomy of old rats partially restores the steroid-induced LH surge (Lu et al., 1983), the role of long-term vs. short-term ovariectomy in attenuating impairments of LH secretion needs to be clarified in rats. The decrease in negative feedback (Gray et al., 1980) is particularly interesting in view of the higher levels of estrogen during the early phase of reproductive senescence in rats (Lu et al., 1980). If estrogen damages negative feedback sensitivity of the neuroendocrine system to estrogen, a destructive positive feedback could drive reproductive senscence. However, in rats it is not known if the decrease in estrogen negative feedback is ovary-dependent.

C. MICE

We undertook studies in C57BL/6J mice to examine if estrogenic hysteresis is a mechanism of reproductive senescence in this species, and to address in detail many of the issues which were not or could not be addressed in rats. In addition, the genetic homogeneity of these mice ensured that ovarian graft studies would not be confounded by graft rejection. Reproductive senescence in female C57BL/6J mice differs in at least two important respects from that in rats. First, female mice rarely develop hyperprolactinemia during reproductive senescence (Mobbs et al., 1985a), nor do they develop pituitary tumors after even extremely high doses of estrogen (Mobbs et al., 1984a). Therefore if estrogenic hysteresis is a mechanism in mice hyperprolactinemia would not be implicated. Second, ovarian failure is at least as important in the loss of reproductive cycles as neuroendocrine impairments (Mobbs et al., 1984b). In this respect reproductive senescence in mice is more similar to that in humans than is reproductive senscence in rats.

Some age-correlated neuroendocrine impairments in mice are ovary-dependent. If ovaries from reproductively senescent (12-18 month-old) C57BL/6J mice are replaced with young ovaries, the older hosts are deficient in maintaining estrous cycles compared with young hosts (Felicio et al., 1983; Mobbs et al., 1984b). These results are consistent with previous work in rats suggesting a neuroendocrine impairment contributes to reproductive senscence, although in mice ovarian impairments are also involved (Gosden et al., 1983; Mobbs et al., 1984b). The deficiency of old mice in supporting estrous cycles with young ovarian grafts is largely, but not completely, prevented if mice are ovariectomized when young (Felicio et al., 1983; Mobbs et al., 1984b). If middle-aged mice are ovariectomized 2 months before being given ovarian grafts, the ability to support estrous cycles is significantly improved, but not as much as if the mice were ovariectomized when young (Mobbs et al., 1984b).

Ovary-dependent impairments in supporting estrous cycles with grafts are directly correlated with ovary-dependent impairments in the estradiol-induced LH surge (Mobbs et al., 1984b). LH surges were decreased by 12-13 months, and further decreased by 18 months (Mobbs et al., 1984b). This decrease was largely, but not quite entirely, attenuated if mice were ovariectomized when young (Mobbs et al., 1984b). The impairment in the LH surge was partially attenuated by ovariectomizing the mice one or two months before inducing the surge; there was no difference between ovariectomizing one or two months before induction, and the attenuation was not as great as ovariectomizing the mice when young (Mobbs et al.,

1984b). These results suggest that ovarian steroids impair neuroendocrine reproductive functions, and that there are short-term reversible and long-term irreversible components to these impairments. Several other age-correlated neuroendocrine impairments in mice are attenuated by ovariectomy when young. One indicator of sensitivity of the neuroendocrine system to estradiol is the suppression of LH secretion; this supression is called negative feedback because LH is stimulatory to estradiol secretion, so loss of LH supression might be expected to cause estradiol to rise. This negative feedback is decreased during reproductive senescence in mice, is the impairment is attenuated by ovariecomy when young (Mobbs et al., 1985a). This study used a series of estradiol implants which caused graded, steady-state increases of estradiol within the physiological range, so the measured ratio of LH/E2 measured in these animals reflected a non-dynamic component of negative feedback. A previous study in rats reported a decrease in the dynamic component of negative feedback with age, although this study did not directly measure steady state levels of estradiol (Gray et al., 1980). Indeed, estradiol does rise during rat reproductive senescence (during early middle age; Lu et al., 1979), but not in mice, perhaps because even in early middle age mice exhibit profound ovarian impairments (Gosden et al., 1983; Mobbs et al., 1984a). Age-correlated impairments in the post-ovariectomy rise in LH are also attenuated by ovariectomy of young mice (Gee et al., 1983), as was reported in rats (Blake et al., 1983). Age-correlated increases in pituitary dopamamine (Telford et al., 1986), pituitary glucose-6-phosphate dehydrogenase (Gordon et al., 1984), and pituitary adenomas (Felicio et al., 1983; Mobbs et al., 1984b) are also attenuated by ovariectomy when young.

Elevated estradiol accelerates age-correlated neuroendocrine impairments in mice. A single injection of 0.05 mg estradiol valerate (EV) given to young cycling female C57BL/6J mice caused permanent loss of estrous cycles (Mobbs et al., 1984a), as reported for rats (Brawer et al., 1978). Implants, in place for 6 weeks, produced sustained but physiological levels of estradiol and stopped estrous cycles only while in place; however, mice given these implants exhibited premature loss of estrous cycles. EV-treated mice did not develop pituitary tumors or hyperprolactinemia, nor was pituitary responsiveness to LHRH impaired, so neither estrogen-induced hyperprolactinemia nor pituitary damage seemed to be responsible for the reproductive impairments (Mobbs et al., 1984a). Ovaries grafted from these acyclic mice into normal mice resumed normal cycling (Mobbs et al., 1984a). However, these mice were unable to support estrous cycles when given normal ovarian grafts (Mobbs et al., 1984a).

Furthermore, the steroid-induced LH surge and the post-ovariectomy LH rise in these mice was impaired (Mobbs et al., 1984b). Subsequent studies confirmed that the single injuection of EV, even if given to young ovariectomized mice, cause permanent impairments in the ability to support estrous cycles, that these effects were dose-related within a physiological range, and that the impairments were additive with impairments which occurred during aging (Mobbs and Finch, in preparation). If a much smaller dose of EV was given to neonatal mice, the mice would begin cycling normally at puberty but would within 1-2 months develop the same impairments observed with EV in the adult or during reproductive senescence (Mobbs et al., 1985b). These studies suggested that elevations of estradiol can cause permanent impairments in the neuronal loci regulating reproductive cycles, without hyperprolactinemia, in mice as well as rats.

Estrogenic hysteresis includes short-term components which are at least partially reversible, and longer-term components which are apparently irreversible. Reproductive senescence involves more than these neuroendocrine impairments, especially in mice, and even the neuroendocrine impairments have an ovary-independent component. At present it is not clear if the permanent impairments develop during normal estrous cycles before the transiton to acyclicity, or if damage develops because of the relatively elevated (progesterone-unopposed) levels of estrogen characteristic of the transition to acyclicity (Lu et al., 1979; Nelson et al., 1981). Although if the latter were true then neuroendocrine impairments might be secondary to ovarian impairments, in either case estrogenic hysteresis is a mechanism for the development of age-correlated impairments.

These studies also suggest that neurohumoral hysteresis arises as a vestigial process related to the mechanism which sets neuroendocrine negative feedback setpoints during development. An important feature of estrogenic hysteresis is the ovary-dependent decrease in negative feedback sensitivity to estrogen during aging (Mobbs et al., 1985a). A similar decrease in negative feedback occurs during pre-pubertal development, appears to regulate the timing of puberty, and appears to be regulated by ovarian estrogen (Ojeda et al., 1980). Therefore estrogen-induced decrease in sensitivity to itself, a form of developmental hysteresis, may be a necessary component in the development of puberty, but when this process is continues in adulthood, it leads to neuroendocrine impairments. This late component of the hysteresis is not selected against because offspring are born before the impairment is expressed. Thus estrogenic hysteresis is consistent with the pleiotropic theory of senscence, which posits that genes leading to senescence are advantageous early in life and are not selected against because they allow offspring to be born before they exert their deleterious effects (Williams, 1957).

D. HUMANS

Estrogenic hysteresis is probably not directly relevant to human female reproductive senescence, since human menopause is probably due almost entirely to ovarian follicular depletion (Davidson et al., 1983). Furthermore, the steroid-induced LH surge is not impaired in post-menopausal women (Odell and Swedloff, 1968). Some loss of negative feedback sensitivity to gonadal steroids may occur in post-menopausal women (Mills and Mahesh, 1978) and older men (Muta et al., 1981), and in rodents a similar loss of negative feedback is steroid-dependent. However, there is no evidence that chronically elevated gonadal steroids increases risk of, for example, pituitary tumors (Hulting et al., 1983). Therefore estrogenic hysteresis is probably best considered as a model to study mechanisms for other forms of neurohumoral hysteresis; estrogenic hysteresis is notable for the ease with which it can be manipulated and the numerous age-correlated impairments which are estrogen-dependent. An example of neurohumoral hysteresis directly relevant to human aging is in the next section.

III. HYPERGLYCEMIC HYSTERESIS

A. HYSTERESIS CRITERIA ARE MET IN HUMANS AND RODENTS

Hyperglycemic hysteresis contributes to age-correlated impairments of glucose regulation in humans and other species by the following criteria:

(1) In humans, rodents, and other species, plasma glucose regulation is progressively impaired with age. This impairment is marked by hypersecretion of insulin and glucagon, insulin resistance, hyperglycemia, and damage to neurons in the ventromedial hypothalamus (VMH) which regulate glucose.

(2) Destruction of these neurons impairs glucose regulation, as indicated by hypersecretion of pancreatic insulin and glucagon, insulin resistance and hyperglycemia.

(3) Hyperglycemia causes irreversible damage to these neurons, impairments in glucose regulation, and numerous other age-correlated impairments.

(4) Caloric restriction delays impairments in glucose regulation and numerous other age-correlated impairments.

Therefore, progressive age-correlated glucose intolerance may be caused by a destructive feedback cycle between VMH neurons and the pancreas, in which mild (postprandial) hyperglycemia episodes when young induce mild suppression or damage to the

VMH, causing hypersecretion of pancreatic insulin and glucagon, leading to insulin resistance and more hyperglycemia, leading to further damage, etc. Eventually hypersecretion can cause pancreatic exhaustion. Age-correlated diabetes, cardiovascular complications, and other diseases could be precipitated by this pancreatic hypersecretion, insulin resistance, and hyperglycemia. The unique ability of caloric restriction to extend maximum lifespan may be due in part to a reduction in hyperglycemic episodes.

B. HYPERGLYCEMIA AND INSULIN RESISTANCE IN HUMAN AGING

Glucose tolerance in humans, rodents, and other species progressively decreases with age, beginning immediately after puberty (for reviews see DeFronzo, 1982; Davidson 1979; see also Zimmet and Whitehouse, 1979). Several studies have demonstrated that age-correlated glucose intolerance in humans is due primarily to a loss of insulin sensitivity, using glucose clamping (Defronzo et al., 1979) or insulin clamping (DeFronzo, 1979; Fink et al., 1984; Rowe et al., 1983; Chen et al., 1985) techniques. In one study moderately old humans (57-82 years) exhibited a 63% loss of insulin sensitivity compared with young (18-36 years) adults (Chen et al., 1985). The moderately old humans in this study were not obese and did not demonstrate abnormal glucose tolerance (Chen et al., 1985); glucose levels in such individuals stay normal only because older individuals also secrete more insulin (Chlouverakis et al., 1967). Insulin levels in response to oral glucose are more than adequate with advancing age (DeFronzo, 1982), although individual pancreatic cells may be deficient (Reaven et al., 1979; Kithara and Adelman, 1979). Despite pancreatic cellular impairments the insulin output in response to oral glucose stays high because the pancreas hypertrophies with age (Reaven et al., 1979; Kithara and Adelman, 1979). Therefore, insulin resistance precedes and may cause age-correlated glucose intolerance.

C. INSULIN RESISTANCE FOLLOWS PANCREATIC HYPERSECRETION

What causes this increased insulin resistance? The decreased sensitivity is independent of age-correlated weight gain (Brancho-Romero and Reaven, 1977) or adiposity (Chen et al., 1985), although these factors exacerbate insulin resistance. The pancreas secretes more insulin with increasing age after stimulation by oral glucose in both rats (Hayashi, 1982; Brancho-Romero and Reaven, 1977) and humans (Hayashi, 1980; Sandberg et al., 1973).. Levels of proinsulin in response to oral glucose also increases in older humans (Duckworth and Kitabchi, 1976). *In vitro* studies with perfused whole pancreas also demonstrate age-correlated increased total secretion (Gold et al., 1976), when the secretion is monitored for several hours. The increase in insulin secretion can be detected in non-diabetic, non-obese individuals and is not a result of the age-correlated hyperglycemia. Statistical analysis by multiple regression reveals that insulin secretion increases with age independent of adiposity, weight, height, or blood sugar (Welborn et al., 1969), although these factors also independently contribute to plasma insulin levels. Hyperinsulinemia can be detected at younger ages than can hyperglycemia in rats (Brancho-Romero and Reaven, 1977) and humans (Chlouverakis et al., 1967). At early middle ages, hyperinsulinemia occurs in the presence of hypo glycemia (Brancho-Romero and Reaven, 1977; Chlouverakis et al., 1967). Furthermore, hyperinsulinemia can be detected even in non-obese older individuals who do not show hyperglycemia (Hayashi, 1980; Chlouverakis et al., 1967).

These studies suggest that age-correlated hyperinsulinemia precedes age-correlated hyperglycemia and insulin resistance. While age-correlated glucose intolerance may depend on increased adiposity (Reaven and Reaven, 1985), clearly hyperinsulinemia (Welborn et al., 1969; Chlouverakis et al., 1967; Hayashi, 1980)) and insulin resistance (Chen et al., 1985) do not depend on adiposity. Using a different approach, non-obese, non-diabetic middle-aged individuals with slightly elevated plasma glucose and insulin responses to oral glucose were given meals similar in composition to the typical American diet; their glucose levels were normal but insulin levels were still elevated (Reaven et al., 1972). Similarly, elderly individuals with normal glucose tolerance tests were given typical mixed meals; glucose levels were elevated 11% in the elderly vs. young controls, whereas insulin levels were elevated 40% (Fink et al., 1984). Therefore hyperinsulinemia appears to precede hyperglycemia and be a more primary feature of senescence.

Pancreatic glucagon plasma levels after oral glucose increase with age in rats (Hayashi, 1982) and humans (Hayashi, 1980), even in older non-diabetic individuals who do not exhibit abnormal glucose tolerance (Hayashi, 1982; Hayashi, 1980). This increase in glucagon is due to a loss of the suppressive effect of glucose on glucagon (Hayashi, 1980; Hayashi, 1982); loss of glucagon suppression is accelerated in diabetics (Hayashi, 1980). In portal vein blood from 12- and 24-month rats, glucagon actually increased 2- to 3-fold in response to gastric glucose, in contrast to the suppressive effect of glucose on glucagon in young rats (Klug et al., 1979).

Elevated physiological levels of insulin cause glucose intolerance (Ramirez and Friedman, 1983) and insulin resistance at both receptor and post-receptor levels (Martin et al., 1983), independent of increased adiposity (Martin et al., 1982), in both humans (Rizza et al., 1985; Nankervis et al., 1985; Soman and DeFronzo, 1980) and rats (Kobayashi and Olefsky, 1978; Martin et al, 1983). Elevated insulin levels also cause desensitization _in vitro_ at both receptor and post-receptor levels (Ercolani et al., 1985; Marshall and Olefsy, 1980; Davidson and Casanello-Ertl, 1979; Chang and Polakis, 1978). Elevated pancreatic glucagon antagonizes the effect of insulin (Christ et al., 1986), is associated with increased hepatic glucose output and consequent hyperglycemia (Henry et al., 1986), and may be a cause of diabetes (Unger and Orci, 1975). Increased glucagon in particular has been emphasized as contributing to age-correlated insulin resistance (Klug et al., 1979; Hayashi 1980, 1982), whereas hypersecretion of insulin has been cited as the proximal cause of insulin resistance and obesity in the Zucker rat (Jenrenaud, 1985). All these data taken together suggest that pancreatic hypersecretion of insulin and glucagon may be a proximal cause of age-correlated insulin resistance.

D. HYPERGLYCEMIA FOLLOWS DAMAGE OF VMH

What causes the age-correlated increased secretion of insulin and glucagon? Lesions of the ventromedial hypothalamus (ventromedial nucleus and arcuate nucleus) cause increased pancreatic secretion (Rohner-Jeanrenaud and Jeanrenaud, 1980; Tokunaga et al., 1986), insulin resistance (Kasuga et al., 1980; Penicaud et al., 1983), and glucose intolerance (Komeda et al., 1980; Matsuo and Shino, 1972; Cameron et al., 1976) similar in some respects to that detected during aging. VMH lesions cause diabetes 10 times more frequently in old than in young rats (Lazaris et al., 1985). VMH lesions in monkeys can also cause a high incidence of diabetes (Hamilton and Brobek, 1963). Lesions of the ventromedial nucleus (VMN) of the hypothalamus cause increased pancreatic secretion of both insulin (Tokunaga et al., 1986; Berthoud et al., 1979) and glucagon (Rohner-Jeanrenaud and Jenrenaud, 1984). The increased secretion of insulin and glucagon after VMH lesion can be detected even using the perfused pancreas in vitro (Rohner-Jeanrenaud and Jeanrenaud, 1980). The pancreatic hypersecretion and glucose intolerance are not a result of hyperphagia and obesity since (i) increased insulin secretion occurs immediately after an electrolytic lesion of the VMH in an anesthetized animal, in the absence of any food ingestion (Tokunaga et al., 1986); (ii) monosodium glutamate, which lesions the VMH in Chinese hamsters, causes glucose intolerance without causing obesity (Komeda et al., 1980); (3) monosodium glutamate lesions induce hyperglycemia in

KK mice at an age before body weight differs from non-lesioned controls (Cameron et al., 1976). Furthermore, hyperphagia and obesity caused by VMH lesion can be prevented by cutting input of the vagus nerve to the pancreas (Inoue and Bray, 1977), or by destroying the pancreas and replacing with a pancreatic graft (Inoue et al., 1977), the direct effect of both of which is to prevent the onset of hyperinsulinemia. Conversely, stimulation of the vagus nerve directly causes immediate hyperinsulinemia and hyperglycemia (Ionescu et al., 1983); the hyperglycemia is presumably due to glucagon secretion. Apparently, increased pancreatic secretion after VMN lesion is due to the loss of the inhibitory influence of the VMN on the vagus nerve, which stimulates pancreatic secretion (Jeanrenaud, 1985). The increased pancreatic secretion preceeds and may cause the glucose intolerance. Therefore, a loss of VMN neurons directly causes increased pancreatic secretion, followed by development of glucose intolerance and hyperglycemia.

E. DAMAGE OF VMH DURING AGING

VMH neurons are progressivly lost with increasing age. Sabel and Stein (1981) reported a progressive age-correlated loss of neurons from rat VMN; by 880 days the neuronal number was reduced by 60% compared with the number in rats 90 days old; in contrast, neurons in the nearby lateral amygdala were reduced only by a non-significant 23%. Hsu and Peng (1978) also reported a loss in neuronal number in the VMH of female rats. Smith-West and Garris (1983) reported a 20% reduction in the VMN nuclear area from 100 to 450 days, as well as a reduction in nuclear density, in the Chinese hamster. These changes were accelerated in the diabetic hamster (Smith-West and Garris, 1983). One study using male rats did not report a loss of VMN neurons (Peng and Hsu, 1982). However, these neurons could become damaged even before an overt loss of neurons is easily detectable. For example, in the VMN of female rats Wise et al. (1984a) detected a loss by 12 months of cytoplasmic estradiol receptors, apparently related to a loss of neurally mediated biochemical and behavioral responses to estradiol (Wise et al., 1984b); at such an early age loss of neurons in the VMN is marginally detectable (Sable and Stein, 1981). Therefore, the progressive age-correlated impairment of VMN neurons, leading to progressive loss of vagal inhibition, could account for the progressive increase in insulin and glucagon secretion, and subsequently an increase in insulin resistance and glucose intolerance.

F. HYPERGLYCEMIA DAMAGES VMH

What causes the age-correlated loss of VMN neurons? Hyperglycemia, due either to pancreatic damage (insulin insufficiency) (Bestetti and Rossi, 1980; Akmayev and Rabkina, 1976) or genetic defects (insulin resistance ?) (Garris et al., 1985; Garris et al., 1982; Smith-West and Garris, 1983), can cause progressive neuronal damage in the VMH. Hyperglycemia resulting from alloxan-induced pancreatic damage reduces the size of neuronal nuclei in the VMN and arcuate specifically, leaving neuronal nuclei in the dorsomedial and periventricular nuclei unaffected (Akmayev and Rabkina, 1976). This reduction is not due to normal responses to blood glucose, since the well-fed state is normally associated with larger VMN nuclei than the fasted state (Pfaff, 1969). In another study, hyperglycemia resulting from streptozotocin-induced pancreatic damage caused many neuropatholgies (fragmented endoplasmic reticulum, loss of organelles, irregular nuclei, etc.) in the arcuate nucleus (Bestetti and Rossi, 1980), although the VMN was not examined. Hyperglycemia from streptozotocin also causes a progressive loss of estradiol receptors in the VMH and a correlated loss of lordosis behavior (Ahdieh et al., 1983), which is controlled by specific neurons in the VMN (Pfaff, 1980). This study was notable in that lowering blood sugar with insulin after several weeks of hyperglycemia did not restore receptors or behavior (Ahdeih etal., 1983). In the genetically diabetic db/db mouse, the number of degenerating neurons progressively increased, and neuronal density decreased, in the VMN and arcuate nucleus compared to controls (Garris et al., 1985). The VMN lesions were a result of hyperglycemia rather than a cause, since hyperglycemia and arcuate nucleus damage were detectable 12 weeks before VMN damage was detectable, and damage to the VMN was progressive. In the genetically diabetic Chinese hamster, the area, number of neurons, and neuronal density of the VMN were all significantly reduced compared with non-diabetic controls (Garris et al., 1982). In another study with the Chinese hamster, diabetes was associated with an acceleration of the normal age-correlated loss of neurons in this nucleus (Smith-West and Garris, 1983).

G. PANCREATIC EXHAUSTION FOLLOWS HYPERSECRETION

Aging, VMH lesions, and certain diabetogenic (hypothalamic?) mutations are all associated with pancreatic hyperplasia and hypersecretion of insulin and glucagon, both in vivo and in vitro (Jeanrenaud, 1985; Coleman, 1982). In the diabetic mutants hyperinsulinemia and glucose intolerance develop early (Jeanrenaud, 1985; Coleman, 1982). However, overstimulation appears to cause eventual exhaustion of the pancreas, leading to a transition from insulin-independent to insulin-dependent glucose intolerance (Coleman, 1982). Furthermore, VMH lesions also cause hyperplasia and apparently eventual exhaustion, which is exaggerated in older individuals and leads to diabetes (Lazaris et al., 1985). The hyperplasia/exhaustion phenomenon following VMH lesions may explain the age-correlated hyperplasia of the pancreas which is accompanied by decreased pacreatic responsiveness to glucose of individual beta cells (Reaven et al., 1979); Kithara and Adelman, 1979).

H. AGING AND DIABETES: A DESTRUCTIVE FEEDBACK CYCLE

These data suggest a mutually destructive feedback cycle, in which normal transient (postprandial) hyperglycemia when young causes a slight decrease in VMN activity (caused by damage or down-regulation of glucose transporters (Karnieli et al., 1981), which in turn (probably via increased glucagon antagonism and hyperinsulinemia-induced desensitization) enhances subsequent insulin resistance and hyperglycemia, in turn causing more damage to the VMN, and so on. This cycle would lead to progressive insulin resistance exacerbated in some individuals by insulin insufficiency (after pancreatic exhaustion). In individuals with genetic predisposition, this insulin resistance could precipitate (perhaps by insulin insufficiency) overt development of clinical diabetes; in other individuals, other hyperglycemia-sensitive impairments might develop. The hyperglycemia could also down-regulate glucose transporters in many cells, amplifying the destructive cycle (Unger and Grundy, 1985).

Hyperglycemic hysteresis suggests that a primary (for definition see Landfield, 1980) lesion during senescence is the hyperglycemia-induced damage to the VMN which results in progressive insulin resistance, in some cases precipitating Type II diabetes. Many other age-correlated impairments could be precipitated by age-correlated hyperinsulinemia, insulin resistance and hyperglycemia. Cardiovascular complications are the major cause of human mortality in aging and are even more common in diabetes (West, 1978; Garcia et al., 1974) and are particularly associated with hyperinsulinemia (Stout, 1981). Decreased fertility in women is the first physiological system to senesce and perhaps the most sensitive to hyperglycemia (Hamel etal., 1986; Tallarigo et al., 1986). All of the following are also associated with senescence and with glucose intolerance: neuropathy (Brown et al., 1982; Clements et al., 1979); immunodeficiencies (Larkin et al., 1985); kidney damage (Steffes and Mauer, 1984); eye damage (Leopold et al., 1978); certain cancers (O'Mara et al., 1985); collagen cross-linking (Schnider and Kohn, 1981); and obesity (Jeanrenaud, 1985). The almost universal age-correlated increase in adiposity has long been a puzzle, in view of the highly regulated manner in which the body defends its set point of adiposity, and it has been proposed that the adiposty set point increases with age (Woods et al., 1985). Hyperglycemic hysteresis suggests that the set point of adiposity increases with age due to the gradual loss or supression of VMN neurons; in turn this increased adiposity exacerbates the hypersecretion-induced insulin resistance. Thus age-related adiposity is characterized by a state similar to the "static" phase of the VMH-lesioned rat, which is not hyperphagic but will defend its new body fat level (Penicaud et al., 1983). Since hyperglycemia can decrease cellular glutathione by more than 60% (Gonzalez et al., 1981), this mechanism could also explain data which suggests that some proteins become denatured during aging because of insufficient intracellular reducing potential (Gafni and Noy, 1985). Hyperglycemia is particularly toxic to neurons in the arcuate nucleus (Bestetti and Rossi, 1980; Akmayev and Rabkina, 1976; Garris et al., Coleman, 1985), and this nucleus, which is necessary for many neuroendocrine functions, is damaged with age (Sable and Stein, 1981; Hsu and Peng, 1978). Furthermore, lesions of the VMN reduce growth hormone release and cause growth retardation (Frohman and Bernardis, 1968) and increase net secretion of glucocorticoids (Dallman et al., 1984). Reduced growth hormone secretion during aging (Sonntag et al., 1980) may cause many other impairments (Sonntag et al. 1983), and altered secretion of glucocorticoids could cause hippocampal damage and other impairments (Landfield, 1980; Sapolsky et al., 1986). Therefore age-correlated damage to the VMH by hyperglycemic hysteresis could cause a cascade of hypothalamic neuroendocrine impairments of the kind postulated by Dilman (1971) and Finch (1976).

IV. GLUCOCORTICOID HYSTERESIS

A. HYSTERESIS CRITERIA MET IN MALE RATS

Glucocorticoid hysteresis is a mechanism in the age-correlated impairment in the regulation of corticosterone secretion in rats by the following criteria:

(1) Secretion of corticosterone, impairments in the suppression of corticosterone after stress, and CA3 hippocampal damage, all increase with age.

(2) Damage to the hippocampus causes impaired ability to suppress corticosterone secretion after stress.
(3) High levels of corticosterone cause hippocampal damage.
(4) Reduction of corticosterone (by adrenalectomy) when young decreases age-correlated hippocampal damage.

B. MALE RATS

Although the effect of aging on glucocorticoid secretion is controversial (Riegle, 1983) Landfield reported that adrenal glucocorticoid secretion increases with age, and that this increase is correlated with hippocampal damage (Landfield et al., 1978). He further showed that high levels of glucocorticoids cause hippocampal damage and adrenalectomy prevented some age-associated hippocampal damage (Landfield, 1980; Landfield, 1983). Based on these studies Landfield was the first to formulate a theory of neurohumoral hysteresis, which he called the "adrenocortical hypothesis of brain and somatic aging" in which adrenal glucorticoid secretion causes hippocampal damage, causing more secretion of adrenal steroids, causing more damage, leading to a "runaway positive feedback loop" (Landfield, 1980; Landfield, 1983).

In a series of elegant studies in male rats Sapolsky et al. confirmed and greatly extended several elements of this hypothesis, focusing particularly on the age-correlated loss of hippocampal glucocorticoid receptors and impaired response to stress (Sapolsky et al., 1986). They showed that the steroid-dependent hippocampal damage was irreversible, specifically entailed loss of neurons and glucocorticoid receptors in the CA3 region of the hippocampus, and that this damage could be accelerated by constant stress, which elevated glucocorticoid levels (Sapolsky et al., 1986). Of particular importance they showed that damage of the hippocampus attenuated negative feedback supression of corticosterone after stress (Sapolsky et al., 1984), and negative feedback after stress was also attenuated during aging (Sapolsky et al., 1983).

C. OTHER SPECIES

There appears to be little evidence that the adrenocortical hypothesis, formulated on the basis of experiments in male rats, directly applies to aging in other species, including humans . Mice do not lose hippocampal glucocortoid receptors during aging (Nelson et al., 1976). In an extensive review, Riegle concludes on the basis of his own work and the work of others that for cattle, goals, and rats there is "no consistent change in basal corticosterone concentrations in aging rats of either sex" (Riegle, 1983). Indeed, Wilson et al. reported a decrease in basal corticosterone levels in 25-month-old male rats (Wilson et al., 1981). Therefore the rise in glucocorticoids during aging may be detectable only after stress or some other stimulus. In humans glucocorticoid regulation after stress does not seem to be impaired during aging (Blickert-Toft, 1975). Furthermore the CA3 region of the human hippocampus, which in the rat is the most vulnerable to glucocorticoid damage and is specifically damaged during aging (Sapolsky et al., 1986), seems specifically spared during human aging (Mani et al., 1986), even in senile dementia (Flood et al., 1987). In contrast the CA4 region, which in rats is not impaired by glucocorticoid toxicity or aging (Sapolsky et al., 1986), is the only hippocampal region to lose a significant number of neurons during human aging (Mani et al., 1986).

D. HYPERGLYCEMIC HYSTERESIS AND CORTISOL IN HUMAN AGING

However, hyperglycemic hysteresis may lead to a net increase in cortisol secretion during human aging. Lesions of the VMN elevate the nadir but not zenith levels of glucorticoids during the circadian cycle (Dallman, 1984). Aging humans may have elevated nadir levels of cortisol (Friedman et al., 1969; Touitou et al., 1983; Stokes et al., 1984), but unimpaired cortisol response to stress (Blichert-Toft, 1978). Aged male rats also have elevated nadir but not zenith corticosterone levels (Klug and Adelman, 1979). This age-correlated increase in basal cortisol levels is exacerbated in diabetic humans (Grad et al., 1971). Similarly, streptozotocin-hyperglycemic rats show elevated nadir levels of glucocorticoids and adrenal hypertrophy, but normal zenith levels of glucocorticoids (Gibson et al., 1985). Chronic hyperglycemia reduces glucocortoid receptors in some tissues (Tornello et al., 1981) and can cause age-like impairments in the regulation of glucocorticoids (L'Age et al., 1974). Therefore hyperglycemic hysteresis may lead to an age-correlated increase of cortisol at the nadir of the circadian rhythm, which in turn could cause secodary impairments in other physiological functions.

IV. OTHER FORMS OF NEUROHUMORAL HYSTERESIS

Elevation of thyroid hormones in the neonatal rat permanently alter the negative feedback sensitivity of the neuroendocrine axis to thyroid hormone (Bakke et al., 1975). This effect of neonatal thyroid hormone appears to be mediated by neurons in the mediobasal hypothalamus (Bakke et al., 1972). Since the circadian rhythm of thyroid hormone is impaired during aging (Klug and Adelman, 1979), some form of thyroid hormone hysteresis may occur during aging. Such a phenomenon would be consistent with the suggestion that neurohumoral hysteresis is a vestigial process related to developmental processes. A recent paper has also reported that after removal of the posterior pituitary, exogenous elevations of vasopressin is toxic to vasopressin neurons in the hypothalamus (Herman et al., 1987), which presumably are involved in the negative feedback regulation of vasopressin. Whether this system is an example neurohumoral hysteresis will require further study of the effects of vasopressin in intact animals.

V. CONCLUSION

Taken together these studies suggest that numerous forms of neurohumoral hysteresis exist, and that neurohumoral hysteresis is a mechanism in the development of age-correlated impairments in several physiological systems. It is still unclear if normal, youthful levels of humoral substances can cause neuroendocrine damage or if this damage occurs only after these substances are increased by some other primary cause. The mechanisms by which hormones and glucose damage the very neurons which regulate these substances will require much further study. However, neurohumoral hysteresis, especially hyperglycemic hysteresis, would appear to merit further study as a mechanism in the aging of humans and other species.

ACKNOWLEDGEMENTS. This work was supported by a post-doctoral fellowship from the NIA, 5F 32A605326-02.

REFERENCES

Ahdieh, H.B., Hamilton, J.M., and Wade, G.N. (1983) Copulatory behavior and hypothalamic estrogen and progestin receptors in chronically insulin-deficient female rats. Physiology and Behavior 31:219-223.

Akmayev, I.G., and Rabkina, A.E. (1976) CNS-pancreas system. The hypothalamic response to insulin deficiency. Endokrinologie 68:211-220.

Ammon, H.P.T., Amm, U., Eujen, R., Hoppe, E., Trier, G., and Verspohl, E.J. (1984) The role of old age in the effects of glucose on insulin secretion, pentosphophate shunt activity, pyridine nucleotides, and glutathione of rat pancreatic islets. Life Sci. 34:247-257.

Arbogast, B.W., Berry, D.L., and Newell, C.L. (1984) Injury of arterial endothelial cells in diabetic, sucrose-fed, and aged rats. Atherosclerosis 51:31-45.

Aschheim, P. (1983) Relation of neuroendocrine system to reproductive decline in female rats. In: (Meites, J., ed.) Neuroendocrinology of Aging, pp. 73-102, Plenum Press, 1983.

Bakke, J.L., Lawrence, N., and Robinson, S. (1972) Late

effects of thyroxine injected into the hypothalamus of the neonatal rat. Neuroendocrinology 10:183-195.

Bakke, J.L., Lawrence, N., Bennett, J., and Robinson, S. (1975) The late effects of neonatal hyperthyroidism upon the feedback regulation of TSH secretion in rats. Endocrinology 97:659-664.

Bestetti, G., and Rossi, G.L. (1980) Hypothalamic lesions in rats with long-term streptozotocin-induced diabetes mellitus. Acta Neuropath. 52:119-127.

Bestetti, G., Hofer, R., and Rossi, G.L. (1987) The preoptic-suprachiasmatic nuclei though morphologically heterogeneous are equally affected by streptozotocin diabetes. Exp. Brain Res. 66:74-82.

Berthoud, H.R., and Jeanrenaud, B. (1979) Acute hyperinsulinemia and its reversal by vegotomy following lesions of the ventromedial hypothalamus in anesthetized rats. Endocrinology 105: 146-151.

Blaha, G.C., and Lamperti, A.A. (1983) Estradiol target neurons in the hypothalamic arcuate nucleus and lateral ventromedial nucleus of young adult, reproductively senescent and monosodium glutamate-lesioned female golden hamsters. J. Gerontol. 35:335.

Blake, C.A., Elias, K.A., Huffmann, L.J. (1983) Ovariectomy of young adult rats has a sparing effect on the suppressed ability of aged rats to release luteinizing hormone. Biol. Reprod. 28:575.

Blickert-Toft, M. (1978) The adrenal glands in old age. Aging (Geriatric Endocrinology) 5:81-102.

Brancho-Romero, E., Reaven, G.M. (1977) Effect of age and weight on plasma glucose and insulin responses in the rat. J. Am. Geri. Soc. 7:299-302.

Brawer, J.R., and Sonnenschein, C. (1975) Cytopathological effects of estradiol on the arcuate nucleus of the female rat. Am. J. Anat. 144:57-87.

Brawer, J.R., Naftolin, F., Martin, J., and Sonnenschein, C. (1978) Effects of a single injection of estradiol valerate on the hypothalamic arcuate nucleus and on reproductive function in the female rat. Endocrinology 103: 501-512.

Brawer, J.R., Schipper, H., and Naftolin, F. (1980) Ovary-dependent degeneration in the hypothalamic arcuate nucleus. Endocrinology 107:274-279.

Brawer, J.R., Ruf, K.I., and Naftolin, F. (1980) Effects of estradiol-induced lesions of the arcuate nucleus on gonadotropin release in response to preoptic stimulation in the rat. Neuroendocrinology 30:144-150.

Brawer, J.R., Schipper, H., and Robaire, B. (1983) Effects of long-term androgens and estradiol exposure on the hypothalamus. Endocrinology 112:510-515.

Brown, M.R., Dyck, P.J., McClearn, G.E., Sima, A.F., Powell,

H.C., and Porte, D. (1982) Central and peripheral nervous system complications. Diabetes 31 (Suppl.): 65-70.
Brown-Grant, K. (1975) On "critical periods" during the postnatal development of the rat. Int. Symposium on Sexual Endocrinology of the perinatal period. INSERM 32j:357-376.
Cameron, D.P., Poon, T.K.-Y., and Smith, G.C. (1976) Effects of monosodium glutamate administration in the neonatal period on the diabetic syndrome in KK mice. Diabetologia 12:621-626.
Cerami, A. (1985) Glucose as a mediator of aging. J. Am. Ger. Soc. 33:626-634.
Cerami, A. (1986) Aging of proteins and nucleic acids - What is the role of glucose? Trends in Biochem. Sci. 11:311-314.
Chang, T.-H., Polakis, S.E. (1978) Differentiation of 3T3-L1 fibroblasts to adipocytes. Effect of insulin and indomethacin on insulin receptors. J. Biol. Chem. 253: 4693-4696.
Christ, B., Probst, I., and Jungermann, K. (1986) Antagonistic regulation of the glucose/glucose 6-phosphate cycle by insulin and glucagon in cultured hepatocytes. Biochem. J. 238:185-191.
Chen, M., Bergman, R.N., Pacini, G., and Porte, D. (1985) Pathogenesis of age-related glucose intolerance in man: Insulin resistance and decreased beta-cell function. J. Clin. Endo. Metab. 60:13-20.
Chlouverakis, C., Jarrett, R.J., Keen, H. (1967) Glucose tolerance, age, and circulating insulin. Lancet 1:806-809.
Clements, R.S. (1979) Diabetic neuropathy - New concepts of its etiology. Diabetes 28:604-611.
Cohen, M.P. (1987) The polyol paradigm and complications of diabetes. Springer-Verlag, NY.
Coleman, D.L. (1982) Diabetes-obesity syndrome in mice. Diabetes 31:1-6.
Dallman, M. (1984) Viewing the ventromedial hypothalamus from the adrenal gland. Am. J. Physiol. 246: R1-R12.
Davidson, M.D. (1979) The effect of aging on carbohydrate metabolism. A review of the English literature and a practical approach to the diagnosis of diabetes mellitus in the elderly. Metabolism 28 :688-705.
Davidson, M.B., and Casanello-Ertl, D. (1979) Insulin antagonism in cultured rat myoblasts secondary to chronic exposure to insulin. Horm. Metab. Res. 11:207-209.
Davidson, J.M., Gray, G.D., and Smith, E.R. (1983) The sexual psychoendocrinology of aging. In: Meites, J., (ed.), Neuroendocrinology of Aging, pp. 221-258, Plenum Press, NY.
DeFronzo, R.A., Tobin, J.D., and Andres, R. (1979) The glucose clamp technique. A method for the quantification of beta cell sensitivity to glucose and of tissue sensitivity to insulin. Am. J. Physiology 237:E214-E223.

DeFronzo, R.A. (1979) Glucose tolerance and aging. Evidence for tissue insensitivity to insulin. Diabetes 28 :1095-1101.
DeFronzo, R.A. (1982) Glucose intolerance and aging. In: Shimke, R. (ed.) Biological Markers of Aging, pp 98-119.
Dilman, V.M. (1971) Age-associated elevation of hypothalamic threshhold to feedback control, and its role in development, aging and disease. Lancet 1:1211-1218.
Duckworth, W.C., and Kitabchi, A. E. (1976) The effect of age on plasma proinsulin-like material after oral glucose. J. Lab. Clin. Med. 88:359-367.
Dudl, R.J., and Ensinck, J.W. (1977) Insulin and glucagon relationships during aging in man. Metabolism 26:33-41.
Ercolani, L., Lin, H.L., and Ginsberg, B.H. (1985) Insulin-induced desensitization at the receptor and post-receptor level in mitogen-activated human T-lymphocytes. Diabetes 34:931-937.
Farooqui, Y.H., and Ahmed, A.E. (1984) Circadian periodicity of tissue glutathione and its relationship with lipid peroxidation in rats. Life Sci. 34:2413-2418.
Felicio, L.S., Nelson, J.F., Gosden, R.G., and Finch, C.E. (1983) Restoration of ovulatory cycles by young ovarian grafts in aging mice: Potentiation by long-term ovariectomy decreases with age. Proc. Soc. Natl. Acad. Sci. USA 80:6076.
Finch, C.E. (1976) The regulation of physiological changes during mammalian aging. Quart. Rev. Biol. 51:49-83.
Fink, R.I., Kolterman, O.G., Kao, M. , Olefsky, J.F. (1984) The role of the glucose transport system in the post-receptor defect in insulin action associated with human aging. J. Clin. Endo. Metab. 58:721-725.
Fink, R., Koleterman, O., and Olefsky, J. (1984) The physiological signficance of the glucose intolerance of aging. J. Gerontol. 39:273-278.
Flood, D.G., Guarnaccia, M., Coleman, P.D. (1987) Dendritic extent in human CA2-3 hippocampal pyramidal neurons in normal aging and senile dementia. Brain Res. 409:88-96.
Friedman, M., Green, M.F., and Sharland, D.E. (1969) Assessment of hypothalamic-pituitary-adrenal function in the geriatric age group. J. Gerontol. 24:292-297.
Frohman, L.A., and Bernardis, L.L. (1968) Growth hormone and insulin levels in weanling rats with ventromedial hypothalamic lesions. Endocrinology 82:1125-1132.
Gafni, A., and Noy, N. (1984) Age-related effects in enzyme catalysis. Mol. Cell. Biochem. 59:113-129.
Garris, D.R., Diani, A.R., Smith, C., and Gerritsen, G.C. (1982) Depopulation of the ventromedial hypothalamic nucleus in the diabetic chinese hamster. Acta Neuropathol 56:63-66.
Garris, D.R., and Coleman, D.L. (1984) Diabetes-associated changes in estradiol accumulation in the aging C57BL/KsJ

mouse brain. Neuroscience Lett. 49:285-290.
Garris, D.R., Coleman, D.L., and Morgan, C.R. (1985) Age- and diabetes-related changes in tissue glucose uptake and estradiol accumulation in the C57Bl/KsJ mouse. Diabetes 34:47-52.
Garris, D.R., West, L.R., and Coleman, D.L. (1985) Morphometric analysis of medial basal hypothalamic neuronal degeneration in diabetes (db/db) mutant C57BL/KsJ mice: Relation to age and hyperglycemia. Dev. Brain Res. 20:161-168.
Gee, D.M., Flurkey, K., and Finch, C.E. (1983) Aging and the regulation of luteinizing hormone in C57BL/6J mice: Impaired elevations after ovariectomy and spontaneous elevations at advanced ages. Biol. Reprod. 28:598.
Gibson, M.J., DeNicola, A.F., Krieger, D.T. (1985) Streptozotocin-induced diabetes is associated with reduced immunoreactive beta-endorphin concentrations in intermediate pituitary lobe and with disrupted circadian periodicity of plasma corticosterone levels. Neuroendocrinology 41:64-71.
Gold, G., Karoly, K., Freeman, C., and Adelman, R.C. (1976) A possible role for insulin in the altered capability for hepatic enzyme adaptation during aging. Biochem. Biophys. Res. Comm. 73:1003-1010.
Gold, P.E., Stone, W.S., and Martin, S.M. (1987) Glucose relationships with memory and sleep in old rats. Soc. Neurosci. Abstr. 13(1): 441.
Gonzales, A.M., Sochor, M., Rowles, P.M., Wilson-Holt, N, and McLean, P. (1981) Sequential biochemical and structural changes occuring in rat lens during cataract formation in experimental diabetes. Diabetes 21:23.
Gordon, M.N., Mobbs, C.V., Morgan, D.G., and Finch, C.E. (1984) Effects of age and estradiol on brain and pituitary glucose-6-phosphate dehydrogenase and 6-phosphogluconate dehydrogenase activities. Soc. Neurosci. Abs. 10:477
Gosden, R.G., Laing, S.C., Felicio, L.S., Nelson, J.F., and Finch, C.E. (1983) Imminent oocyte exhaustion and reduced follicular recruitment mark the transition to acyclicity in aging C57BL/6J mice. Biol. Reprod.
Grad, B., Rosenberg, G.M., Liberman, H., Trachtengerg, J., and Kral, V.A. (1971) Diurnal variation of the serum cortisol level of geriatric subjects. J. Gerontol. 26:351-357.
Gray, G.D., Wexler, B.C. (1980) Estrogen and testosterone sensitivity of middle-aged female rats and the regulation of LH. Exp. Gerontol. 15:201.
Haffner, S.M. (1987) Hyperinsulinemia as a possible etiology for the high prevalence of non-insulin-dependent diabetes in mexican americans. Diabete and Metabolisme 13:337-344.
Hamel, E.E., Santisteban, G.A., Ely, J.T.A., and Read, D.H.

(1986) Hyperglycemia and reproductive defects in non-diabetic gravidas: A mouse model test of a new theory. Life Sci. 39:1425-1428.
Hamilton, C.L., and Brobeck, J.R. (1963) Diabetes mellitus in hyperphagic monkeys. Endocrinology 73:512-515.
Hayashi, K. (1982) Insulin insensitivity and hyposupressibility of glucagon by hyperglycemia in aged Wistar rats. Gerontology 28:10-18.
Hayashi, K. (1980) Glucose tolerance in the elderly with special reference to insulin and glucagon responses. Wakayama Med. Rep. 23:29-39.
Henry, R.R., Wallace, P., and Olefsky, J.M. (1986) Effects of weight loss on mechanisms of hyperglycemia in obese non-insulin-dependent diabetes mellitus. Diabetes 35:990-998.
Herman, J.P., Marciano, F.F., Wiegand, S.J., and Gash, D.M. (1987) Selective cell death of magnocellular vasopressin neurons in neurohypophysectomized rats following chronic administration of vasopressin. J. Neurosci. 7:2564-2575.
Hsu, H.K., and Peng, M.T. (1978) Hypothalamic neuron number in old female rats. Gerontology 24:434-440.
Huang, H.H., Marshall, S., Meites, J. (1976) Capacity of old vs. young female rats to secrete LH, FSH, and prolactin. Biol. Reprod. 14:538.
Ingram, D.K., London, E.D., and Reynolds, M.A. (1982) Circadian rhythmicity and sleep: effects of aging in laboratory animals. Neurobiol. Aging 3:287-292.
Inoue, S., and Bray, G.A. (1977) The effect of subdiaphragmatic vagotomy in rats with ventromedial hypothalamic lesions. Endocrinology 100:108-114.
Inoue, S., Bray, G.A., Mullen, Y.S. (1977) Effect of transplantation of pancreas on development of hypothalamic obesity. Nature 266:742-744.
Ionescu, E., Rohner-Jeanrenaud, F., Berthoud, H.R., and Jeanrenaud, B. Increases in plasma insulin levels in response to electrical stimulation of the dorsal motor nucleus of the vagus nerve. Endocrinology 112: 904-910.
Ionescu, E., Sauter, J.F., and Jeanrenaud, B. (1985) Abnormal oral glucose tolerance in genetically obese (fa/fa) rats. Am. J. Phys. 248:E500-E506.
Jeanrenaud, B. (1985) An hypothesis on the aetiology of obesity: dysfunction of the central nervous system as a primary cause. Diabetologia 28:502-513.
Karnieli, E., Hissin, P.J., Simpson, I.A., Salans, L.B. (1981) A possible mechanism of insulin resistance in the rat adipose cell in streptozotocin-induced diabetes mellitus. Depletion of intracellular glucose transport systems. J. Clin. Invest. 68:811-814.
Kasuga, M., Inoue, S., Akanuma, Y., and Kosaka, K. (1980)

Insulin receptor function and insulin effects on glucose metabolism in adipocytes from ventromedial hypothalamus-lesioned rats. Endocrinology 107:1549-1555.

Katayama, S., Brownscheidle, C.M., Wootten, V., Lee, J.B., and Shimaoka, K. (1984) Absent or delayed preovulatory luteinizing hormone surge in experimental diabetes mellitus. Diabetes 33:324-327.

Kawashima, S. (1960) Influence of continued injections of sex steroids on the estrous cycle in the female rat. Annot. Zool. JPN. 33:226.

Kithara, A., and Adelman, R.C. (1979) Altered regultion of insulin secretion in isolated islets of different size in aging rats. Biochem. Biophys. Res. Comm. 87:1207-1213.

Klug, T.L., and Adelman, R.C. (1979) Altered hypothalamic-pituitary regulation of thyrotropin in male rats during aging. Endocrinology 104:1136-1142.

Klug, T.L., Freeman, C., Karoly, K., and Adelman, R.C. (1979) Altered regulation of pancreatic glucagon in male rats during aging. Biochem. Biophys. Res. Comm. 907-912.

Kobayashi, M., and Olefsky, J.M. (1978) Effect of experimental hyperinsulinemia on insulin binding and glucose transport in isolated rat adipocytes. Am. J. Phys. 235: E53-E62.

Komeda, K., Yokote, M., and Oki, Y. (1980) Diabetic syndrome in the Chinese hamster induced with monosodium glutamate. Experientia 36:232-234.

Kow, L.-M., and Pfaff, D.W. (1985) Actions of feeding-relevant agents on hypothalamic glucose-responsive neurons in vitro. Brain Res. Bull. 15:509-513.

L'Age, M., Langholz, J., Fechner, W., Salzman, H. (1974) Disturbances of the hypothalamo-hypopysial-adrenocortical system in the alloxan diabetic rat. Endocrinology 95:760-765.

Landfield, P., Waymire, J., and Lynch, G. (1978) Hippocampal aging and adrenocorticoids: A quantitative correlation. Science 202:1098-1102

Landfield, P. (1980) Adrenocortical hypothesis of brain and somatic aging.
In: Shimke, R.T. (ed.) Biological Mechanisms in Aging, NIH Publication No. 81-2194, Washington, DC, pp. 658-672.

Landfield, P.W., Sundberg, D.K., Smith, M.S., Eldridge, J.C., and Morris, M. (1980) Mammalian aging: Theoretical implications of changes in brain and endocrine systems during mid- and late-life in rats. Peptides 1 (Suppl. 1): 185-196.

Larkin, J.G., Frier, B.M., and Ireland, J.T. (1985) Diabetes mellitus and infection. Postgraduate Medical Journal 61:233-237.

Lazaris, J.A., Goldberg, R.S., and Kozlov, M.P. (1985) Studies on diabetes mellitus after ventromedial hypothalamic lesions in adult and aged rats. Endocrinologia Experimentalis 19:67-76.

Leopold, I.H., Mosier, M.A. (1978) Cataracts in diabetes mellitus. Geriatrics 33:33-41.

Liebelt, R.A., and Perry, J.H. (1957) Hypothalamic lesions associated with gold thioglucose-injected obesity. Proc. Soc. exp. Biol. 95:774-777.

Lu, J.H.K., Hopper, B.R., Vargo, T.M., and Yen, S.S.C. (1979) Chronological changes in sex steroid, gonadotropin, and prolactin secretion in aging female rats displaying different reproductive states. Biol Repro. 21:193-203.

Lu, J.H.K., Gilman, D.P., Meldrum, D.R., Judd, H.L., and Sawyer, C.H. (1981) Relationship between circulating estrogens and the central mechanisms by which ovarian steroids stimulates luteinizing hormone secretion in aged and young female rats. Endocrinology 108:836-841.

Mani, R.B., Lohr, J.B., Jeste, D.V. (1986) Hippocampal pyramidal cells and aging in the human: A quantitative study of neuronal loss in sectors CA1 to CA4. Exp. Neurol. 94:20-40.

Marshall, S., and Olefsky, J.M. (1980) Effects of insulin incubation on insulin binding, glucose transport, and insulin degradation by isolated rat adipocytes. J. Clin. Invest. 66:763-772.

Martin, C., Desai, K.S., and Steiner, G. (1983) Receptor and post-receptor insulin resistance induced by in vivo hyperinsulinemia. Can. J. Pharm. 61:802-807.

Matsuo, T., and Shino, A. (1972) Induction of diabetic alterations by goldthioglucose-obesity in KK, ICR, and C57Bl mice. Diabetelogia 8:391-397.

Mills, T.M., and Mahesh, V.B. (1978) Pituitary function in the aged. Aging (Geriatric Endocrinology) 5:1-11.

Mobbs, C.V., Flurkey, K., Gee, D.M., Yamamoto, K., Sinha, Y.N., and Finch, C.E. (1984a) Estradiol-induced adult anovulatory syndrome in female C57BL/6J mice: Age-like neuroendocrine, but not ovarian, impairments. Biol. Repro. 30:556-563.

Mobbs, C.V., Gee, D.M., and Finch, C.E. (1984) Reproductive senescence in female C57BL/6J mice: Ovarian impairments and neuroendocrine impairments that are partially reversible and delayable by ovariectomy. Endocrinology 115:1653-1662.

Mobbs, C.V., Cheyney, D., Sinhan, Y.N., and Finch, C.E. (1985a) Age-correlated and ovary-dependent changes in relationships between plasma estradiol and luteinizing hormone, prolactin, and growth hormone in femal C57BL/6J mice. Endocrinology 116:813-820.

Mobbs, C.V., Kannegieter, L., Sinha, Y.N., and Finch, C.E. (1985b) Delayed anovulatory syndrome induced by estradiol in female C57BL/6J mice: Age-like neuroendocrine, but not ovarian, impairments. Biol. Reprod. 32:1010-1017.

Mordes, J.P., and Rossini, A.A. (1981) Animal models of diabetes. Am. J. Med. 70:353-360.
Morgan, L.O., Vonderahe, A.R., and Malone, E.F. (1937) Pathological changes in the hypothalamus in diabetes mellitus. J. Nerv. Ment. Dis. 85:125-138.
Muta, K., Kato, K., Akamine, Y., Ibayishi, H. (1981) Age-related changes in the feedback regulation of gonadotropin secretion by sex steroids in men. Acta Endocrinol. 96:154-167.
Nakano, K., Yoshida, T., Kondo, M., and Muramatsu, S. (1984) Immune responsiveness and phagocytic activity of macrophages in streptozotocin (SZ)-induced diabetic mice. Endocrinol. Japon. 15-22.
Nankervis, A., Proietto, J., Aitken, P., and Alford, F. (1985) Hyperinsulinemia and insulin insensitivity: studies in subjects with insulinoma. Diabetologia 28:427-431.
Nelson, J.F., Holinka, C.F., Latham, K.R., Allen, J.K., and Finch, C.E. (1976) Corticosterone binding in cytosol from brain regions of mature and senescent male C57Bl/6J mice. Brain Res. 115:345-351.
Nelson, J.F., Felicio, L.S., Osterburg, H.H., Finch, C.E. (1981) Altered profiles of estradiol and progesterone associated with prolonged estrous cycles and persisten vaginal cornification in aging C57BL/6J mice. Biol Reprod. 24: 784.
Odell, W.D., Swerdloff, R.S. (1968) Progesterone-induced luteinizing and follicle-stimulating hormone surge in post-menopausal women: a simulated ovulatory peak. Proc. Natl. Acad. Sci. USA 61:529.
O'Mara, B.A., Byers, T., Schoenfeld, E. (1985) Diabetes mellitus and cancer risk: A multisite case-control study. J. Chronic Dis. 38:435-441.
Peng, M.T., and Hsu, H.K. (1982) No neuron loss from hypothalamic nuclei of male rats in old age. Gerontology 28:19-22.
Penicaud, L., Larue-Achagiotis, C., and Le Magnen, J. (1983) Endocrine basis for weight gain after fasting or VMH lesion in rats. Am. J. Physiol. 245: E246-E252.
Pfaff, D.W. (1972) Histological differences between ventromedial hypothalamic neurons of well-fed and underfed rats. Nature 223:77-79.
Pfaff, D.W. (1980) Estrogens and brain function. Springer-Verlag, New York.
Ramirez, V.D., and Sawyer, C.H. (1965) Advancement of puberty in the female rat by estrogen. Endocrinology 76:282.
Ramirez, I., and Friedman, M.I. (1982) Suppression of food intake by intragastric glucose in rats with impaired glucose tolerance. Physiology and Behavior 31:39-43.
Reaven, G.M., Olefsky, J., and Farquar, J.W. (1972) Does

hyperglycemia or hyperinsulinemia characterize the patient with chemical diabetes? The Lancet 2:1247-1249.
Reaven, E.P., Gold, G., and Reaven, G.M. (1979) Effect of age on glucose-stimulated insulin release by the beta cell of the rat. J. Clin. Invest. 64:591-599.
Reaven, E., Wright, D., Mondon, C.E., Solomon, R., Ho, H., and Reaven, G.M. (1983) Effect of age and diet on insulin secretion and insulin action in the rat. Diabetes 32:175-180.
Reaven, G.M., and Reaven, E.P. (1985) Age, glucose intolerance, and non-insulin-dependent diabetes mellitus. J. Am. Ger. Soc. 33:286-290.
Rizza, R.A., Mandarino, L.J., Genest, J., Baker, B.A., and Gerich, J.E. (1985) Production of insulin resistance by hyperinsulinemia in man. Diabetologia 28:70-75.
Rohner-Jeanrenaud, F., and Jeanrenaud, B. (1980) Consequences of ventromedial hypothalamic lesions upon insulin and glucagon secretion by subsequently isolated perfused pancreases in the rat. J. Clin. Invest. 65:902-910.
Rohner-Jeanrenaud, F., Hockstrasser, A.-C., and Jeanrenaud, B. (1983) Hyperinsulinemia of preobese and obese fa/fa rats is partly vagus mediated. Am. J. Phys. 244: E317-E322.
Rohner-Jeanrenaud, F., and Jeanrenaud, B. (1984) Oversecretion of glucagon by pancreases of hypothalamic-lesioned rats: a re-evaluation of a controversial topic. Diabetologia 27:535-539.
Rowe, J.W., Minaker, K.L., Pallota, J.A., Flier, J.S. (1983) Characterization of the insulin resistance of aging. J. Clin. Invest. 71:1581-1589.
Sabel, B.A., and Stein, D.G. (1981) Extensive loss of subcortical neurons in the aging rat brain. Exp. Neurol. 73:507-516.
Sandberg, H., Yoshimine, N., Maeda, S., Symons, D., and Zavodnick, J. (1973) Effects of an oral glucose load on serum immunoreactive insulin, free fatty acid, growth hormone, and blood sugar levels in young and elderly subjects. J. Am. Ger. Soc. 10:433-438.
Sapolsky, R.M., Krey, L., and McEwen, B.S. (1983) The adrenocortical stress-response in the aged male rat: impairment of recovery from stress. Exp. Gerontol. 18:55.
Sapolsky, R.M., Krey, L.C., and McEwen, B.S. (1984) Glucocorticoid-sensitive hippocampal neurons are involved in terminating the adrenocortical stress response. Proc. natl. Acad. Sci. USA 81:6274.
Sapolsky, R.M., Krey, L.C., and McEwen, B.S. (1986) The neuroendocrinology of stress and aging: The glucocorticoid cascade hypothesis. Endocrine Rev. 7:284-301.
Schnider, S.L., and Kohn, R.R. (1981) Effects of age and diabetes mellitus on the solubility and non-enzymatic

glycosylation of human skin collagen. J. Clin. Invest. 67:1630-1635.

Seals, D.r., Hagber, J.M., Allen, W.K., Hurley, B.F., Dalsky, G.P., Ehsani, A.A., and Holloszy, J.O. (1984) Glucose tolerance in young and older athletes and sedentary men. J. Appl. Physiol : 56(6):1521-1525.

Smith, R.E., and Horwitz, B.A. (1969) Brown fat and thermogenesis. Physiol. Rev. , 49:330-425.

Smith-West, C., and Garris, D.R. (1983) Diabetes-associated hypothalamic neuronal depopulation in the aging Chines hamster. Dev. Brain Res. 9:385-389.

Soman, V.R., and DeFronzo, R.A. (1980) Direct evidence for downregulation of insulin receptors by physiologic hyperinsulinemia in man. Diabetes 29_159-163.

Sontag, W.E., Steger, R.W., Forman, L.J., and Meites, J. (1980) Decreased pulsatile release of growth hormone in old males. Endocrinology 107: 1875-1881.

Sontag, W.E., Forman, L.J., and Meites, J. (1983) Changes in growth hormone secretion in aging rats and man, and possible relation to diminished physiological functions. In: Meites, J. (ed.) Neuroendocrinology of Aging, pp. 275-308, Plenum Press, New York.

Spindler-Vomachka, M., and Johnson, D.C. (1985) Altered hypothalamic-pituitary function in the adult female rat with streptozotocin-induced diabetes. Diabetologia 28:38-44.

Sridaran, R., and Blake, C.A. (1978) Effects of neonatal monosodium glutamate (MSG) on pulsatile plasma LH and estrogen feedback on LH release in ovariectomized (OVX) rats. Physiologist 21:114.

Steffes, M.W., and Mauer, S.M. (1982) Diabetic glomerulopathy in man and experimental animal models. In: Richter, G.W., Epstein, M.A. (eds.): International Review of Experimental Pathology, pp 147-159, Academic Press, New York.

Stokes, P.E., Stoll, P.M., Koslow, S.H., Maas, J.W., Davis, J.M., Swann, A.C., and Robins, E. (1984) Pretreatment DST and hypothalamic-pituitary-adrenocortical function in depressed patients and comparison groups. Arch. Gen. Psych. 41:257-267.

Stout, R.W. (1981) The role of insulin in atherosclerosis in diabetics and non-diabetics. A review. Diabetes 30 (Supp. 2): 54-57.

Tallarigo, L., Giampietro, O., Penno, G., Miccoli, R., Gregori, G., and Navalesi, R. (1986) Relation of glucose tolerance to complications of pregnancy in nondiabetic women. N. Engl. J. Med. 315:989-992.

Telford, N., Mobbs, C.V., Sinha, Y.N., and Finch, C.E. (1986) The increase of anterior pituitary dopamine in aging C57BL/6J mice is caused by ovarian steroids, not intrinsic pituitary aging. Neuroendocrinology 43:135-142.

Tokunaga, K., Fukushima, M., Kemnitz, J.W., and Bray, G.A. (1986) Effect of vagatomy on serum insulin in rats with paraventricular or ventromedial hypothalamic lesions. Endocrinology 119:1708-1711.
Tornello, S., Fridman, O., Weisenberg, L., Coirini, H., De Nicola, A. (1981) Differences in corticosterone binding by regions of the central nervous system in normal and diabetic rats. J. Steroid Biochem. 14:77-84.
Touitou, Y., Sulon, J., Bogdan, A., Reinberg, A., Sodoyez, J.-C., and Demey-Ponsart, E. (1983) Adrenocortical hormones, ageing, and mental condition: seasonal and circadian rhythms of plasma 18-hydroxy-11-deoxycorticosterone, total and free cortisol, and urinary corticosteroids. J. Endo. 93:53-64.
Unger, R.H., and Orci, L. (1975) The essential role of glucagon in the pathogenesis of diabetes mellitus. Lancet 1:14-16.
Unger, R.H., and Grundy, S. (1985) Hyperglycemia as an inducer as well as a consequence of impaired islet cell function and insulin resistance: implications for the management of diabetes. Diabetologia 28: 119-121.
Welborn, T.A., Stenhouse, N.S., and Johnstone, C.G. (1969) Factors determining serum insulin response in a population sample. Diabetologia 5:263-266.
West, K.M. (1978) Epidemiology of diabetes and its vascular lesions. Elsevier North-Holland, New York.
West, D.B., Seino, Y., Woods, S.C., and Porte, D. (1980) Ventromedial hypothalamic lesions increase pancreatic sensitivity to streptozotocin in rats. Diabetes 29:948-951.
Williams, G.C. (1957) Pleiotropy, natural selection, and the evolution of senescence. Evolution 11:398.
Wise, P.M., Ratner, A., and Penke, G.T. (1976) Effect of ovariectomy on serum prolactin concentrations in old and young rats. J. Reprod. Fert. 47:363.
Wise, P.M., and Parsons, B. (1984) Nuclear estradiol and cytosol progestin receptor concentrations in the brain and pituitary gland and sexual behavior in ovariectomized estradiol-treated middle-aged rats. Endocrinology 115:810-816.
Wise, P.M., McEwen, B.S., Parsons, B., and Rainbow, T.C. (1984) Age-related changes in cytoplasmic estradiol receptor concentrations in microdissected brain nuclei: Correlations with changes in steroid-induced sexual behavior. Brain Res. 321:119-126.
Wise, P.M. (1983) Aging of the femal reproductive system. In: Rothstein, M., (ed.) Review of Biological Research in Aging, Vol. 1, pp. 195-222. Liss, New York.
Woods, S.C., Porte, D., Bobbioni, E., Ionescu, E., Sauter, J.-F., Rohner-Jeanrenaud, F., and Jeanrenaud, B. (1985) Insulin: Its relationship to the central nervous system and to the control of food intake and body weight. Am. J. Clin.

Nutr. 42:1063-1071.

Zarco de Coronado, I., Yepez Chamorro, M.C. (1987) Increase of circulating levels of glucose after the electrolytic lesion of the arcuate hypothalamic nucleus. Soc. Neurosci. Abstr. 13: p. 403.

Zimmet, P. and Whitehouse, S. (1979) The effect of age on glucose tolerance. Diabetes 28:617-628.

13

SALMONIDS AND ANNUAL FISHES: DEATH AFTER SEX

Walton W. Dickhoff[1]

School of Fisheries
University of Washington
Seattle, Washington 98195
and
Northwest and Alaska Fisheries Center
National Marine Fisheries Service
Seattle, Washington 98112
U.S.A.

Then seized with feverish passion, we gave play
To acts of love that wore our love away.
Tho' death was gaining on us, longing bade
Our beings mingle in that evening shade.
Our passions were unchecked by thoughts of harm;
Our kindled looks were quenched by no alarm;
Each for the other, at the point of death...

from "At the Water's Side"
Guy de Maupassant, 1880

I. INTRODUCTION

The rapid aging and death of the Pacific salmon after a long migration and subsequent spawning in its natal stream is an intriguing end to a

[1] The author was supported by grants from the National Science Foundation (PCM 86-15521) and Washington Sea Grant (R/A-49) during preparation of this chapter. Work on Atlantic salmon was supported by U.S. Department of Agriculture (S & E-Aq) under Agreement No. 85-CRSR-2-2603.

dramatic life cycle. The rapidity and precise timing of senescence and death of Pacific salmon is an often cited example for a functional role of aging and death. Furthermore, this phenomenon serves as a good subject for investigation of the general mechanisms of aging and death in animals. This review will address some of those mechanisms in salmon and lamprey and point out promising areas for future research.

The death of Pacific salmon after a single reproductive event is an example of a semelparous mode of reproduction (Cole, 1954). Animals that produce multiple offspring or litters over an extended reproductive life-span are classified as iteroparous. Among fishes, the best known examples of semelparous species include the teleosts, Pacific salmon and eels, and an agnathan, the lamprey, although semelparity has been observed in other species, for example, some gobiid fishes (Miller, 1973). Undoubtedly semelparity will be found in other teleost species when more is known about their life histories. It could be concluded that the occurrence of semelparity as a reproductive strategy of fishes is not uncommon. There may be several reasons for the existence of semelparity in fishes. Consideration of factors that contribute to the rate of population growth led Cole (1954) to the conclusion that iteroparity is favored when fecundity is relatively low or when the pre-reproductive period is prolonged. Semelparity would be favored when animals produced a large number of viable offspring per mating and begin reproducing at an early age. Female Pacific salmon typically produce between two and six thousand eggs and may mature as early as two years of age. According to Cole, once semelparity had been established, there would be little selective pressure or advantage for Pacific salmon to change from semelparous to iteroparous reproduction. Although a species that is iteroparous would always have a greater rate of population growth (assuming equal survivorship) than if it was semelparous, a short generation time and high fecundity would minimize the difference in population growth.

Although semelparous reproduction is quantitatively reasonable for population growth in Pacific salmon, other biological advantages of semelparity in these species are not as immediately apparent. One suggestion is that this mode of reproduction favors exploitation of remote spawning sites. It is reasonable to speculate that the energetic demands of the long upstream migration of adult Pacific salmon is so exhaustive that there is not sufficient energy remaining for subsequent downstream and oceanic migration. This may be an example of selection for antagonistic pleiorophic genes (Williams, 1957, Sacher, 1982), genes which have both favorable and unfavorable actions. For example genes which favor mobilization of tissue stores of energy during the upstream migration would allow adult Pacific salmon to reach favorable spawning sites. However, once these sites are reached and spawning is completed, there is only enough energy remaining for a brief period of nest guarding. Another possible advantage for semelparity in Pacific salmon is that the decaying bodies of the spawned-out adults fertilize the nursery areas of the young. The glacial rivers and streams of the Northwest and Alaska are often poor in nutrients. Bacteria and zooplankton feeding on the decaying bodies of the adults in the streams may eventually provide food for the salmon alevins after they emerge from the gravel beds. Clearly, the rapid aging and senescence of Pacific salmon is an example of a genetically programmed death. As such, Pacific salmon and other semelparous fish are useful for investigation of the mechanisms of aging and death. However, it is difficult to say to what extent information obtained from semelparous species may be applicable to senescence and death of iteroparous species.

II. NEUROENDOCRINE FACTORS IN SEMELPAROUS FISH

Many of the recognized theories on senescence and death in vertebrates (Hayflick, 1985) have some basis of fact in fishes. Both the immune system and the neuroendocrine system have been implicated in the death of semelparous fish. In general, mechanisms contributing to the death of semelparous fishes include: 1) reproductive hormones, 2) corticosteroids, 3) depletion of metabolizable energy stores, and 4) disease due to a suppressed immune response. There is evidence to suggest that most of these mechanisms are interdependent, and they appear to act simultaneously. The challenge is to determine whether one particular factor is primary.

The neuroendocrine basis of senescence and death in fishes has been reviewed by Larsen (1973, 1985). Her recent review compared mechanisms of death in eels, Pacific salmon and lampreys. In eels, (*Anguilla anguilla*) administration of pituitary extracts or human chorionic gonadotropin induces ovulation in females and spermiation in males. However, females die within weeks after ovulation whereas males may live through another cycle of reproduction. Such results might support the view that the male is the hardier sex in that energetic demands of ovarian maturation and ovulation are greater than the energetic demands of spermatogenesis and spermiation. However, female eels which have been induced to spawn still have significant mobilizable energy reserves (Boëtius and Boëtius, 1967). Furthermore, there is evidence to suggest that eels should have sufficient energy reserves for migration and spawning, and that energy depletion should not be considered to be a life-limiting factor (Boëtius and Boëtius, 1980). Therefore, the evidence suggests that the reproductive hormones are responsible for death of female eels; the mechanism of death in male eels is not apparent. It is interesting with regard to male-female longevity that in the iteroparous marine Norway goby (*Pomatoschistus norvegicus*) the female dies within a few months after spawning for a second time, but the male remains alive for about six months (Gibson and Ezzi, 1981).

A. Lamprey

Larsen's research (Larsen, 1973, 1980) on the cause of death in river lampreys has shown that gonadectomy prolongs life for up to four months and hypophysectomy prolongs life for up to 13 months after the normal time of spawning. Estradiol treatment of intact lampreys prolonged life for up to four months (Larsen, 1974), whereas testosterone treatment either had no effect or caused a slight prolongation of life (Larsen, 1987). Prolongation of life by reproductive steroids would appear to be contradictory to the prolongation of life observed after removal of the gonads or pituitary, since these tissues are usually assumed to be responsible for steroid production and gonad maintenance, respectively. One hypothesis to explain the apparent contradiction would involve pituitary gonadotropin acting both directly and through a gonadal factor (other than estradiol or testosterone) as the agents of mortality. When the pituitary is removed, prolongation of life is maximal since both the source of gonadotropin is removed and the production of the pituitary-dependent gonadal factor is reduced. When the gonad is removed, pituitary gonadotropin may be the only factor inducing mortality, and prolongation of life is intermediate. When estradiol is given to intact, maturing lampreys, it may reduce gonadotropin production through negative feedback and life is prolonged. Although this hypothesis might be plausible, recent work indicates that estradiol and testosterone may be produced extragonadally in *Lampetra fluviatilis* (Kime and Larsen, 1987). Thus, the neuroendocrine relationships in maturing lampreys may be complex and not typical of such known systems in teleosts. Larsen's most recent speculation includes the possibility that corticosteroids from the pronephroi (adrenal homolog) may be involved in the post-spawning death of the lamprey (Larsen, 1987).

B. Salmon

Although it is often stated that all Pacific salmon (*Oncorhynchus* spp.) are semelparous whereas other salmonids (*Salmo, Salvenlinus)* are iteroparous, there are some exceptions. Precociously mature male sockeye (O. nerka), chinook (*O. tshawytscha*), and pink (O. gorbuscha) salmon do not die after spawning (Fraser, 1918; Robertson, 1957; Funk and Donaldson, 1972). Furthermore, significant proportions of iteroparous anadromous Atlantic salmon (*Salmo salar*) and steelhead trout (*S. gairdneri*) will die after their first spawning as adults. Non-migratory rainbow trout (*S. gairdneri)* usually do not die after their first spawning.

A contributing factor to the death of anadromous salmonid fishes is the depletion of available energy stores that results from the physical effort expended during the long migration to the spawning grounds. Increased energy demands may also be incurred as a result of gametogenesis and the need to change osmo/ionoregulatory systems to adapt from seawater to fresh water. It can be argued that the energy demands of long migrations and freshwater adaptation are not primary causes of death in Pacific salmon, since land-locked sockeye salmon (kokanee) also die after spawning.

Maturing Pacific salmon in the wild enter a period of partial or complete fasting prior to spawning. At this time there is observable atrophy of the gastro-intestinal tract. Forced feeding of maturing or spawned sockeye salmon reverses stomach atrophy and induces volitional feeding within a few weeks and may prolong life for as much as 10 weeks compared to unfed control fish (McBride *et al.*, 1965). Although feeding may help survival of Pacific salmon, it does not prolong life for an extended period. Salmon held in captivity without food can remain alive for several months (Robertson *et al.*, 1963). However, forced feeding of Atlantic salmon after spawning is important for their long term survival and successful completion of reproduction in the subsequent year (Hill and Semple, pers. comm.). Presumably, starvation causes a reduction in blood insulin levels in adult salmon as it does in juveniles (Plisetskaya, *et al.*, 1986). Feeding may restore insulin levels and favor repletion of metabolizable energy stores.

Disease can be considered another contributing factor to the death of maturing salmonids. After entry into freshwater, salmon are highly susceptible to fungal (*Saprolegnia* spp.) and bacterial infections (furunculosis; *Aeromonas* spp.). Immunological resistance to infection is probably reduced due to elevated blood levels of cortisol during final maturation. Cortisol has been shown to decrease the number of lymphocytes and reduce antibody production in juvenile salmon and in brown trout (Pickering, 1984; Maule *et al.*, 1987). Disease can be a major cause of either pre- or post-spawning death in wild salmon if their migration to the spawning grounds is delayed. However, disease cannot be a primary mechanism of death in salmon, since post-spawning mortality occurs in apparently healthy fish.

The marked hyperplasia of corticosteroid-secreting interrenal tissue accompanying sexual maturation and spawning of Pacific salmon was first described by Robertson and Wexler (1957). Interrenal hyperplasia was found to be comparable in Pacific salmon and migratory rainbow trout (steelhead), but was very much less pronounced in non-migratory rainbow trout (Robertson and Wexler, 1959). Interrenal hyperplasia has been associated with elevated blood levels of 17-hydroxycorticosteroids (Hane and Robertson, 1959). Histological studies of organs of migrating and spawning Pacific salmon revealed degeneration with or without atrophy in the stomach, liver, spleen, thymus, thyroid, gonads, pituitary, kidney and cardiovascular system (Greene, 1926; Robertson and Wexler, 1960). This pattern of degeneration was reminiscent of Cushing's syndrome, experimental hyperadrenocorticism and aging. Accordingly, it was suggested that the death of Pacific salmon was due to activation of the interrenals and subsequent degenerative effects of high circulating levels of corticosteroids. On the other hand, not all of the evidence fit this hypothesis. The similarity in the degree of interrenal hyperplasia in all of the Pacific salmon and migratory trout did not explain the fact that a significant proportion of the migratory trout survive after spawning (Robertson and Wexler, 1959). Furthermore, the levels of circulating corticosteroids did not completely correspond with survival or the extent of degenerative changes in tissues. Although Pacific salmon had the highest levels of 17-hydroxycorticosteroids, non-migratory trout had higher levels than did the migratory trout (Robertson *et al.*, 1961). Robertson and Wexler (1960) stated that except for the cardio-vascular changes the histologically observable tissue degeneration in Pacific salmon was "more pronounced and generalized than are the corresponding ones found in Cushing's syndrome, experimental hyperadrenocorticism or aging." Finally, treatment of immature rainbow trout with cortisol implants for several months did not uniformly induce death, although they did cause degenerative changes in tissues (Robertson *et al.*, 1963). A most important piece of the puzzle was supplied with the discovery that gonadectomy of sockeye salmon prevented interrenal hyperplasia and prolonged survival of the fish for up to four to five years (Robertson, 1961).

Robertson's experiments on surgically castrated sockeye salmon indicated that gonadectomy had to be performed on immature fish in order to show that survival could be prolonged for several years (Robertson, 1961). Gonadectomy of mature fish has only a small, although significant, effect since interrenal hyperplasia and degeneration of the tissues had already taken place. Robertson (1961) speculated that some factor from the gonads stimulated the interrenals of maturing salmon. The idea that estrogen may stimulate the interrenal found some support by the work of Hane and colleagues (Hane and Robertson, 1959; Hane *et al.*, 1966) who found higher blood levels of 17-hydroxycorticosteroids in female compared to male chinook salmon. Experimental demonstration of the influence of gonadal factors on interrenal function was provided by Fagerlund and Donaldson (1969) who showed that 11-ketotestosterone and 17-alphamethyltestosterone increased both the distribution volume and secretion rate of cortisol when injected into gonadectomized male sockeye salmon. Evidence that estradiol treatment can also induce interrenal hyperplasia in salmon has been provided by van Overbeeke and McBride (1971). Based on histological studies of the maturing sockeye salmon, it has been concluded that the corticotrophs of the pituitary gland are not markedly changed during the development of interrenal hyperplasia (McBride and van Overbeeke, 1969).

Based on the studies cited above a hypothetical model for the primary mechanism of post-spawning death could be proposed as shown in figure 1. Increasing levels of gonadotropin during sexual maturation stimulate the production of gonadal steroids (androgens in the male; androgens and estrogen in the female) which stimulate the production of cortisol. The long term effects of elevated cortisol result in tissue degeneration, immune suppression and general loss of the ability to maintain homeostasis. Death results from degeneration of vital tissues.

Although the proposed model (Fig. 1) basically explains the majority of evidence regarding the causes of death, there are several issues that can be raised to suggest that the model is incomplete. It has been argued that the elevations in blood corticosteroids observed in Pacific salmon during sexual maturation may be due to stress associated with handling, physical exercise or disease. Fagerlund (1967) has shown that sexual maturation and spawning did not induce sustained increases in blood cortisol in healthy

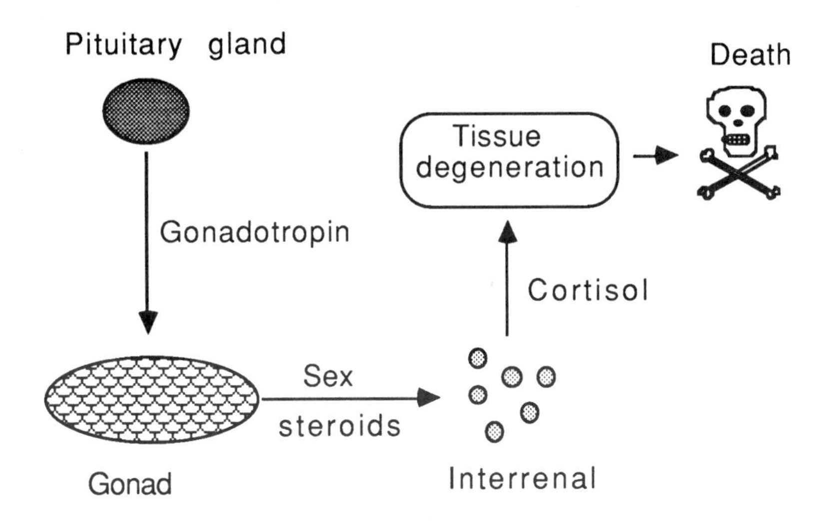

Fig. 1. Schematic diagram of the endocrine factors contributing to death of Pacific salmon after spawning.

sockeye salmon. Relatively low levels of plasma cortisol were found in spawning and spent wild sockeye sampled on the spawning grounds (Fagerlund, 1967). High levels of plasma cortisol were found primarily in dying or diseased fish.

The relationship between plasma cortisol levels and death after spawning is not entirely clear. Although many reports (see above) show higher levels of cortisol in semelparous versus iteroparous salmonids, high levels of corticosteroids can be observed in non-migratory trout (Robertson *et al.*, 1961) or in brown trout (*Salmo trutta*) which are infected with fungus (*Saprolegnia*) at the time of sexual maturation (Pickering and Christie, 1981). This point may not be important, for example, if it could be shown that semelparous species are more sensitive to circulating cortisol.

The relationship between sex steroid and cortisol changes are not entirely clear. In salmonids plasma levels of estradiol increase early on during vitellogenesis and may reach peak levels several weeks or months before spawning(Ueda *et al.*, 1984). Estradiol levels usually decrease at the time of final maturation (Sower and Schreck, 1982, Scott *et al.*, 1983; Sower *et al.* 1984). We have observed that plasma levels of androgens in precociously maturing male salmonids may increase at six to nine months before final maturation (Stuart-Kregor *et al.*, 1981). Plasma androgen levels either remain unchanged or tend to decrease at the time of final maturation. These observations suggest that the relative timing of changes in sex steroids and interrenal hyperplasia are not precisely regulated. The studies on the effects of sex steroids on cortisol dynamics did not show a significant elevation in plasma levels of cortisol, and gonadectomy did not decrease plasma cortisol levels (Donaldson and Fagerlund, 1968; Fagerlund and Donaldson, 1969). Furthermore, treatment of gonadectomized fish with methyltestosterone did not increase fatality (Fagerlund and Donaldson, 1969).

Finally, the mechanism proposed in figure 1 does not seem to be a very precise one. If the cause of mortality is tissue degeneration due to corticosteroids, then the clock of death is set in motion early on. There may be little room for adjustments in the timing of the spawning run. There can be significant year to year variation in such factors important in spawning migration as river level and flow, temperature, rainfall, the strength of natal stream olfactory cues and availability of appropriate sites for building a redd. Although significant pre-spawning mortality occurs when environmental conditions are markedly unfavorable, I would expect greater pre-spawning mortality under normal conditions than that usually observed if the mechanism proposed in figure 1 was the primary one. A teleologically more appealing model would include a mechanism whereby death could be delayed until spawning is completed. A possible mechanism would involve the ovulatory surge of gonadotropin as the immediate instrument of death. Some evidence is available to support this possibility.

For the last seven years the National Marine Fisheries Service (NMFS)in Seattle has maintained several stocks of Atlantic salmon (*S. salar*) as captive broodstock. During the last five years some of these fish have been induced to spawn using a gonadotropin releasing-hormone analogue (GnRHa; D-Ala^6, desGly^{10} LHRH ethylamide). A consistent observation in our induced spawning studies has been that the GnRHa-treated fish appear to die soon after spawning at a higher rate than observed for either the saline-injected controls or untreated fish. Attempts to maintain the females alive after their first spawning and recondition them to reproduce again is usually successful in 10 to 40 % of the naturally spawned fish. None of the GnRHa-injected fish can be maintained alive for more than a month. This observation suggests that GnRHa injections may be lethal. This possibility was examined in an experiment designed to evaluate the dose-related induction of ovulation in Atlantic salmon.

Adult Atlantic salmon (Union River stock) that had been reared throughout their life cycle in captivity in the NMFS facilities were transferred from seawater net-pens to fresh water tanks during the first week of October. After one week of acclimation to fresh water, 120 females (ten groups of 12) were injected intraperitoneally with either 0.7 % saline or varying doses of GnRHa. Two injections were given three days apart. The total doses injected were 2, 11, 20, 21, 30 and 40 μg/kg. Ovulation was significantly advanced in the fish receiving the doses of 30 and 40 μg/kg. The fish were individually marked with passive integrated transponder (PIT) tags (Identification Devices, Inc., Boulder, CO), housed together in

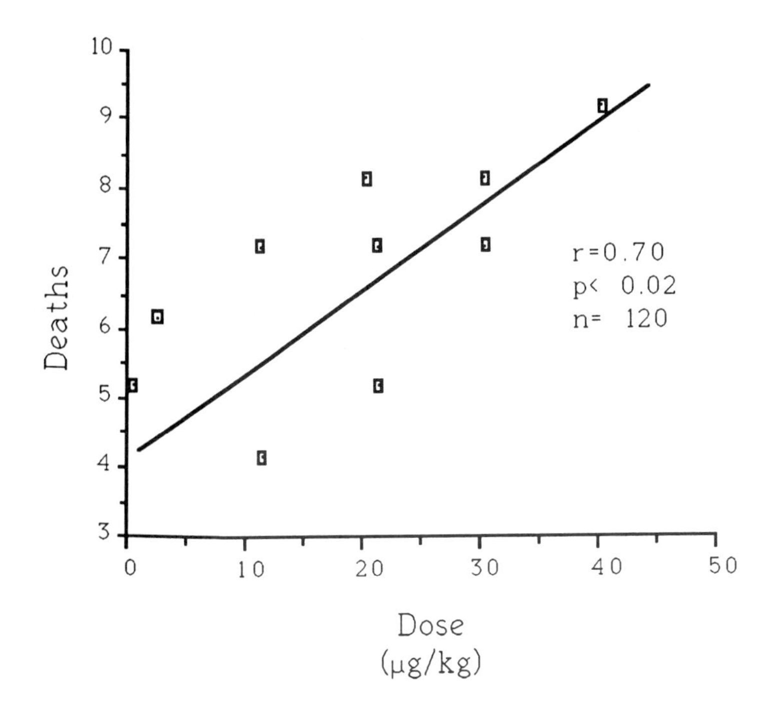

Fig. 2. The relationship between dose of injected GnRHa and mortality of mature Atlantic salmon during 14 days after the first of two injections. Deaths indicate the number of fish dying in each group of 12 fish. (From a previously unpublished study by E. Mooney and W. Dickhoff).

three cylindrical fiberglass tanks (3 m diameter) with flow-through water (13-15°), and checked daily for 14 days after the first injection. The mortality for each injected group is shown in figure 2. A statistically significant relationship was observed between mortality and the dose of GnRHa injected. The majority of the fish receiving the two highest doses spawned by day 14. Most of the remaining fish completed ovulation during the third week after injection. A higher than usual mortality was encountered, probably due to the unusually high water temperature.

The significant relationship between injected GnRHa and the mortality of Atlantic salmon invites speculation on the possible mechanism of the mortality. It is possible that GnRHa or pituitary gonadotropin released in response to the GnRHa may cause additional activation of the interrenal and produce a surge in cortisol. Such a surge in cortisol might be enough to result in the final wave of degeneration to kill the fish. Alternatively, Gn-

RHa itself or gonadotropin may be direct agents of death. In support of this notion, it has been observed that ovarian fluid collected from GnRHa-injected Atlantic salmon frequently has a cytotoxic effect on fish cells *in vitro* (Bronson, pers. comm.). In routine assays for infectious hematopoetic necrosis (IHN) virus it was observed that ovarian fluid from GnRHa-treated fish caused cell lysis similar to cell lysis in response to IHN. However, there was no measurable IHN virus in these cultures. The agent in ovarian fluid that was responsible for the cell lysis has not been identified. It is possible that proteolytic enzymes or enzyme activators may have been present in the ovarian fluid. There is evidence that ovulation in fish is induced, at least in part, by proteolytic enzymes (Oshiro and Hibiya, 1975, 1982). Recently, Berndtson and Goetz (1987) have identified both metallo and serine proteases in the follicle wall of brook trout and goldfish. In mammals, luteinizing hormone stimulates the production of plasminogen, a serine protease, and plasminogen activator by granulosa cells of maturing Graafian follicles (Strickland and Beers, 1976; Reich, 1978). Activated plasminogen (plasmin) degrades proteins in the follicular wall to release the ovum during ovulation. If such a system were operating in salmonids, then large preovulatory increases in gonadotropin may induce large releases of proteolytic enzymes or enzyme activators which could accumulate in the ovarian fluid or in the circulation (Figure 3). Such an occurrence would have a particular damaging effect in salmonids since the ovary is not encased in mesenteries, but is open to the peritoneum. High concentrations of

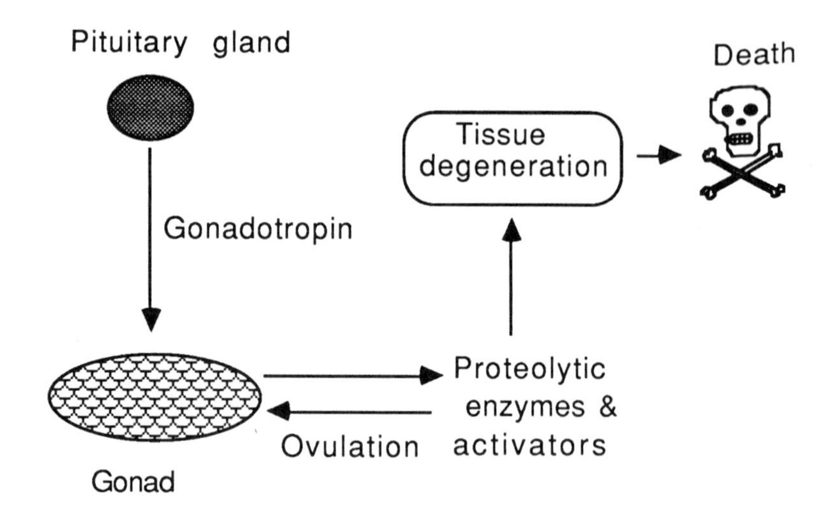

Figure 3. Hypothetical model involving proteolytic enzymes as the cause of death in Pacific salmon.

proteolytic enzymes in the circulation of ovulating salmon could explain the observation that tissue degeneration in spawning Pacific salmon is more pronounced and generalized than in hyperadrenocorticism (Robertson and Wexler, 1960). A large release of proteolytic enzymes might also explain the death induced by injection of human chorionic gonadotropin in Pacific halibut (*Hippoglossus stenolepis*) that we have seen (Dickhoff, unpublished) and in Atlantic halibut (*Hippoglossus hippoglossus*) observed in Norway (Huse, pers comm). This hypothesis needs experimental testing.

Undoubtedly, the factors that influence death of Pacific salmon are working in concert, and depending on the specific environmental conditions present and the genetic background of a particular stock, some factors will predominate. Thus, the relative importance of the factors may vary from year to year. However, as discussed above, their relative importance can be categorized approximately as either primary or secondary (Table I). The primary factors include both activation of the reproductive system and interrenal tissue. Energy required for the morphological, physiological and behavioral changes during reproductive maturation put a large demand on the fish. The damaging effects of activation of the interrenal tissue (dependent on reproductive hormones) sets the stage for death. Final maturation and spawning is the closing act.

Other possible endocrine agents of mortality in Pacific salmon need to be explored. Some purified preparations of prolactin have been shown to induce mortality in several teleosts (Gona, 1979). It is not clear what the mechanism of death could have been or whether this observation has any

Table I. Factors affecting death of Pacific salmon.

Factor	Primary	Secondary
Energy demands		
production of gametes		-
seawater to freshwater acclimation		+
migration		+
lack of feeding		+
Disease		+
Hyperadrenocorticism		
immune suppression	+	+
catabolism	+	+
Reproduction		
hyperadrenocorticism	+	+
reproductive hormones	+	+

physiological significance. Increases in blood levels of prolactin during sexual maturation have been demonstrated at least in female salmon (Hirano *et al.*, 1985).

A good model animal for the studies of factors affecting maturation and death of Pacific salmon is the hormonally sterilized salmon. Treatment of salmonids with androgens near the time of hatching and during the first two to three months of feeding is an efficient technique for producing large numbers of fish in which the gonads do not develop (Hunter and Donaldson, 1983). Additional insights could be made by comparing endocrine changes in both precociously maturing male chinook salmon, which continue living after reproductive maturation, and normal male chinook that die after spawning.

ACKNOWLEDGMENTS

The author wishes to thank Craig V. Sullivan, Elizabeth Mooney, F. William Waknitz, Erika Plisetskaya, Melinda Bernard and Penny Swanson for their comments and assistance during the preparation of this chapter.

REFERENCES

Berndtson, A.K. and Goetz, F. (1987). Proc. IIIrd Int. Symp. Reprod. Physio. Fish, St. John's, Newfoundland, Canada. p.55.
Boëtius, I, and Boëtius, J. (1967). Meddr. Danm. Fiskeri-og Havunders. N.S. **4**, 339.
Boëtius, I, and Boëtius, J. (1980). Dana **1**, 1.
Cole, L. C. (1954). Quart. Rev. Biol. **29**, 103.
Donaldson, E.M. and Fagerlund, U. H. M. (1968). Gen. Comp. Endocrinol. **11**, 552.
Fagerlund, U.H.M. (1967). Gen. Comp. Endocrinol. **8**, 197.
Fagerlund, U.H.M. and Donaldson, E.M. (1969). Gen. Comp. Endocrinol. **12**, 438.
Fraser, M.C. (1918). Contrib. Canad. Biol. No. 38a, 105.
Funk, J.D. and Donaldson, E.M. (1972). Can. J. Zool. **50**, 1413.
Gibson, R.N. and Ezzi, I. A. (1981). J. Fish Biol. **19**, 697.
Gona, O. (1979). Gen. Comp. Endocrinol. **37**, 468.
Greene, C.W. (1926). Physiol. Rev. **6**, 201.
Hane, S. and Robertson, O.H. (1959). Proc. Nat. Acad. Sci. USA **45**, 886.

Hane, S., Robertson, O.H., Wexler, B.C. and Krupp, M.A. (1966). Endocrinology **78**, 791.
Hayflick, L. (1985). Exp. Geront. **20**, 145.
Hirano, T., Prunet, P., Kawauchi, H., Takahashi, A., Ogasawara, T., Kubota, J., Nishioka, R.S., Bern, H. A., Takada, K. and Ishii, S. (1985). Gen. Comp. Endocrinol. **59**, 266.
Hunter, G.A. and Donaldson, E.M. (1983). In: Fish Physiology, Hoar, W.S., Randall, D.J. and Donaldson, E.M., eds., Academic Press, New York, p. 223.
Kime, D.E. and Larsen, L.O. (1987). Gen. Comp. Endocrinol. **68**, 189.
Larsen, L.O. (1973). Thesis. University of Copenhagen, Copenhagen. 172 pp.
Larsen, L.O. (1974). Gen. Comp. Endocrinol. **24**, 305.
Larsen, L.O. (1980). Can. J. Fish. Aquat. Sci. **37**, 1762.
Larsen, L.O. (1985). Current Trends in Comparative Endocrinology, Lofts, B. and Holmes, W.N., eds. Hong Kong University Press, p. 613.
Larsen, L.O. (1987). Gen. Comp. Endocrinol. **68**, 197.
Maule, A., Schreck, C.B. and Kaattari, S.L. (1987). Can. J. Fish. Aquat. Sci. **44**, 161.
McBride, J.R., Fagerlund, U.H.M., Smith, M. and Tomlinson, N. (1965). J. Fish. Res. Bd. Canada **22**, 775.
McBride, J.R. and van Overbeeke, A.P. (1969). J. Fish. Res. Bd. Canada **26**, 1147.
Miller, .J. (1973). J. Fish Biol. **5**, 353.
Oshiro, T. and Hibiya, T. (1975). Bull. Japan. Soc. Sci. Fish. **41**, 115.
Oshiro, T. and Hibiya, T. (1982). Bull. Japan. Soc. Sci. Fish. **48**, 623.
Pickering, A.D. (1984). Gen. Comp. Endocrinol. **53**, 252.
Pickering, A.D. and Christie, P. (1981). Gen. Comp. Endocrinol. **44**, 487.
Plisetskaya, E.M., Dickhoff, W.W., Paquette, T.L. and Gorbman, A. (1986). Fish Physiol. Biochem. **1**, 35.
Reich, E. (1978). Molecular basis of biological degradative processes. p. 155.
Robertson, O.H. (1957). Am. Philos. Soc. Year Book 1956, p. 215.
Robertson, O.H. (1961). Proc. Nat. Acad. Sci. USA **47**, 609.
Robertson, O.H., Hane, S, Wexler, B.C. and Rinfret, A.P. (1963). Gen. Comp. Endocrinol. **3**, 422.
Robertson, O.H., Wexler, B.C., and Miller, B.F. (1961). Circ. Res. **9**, 826.
Robertson, O.H. and Wexler, B.C. (1957). Science **125**, 1295.
Robertson, O.H. and Wexler, B.C. (1959). Endocrinology **65**, 225.
Robertson, O.H. and Wexler, B.C. (1960). Endocrinology **66**, 222.
Sacher, G.A. (1982). Perspec. Biol. Med. **25**, 339.
Scott, A.P., Sumpter, J.P. and Hardiman, J.P. (1983). Gen. Comp. Endocrinol. **49**, 128.
Sower, S.A. and Schreck, C.B. (1982). Gen. Comp. Endocrinol. **47**, 42.

Sower, S.A., Iwamoto, R.N., Dickhoff, W.W. and Gorbman, A. (1984). Aquaculture **43**, 35.
Strickland, S. and Beers, W. (1976). J. Biol. Chem. **251**, 5694.
Ueda, H., Hiroi, O., Hara, A., Yamauchi, K. and Nagahama, Y. (1984). Gen. Comp. Endocrinol. **53**, 203.
van Overbeeke, A.P. and McBride, J.R. (1971). J. Fish. Res. Bd. Canada **28**, 477.
Williams, G.C. (1957). Evolution **11**, 398.

III
Neuroendocrinology and the Environment

14

DEVELOPMENT AND SENESCENCE OF THE NEUROENDOCRINE SYSTEMS CONTROLLING GROWTH AND RESPONSES TO THE ENVIRONMENT: AN INTRODUCTION

Colin G. Scanes

Department of Animal Sciences
Rutgers University
New Brunswick, NJ 08903

The hormones of the thyroid, adrenal cortex, and of the hypothalamo-pituitary growth hormone - somatomedin axis share various features. These include: -

(a) their secretion and roles show common features (figure 1, also see below),

(b) their roles in both development/growth and in the response to environmental pertebations and

(c) their influence on the immune system. The neuro-endocrine endocrine control of the immune system obviously also involves hormones other than those of the hypothalamo-pituitary axis.

For the hypothalamic-pituitary, growth hormone (GH)-Somatomedin axis the releasing factors controlling GH secretion are growth hormone releasing factor (GRF, somatocrinin) (stimulatory), somatostatin (SRIF) (inhibitory) and in some species, thyrotropin releasing hormone (TRH) (stimulatory). The tropic hormone is GH (somatotropin) itself and the target gland is the liver. This produces somatomedin C (insulin-like growth factor I, IGFI) which affects the growth and metabolism of various tissues including skeleton (cartilage). It should be noted that the situation does not

Development, Maturation, and Senescence of
Neuroendocrine Systems: A Comparative Approach

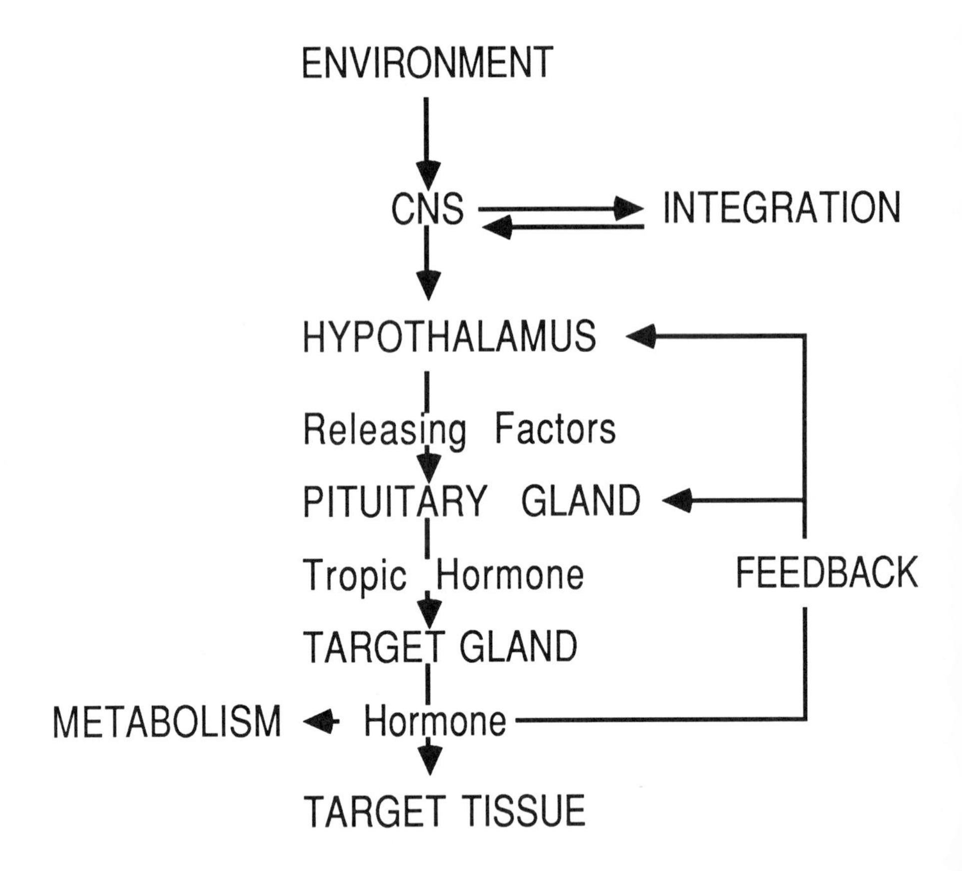

Figure 1 Control and Role of Pituitary Hormones Environment

appear to be quite as simple, in fact. In addition to having its effects mediated by IGFI, GH has direct effects on various tissues. GH modifies the rate of both lipolysis (Goodman, 1981; Campbell and Scanes, 1985) and lipogenesis (Schaffer, 1985, Etherton *et al.*, 1987) in adipose tissue and influences immune function (Edwards *et al.*, 1988). Moreover the tight coupling of GH and somatomedin release can be reduced or abolished by various factors which act, for instance, to down regulate GH receptors (Picard and Postel-Vinay, 1984). In the case of IGFI, it is uncertain the relative importance of circulating IGFI to locally produced IGFI. Indeed, there is strong evidence that target tissues can produce IGFI (at least partially independently of GH), which then exerts both paracine or autocrine effects.

The hypothalamic-hypophyseal-adrenocortical axis is the 'classical' mediator of stress. The neurosecretory neurons in the hypothalamus control the secretion of adrenocorticotropic hormone (ACTH) by the release of releasing factors into the hypophyseal portal blood vessels. The more releasing factor for ACTH is corticotropin releasing factor. However other hypothalamic peptides and also neurotransmitters also influence ACTH secretion. These include arginine vasopressin and epinephrine which can act synergistically with CRF to increase ACTH release and epinephrine (Rivier and Vale, 1983; Vale _et al_., 1983). Secretion of glucocorticoids (cortisol in most mammals including humans, and sheep and corticosterone in rats and chickens) is predominantly controlled by circulating concentrations of ACTH. Other factors can influence glucocorticoid secretion. For instance, prolactin inhibits adrenal conversion of glucocorticoid to inactive metabolites (Carsia _et al_., 1984; 1987). Moreover, total secretion of glucocorticoids is dependent on the size and development of adrenal cortical cells (specifically those of the zona fascienlata). The glucocorticoids act on many tissues. In a stressed animal, glucocorticoids are catabolic in muscle and gluconegenic in the liver. The immune system can also be inhibited by glucocorticoids. Moreover, glucocorticoids are involved in development. In particular, the pre-partum increase in plasma concentrations of glucocorticoids appear responsible for the induction of various enzymes required for post natal life (eg Kuhn _et al_., 1984), and for the initiation of parturitation itself in at least some species.

The hypothalamic-pituitary-thyroid axis appears to be relatively simple. The principal hypothalamic hypophysiotropic factor which stimulates the secretion of thyrotropin (thyroid stimulating hormone, TGH) is thyrotropin releasing hormone (TRH). In addition, SRIF inhibits release of TSH. In turn, TSH stimulates the secretion of thyroxine (T_4) for the thyroid gland. Unlike the other hormones from pituitary controlled target endocrine glands, T_4 appears to be a pre-hormone. Monodeiodination of T_4 to triiodothyronine (T_3) occurs in the liver and also in some target tissues (eg. the pituitary gland). Monodeiodination in the liver is thought to be under independent control to the release of T_4 from the thyroid. It is dependent on stage of liver development (discussed in detail by McNabb, 1988), and may also be under endocrine and/or nervous control. Thyroid hormones are involved in the control of tissue (eg. muscle metabolism) and also in development/differentiation.

Environment The secretion of the tropic hormones and of the target gland hormones is markedly affected by various environmental factors. Obviously this is integrated by the central nervous system and moderated by the release of peptide releasing factors from hypothalamic neurosecretory terminals (in the median eminence). The ability of the various hypothalamo-pituitary-target endocrine gland to respond to environmental influences is age dependent. Not only must the requisite developmental sequences have taken place but also the response to environmental factors may be influenced by senescence or any differences in the role of a hormone in the stage of development. These are discussed in detail in the following chapters.

In the case of the GH-somatomedin C axis, GH secretion is profoundly influenced by nutrition in a variety of species. For instance, in rats plasma concentrations of GH are reduced by fasting (Takahashi et al., 1971) while in human (Glick, 1969) and chickens (Harvey et al., Scanes et al., 1981), GH secretion is elevated by nutritional deprivation (particularly of protein and/or energy). While direction of the effect of inadequate nutrition varies from species, an effect is consistently observed. Moreover in all species examined, protein/energy deprivation is almost uniformly accompanied by decreases in the circulating concentrations of IGFI (rats Prewitt et al., 1982, chicken reviewed Lauterio and Scanes, 1987). This paradox can be explained by nutritional inadequacy (independently) affecting both GH secretion and release of IGFI. The neuroendocrine control of GH release can be modified by other environmental factors. Various stressors affect GH secretion. Again the directionally of the shift is species dependent. Plasma concentrations of GH are reduced by ether, hatching and insulin induced hypoglycemic in chickens (reviewed Scanes, 1987) but the 'classical' environmental factor which activates the hypothalamo-pituitary-adrenal cortical axis is stress (Selye, 1950). A problem with this includes the different use of stress in scientific and colloguial English, the difficulty in specifically defining the term stress, and the circulate argument that 'stress increases glucocorticoid secretion' and if glucocorticoid secretion is elevated the animal must have been stressed. A variety of environmental conditions have been observed to increase the secretion of various glucocorticoids with age related differences in the response to stimuli, the magnitude of response and even which glucocorticoids is release. The ontogeny and senescence of the adrenal cortex and its control are discussed by Carsia and Malamed (this volume).

References

Campbell, R.M. and Scanes, C.G. (1985) Proc. Soc. Exp. Biol. Med. 180, 513-517.
Carsia, R.V., Scanes, C.G. and Malamed, S. (1984) Endocrinology 115, 2464-2472.
Carsia, R.V., Scanes, C.G. and Malamed, S. (1987) Comp. Biochem. Physiol. 88A, 131-140.
Edwards, C.K., Ghiasudden, S.M., Schepper, J.M., Yunger, L.M. and Kelley, K.W. (1988) Science 239, 769-771.
Etherton, T.D., Evock, C.M. and Kensinger, R.S. (1987) Endocrinology 121, 699-703.
Goodman, H.M. (1981) Endocrinology 109, 120-129.
Glick, S.M. (1969) In: Frontiers in Neuroendocrinology (pp 141-182) Eds: W.F. Ganong and L. Martin, London University Press.
Harvey, S., Scanes, C.G., Chadwick, A. and Bolton, N.J. (1978) J. Endocrinol. 76, 501-506.
Kuhn, E.R., Decuypere, E. and Rudas, P. (1984) J. Exp. Zool. 232, 653-658.
Lauterio, T.J. and Scanes, C.G. (1987) Proc. Soc. Exp. Biol. Med. 185, 420-426.
Picard, F. and Postel-Venay, M-C. (1984) Endocrinology 114, 1328-1333.
Prewitt, T.E., D'Ercole, A.J., Switzer, B.R. and VanWyk, J.J. (1982) J. Nutr. 112, 144-150.
Rivier, C. and Vale, W. (1983) 113, 939-942.
Scanes, C.G. (1987) CRC Critical Reviews in Poultry Biology 1, 51-105.
Scanes, C.G., Griminger, P. and Buonomo, F.C. (1981) Proc. Soc. Exp. Biol. Med. 168, 334-337.
Schaffer, W.T. (1985) Am. J. Physiol. 248, E719.
Selye, H. (1950) The Physiology and Pathology of Expure to Stress. Acta Inc., Montreal.
Takahashi, K., Daugherday, W.H. and Kipnis, D.M. (1971) Endocrinology 88, 909-919.
Vale, W., Vaughan, J., Smith, M., Yamamoto, G., Rivier, J. and Rivier, C. (1983) Endocrinology 113, 1121-1131.

15

NEUROENDOCRINE MODELS REGULATING LIFESPAN

Paola S. Timiras

Department of Physiology-Anatomy
University of California
Berkeley, California

I. INTRODUCTION

Development, maturation and senescence of the neuroendocrine system are important and involve several key molecular, cellular and organismic functions necessary for homeostasis, adaptation and survival. Neuroendocrine functions are also important because their sequential development may direct the passage from one period of the lifespan to the next. Starting with fertilization and terminating with death, development, maturation and senescence of the whole organism parallel, indeed depend upon neural and endocrine regulation.

This communication deals with various issues that are relevant to the subject of this symposium and are the concern of research in my laboratory. While these data are still fragmentary, they are supportive of a key role for the neuroendocrine system in the timetable of age-related changes in brain neurotransmitters, brain hormone receptors and neuronal pathology.

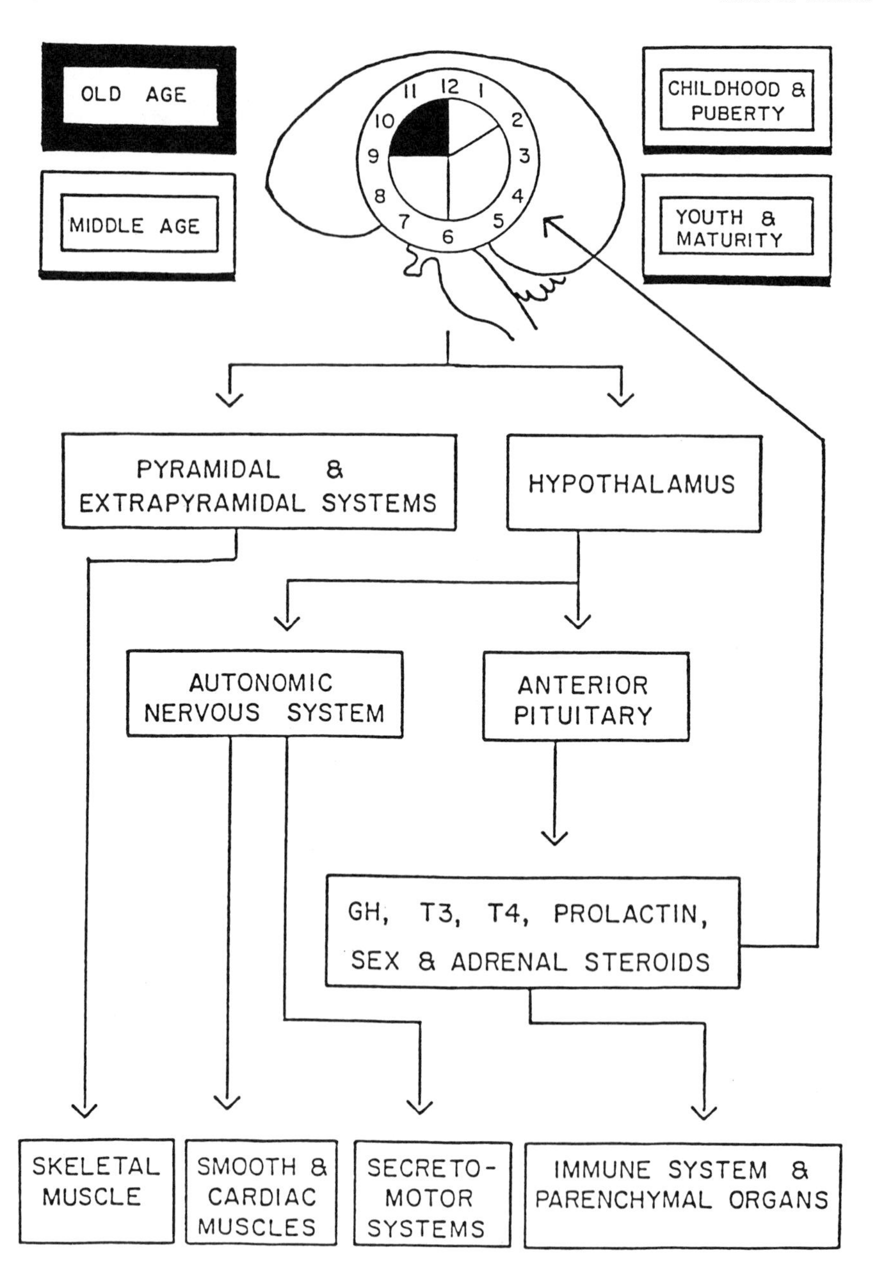

Fig. 1 Neural and hormonal signals regulate the various phases of the lifespan and establish the neuroendocrine theory of development, maturation and aging (Sharma, 1988).

II. A NEUROENDOCRINE PACEMAKER MAY REGULATE DEVELOPMENT, MATURATION AND SENESCENCE OF CELLS, ORGANS AND ORGANISMS.

A composite theory views development and aging as a continuum of age-related changes in the neural and endocrine regulation of body functions (see reviews by Everitt, 1982; Finch, 1976, 1977; Finch and Landfield, 1985; Meites et al., 1987; Timiras, 1978, 1982a, b; 1985a, b; Walker and Timiras, 1982). According to this theory, command neurons, in higher brain centers, possibly in the hypothalamus, act as pacemakers that control the "biological clock" governing this continuum (Figure 1). As shown at the top of the diagram, the neuroendocrine clock moves, after fertilization and prenatal development, through childhood and puberty, to youth and maturity, and on to middle and finally old age. As shown in the center portion of the diagram, signals from the pacemaker or command hypothalamic neurons would be relayed by neurotransmitters, acetylcholine, monoamines and peptides, directly through the pyramidal and extrapyramidal systems to the skeletal muscles to regulate motor activity; or they would be relayed to the autonomic and neuroendocrine cells. Autonomic stimulation mediates the peripheral sympathetic and parasympathetric responses of the heart, smooth muscle and exocrine glands. Hypothalamic hormones regulate the secretion of the anterior pituitary hormones, and these, in turn, stimulate the peripheral endocrine to secrete their respective hormones. Finally, at the bottom of the diagram, several hormones -- growth hormone (GH) thyroid hormones, T_3 and T_4, prolactin and adrenal and sex steroids -- act on specific receptors in target tissues to regulate metabolic, immunologic, reproductive and adaptive functions.

Age-related changes can result from the unfolding of a program genetically encoded at fertilization or from the switching on and off of certain essential genes. With time, aging may result from deterioration of the program or, alternatively, from cessation of the program or possibly from activation of some indesirable genes (Sharma, 1988). In any case, aging would be manifested through a slowing down, or imbalance of neuroendocrine signals. As a consequence, neurotransmitter and hormonal signals would be altered with repercussions on neural, muscular and secretory functions, e.g. involution of reproductive organs, loss of fertility, diminished muscular strength, lessened ability to confront stress, and impaired metabolic, secretory, cardiovascular and respiratory functions. So it is possible that programmed changes in gene activity are not restricted to fertilization but are triggered by neural and endocrine signals at any point throughout the lifespan.

III. AGE-RELATED CHANGES IN NEUROTRANSMITTER SIGNALLING DIFFER WITH THE BRAIN REGION.

In attempting to identify the pacemaker, we screened several neurotransmitters in discrete brain areas. Earlier work revealed that most neurotransmitter systems undergo changes not only during development and maturation of the brain but also extending into senescence. The cholinergic system is one of the most extensively studied because of its putative relationship to several important functions, such as memory, and its possible involvement in brain pathology such as the catastrophic loss of cholinergic neurons in the nucleus of Meynert in the senile dementia of the Alzheimer type (Timiras, 1988). In a long-term study, mice were fed a normal or a choline-enriched diet (Mizumori *et al.*, 1985). Several behavioral tests for memory as well as measurements of the enzyme choline acetyltyransferase, CAT, involved in acetylcholine synthesis were conducted at young and old ages (Figure 2). Long-term memory transfer took longer in old mice as compared to young, but when the old animals were fed a choline-enriched diet, the age-related lengthening of transfer time was reduced. However, neither age or diet significantly affected CAT activity in any of the three brain areas studied (Figure 3). Regional differences in activity, with the highest levels in the corpus striatum, followed by the hippocampus and, much lower, the cerebral cortex are in agreement with other previous investigations. The dichotomy between behavioral decrements with aging and the beneficial

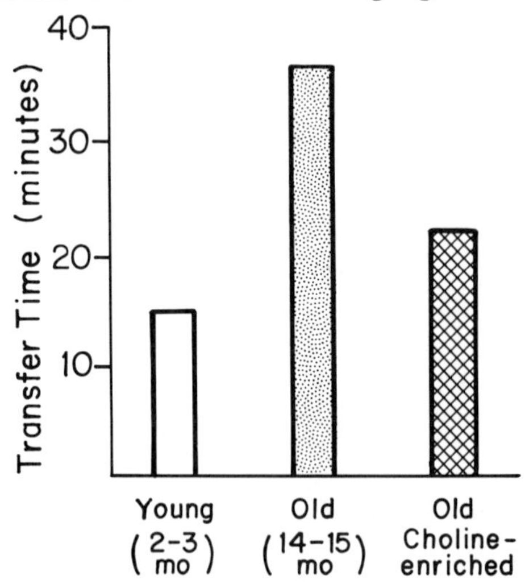

Fig. 2 Age-related changes for long-term memory transfer in mice: effect of choline-enriched diet (adapted from Mizumori *et al.*, 1985).

effects of an enriched diet on one side and the absence of alterations in the levels and activity of cholinergic inputs on the other is not new. It may, in fact, explain the failure, in humans, of treating or at least improving senile dementia by the administration of acetylcholine.

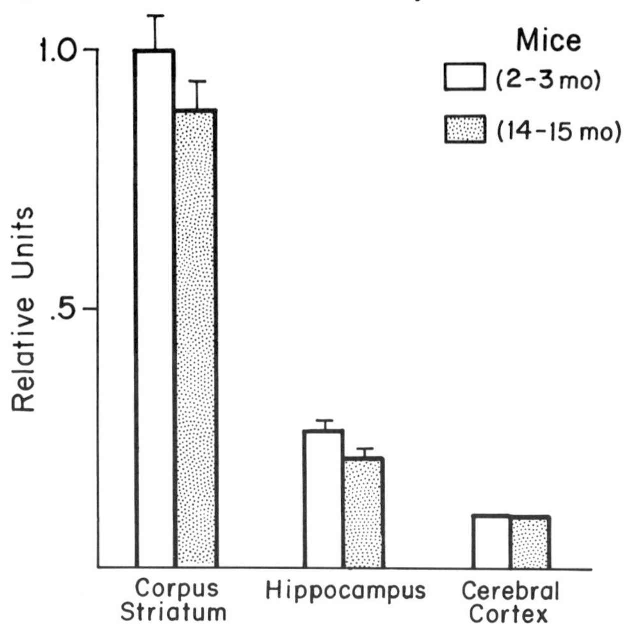

Fig. 3 Age-related changes in choline acetyl transferase (CAT) activity in three brain regions of mice (adapted from Mizumori et al., 1985).

A more promising identification of the pacemaker or pacemaker site is provided by studying the monoamines. A definite imbalance -- increased serotonin and decreased catecholamines -- is set up in the hypothalamus with aging (Timiras et al., 1982a, b, 1984). Likewise a shift occurs in the levels of excitatory and inhibitory amino acids with aging in various areas of the central nervous system (Timiras et al., 1973a, b). In the case of gamma-amino-butyric acid, GABA, levels of this inhibitory amino acid increase with aging in whole spinal cord and brain. To determine whether discrete regional differences could be identified, the activity of glutamic acid decarboxylase, GAD, the synthesizing enzyme of GABA, was measured in microdissected hypothalamic nuclei and, for comparison, in some brain areas of the rat (Figure 4). Within the hypothalamus, some nuclei are not affected by age (e.g. basal, ventromedial and arcuate nuclei) while others such as the medial preoptic and anterior nuclei (related to the secretion of gonadotropin-releasing hormones and to the regulation of rhythmic activities) show a significant increase in GAD activity (Sternberg et al., 1987). Regional differences are also evident when one compares enzyme activity in the hypothalamus with other brain areas: progressively lower than in the hypothalamus in the septum, caudate-putamen nucleus and cerebral cortex. Note that enzyme activity increases with age only in the two areas where it is highest. Taken together, these data support the involvement of specific neurotransmitter with age and their discrete regional distribution, evocative of a pacemaker action.

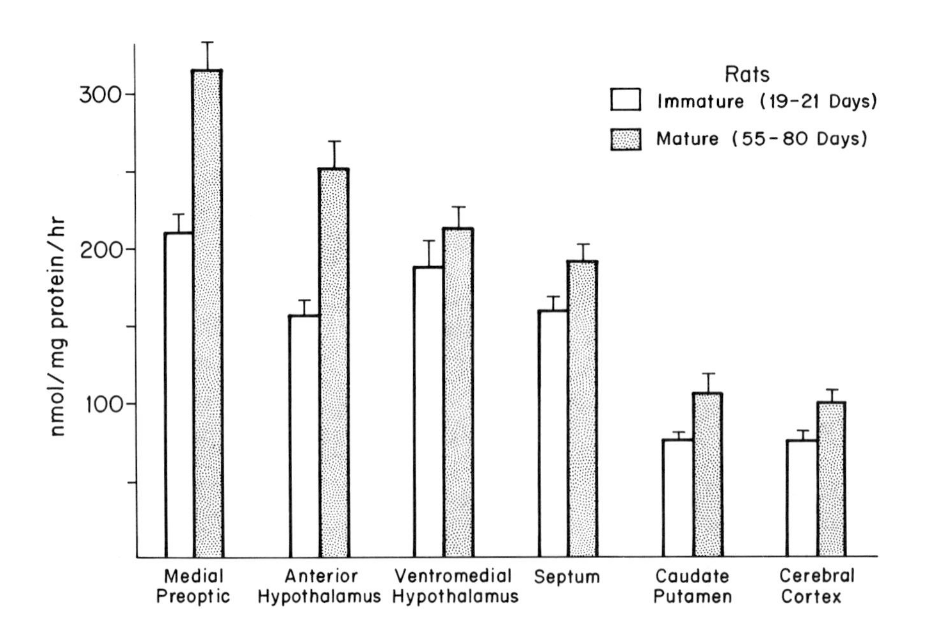

Fig. 4 Brain region and age influence developmental changes in glutamic acid decarboxylase (GAD) activity in rats (adapted from Sternberg *et al.*, 1987)

IV. CORTICOSTEROID BINDING DIFFERS WITH AGE, BRAIN REGION, TEMPERATURE AND CALCIUM.

Age and regional differences in neuroendocrine signals are not limited to enzyme activity but extend to hormonal receptors. Glucocorticoid receptors are present in various brain areas but their number varies with the area, age and levels of hormones. Thus, glucocorticoid binding

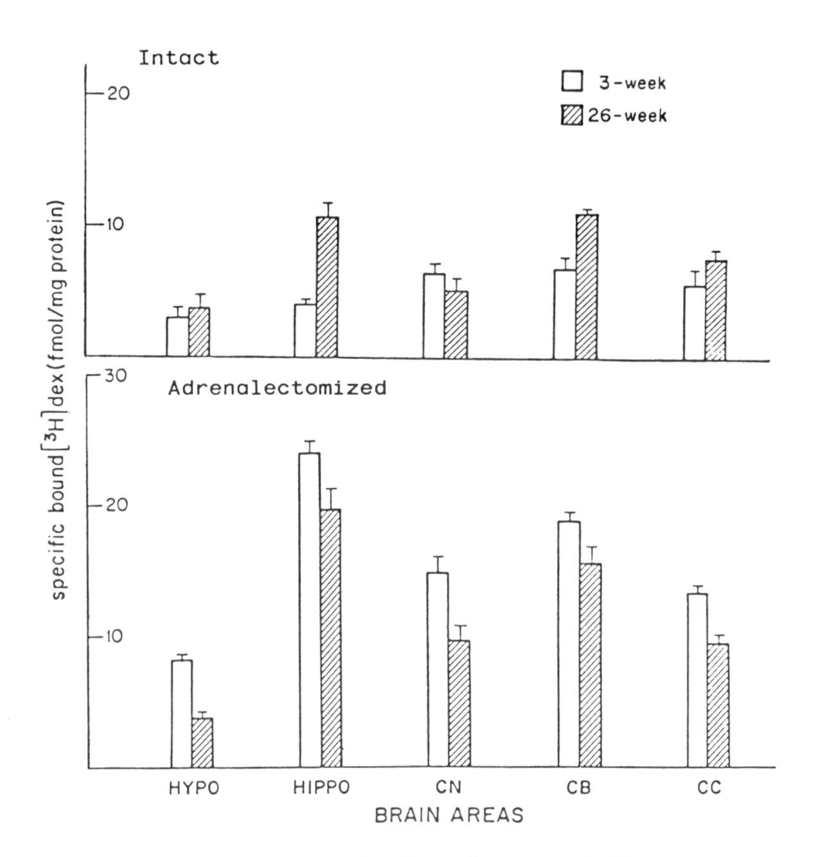

Fig. 5 Brain region, age and adrenalectomy influence cytosolic binding corticosteroids in rats (Sharma and Timiras, 1986).

(represented by binding of the synthetic hormone dexamethasone) is highest in the hippocampus and cerebellum of adult intact rats (Figure 5). When the animals are adrenalectomized, binding is up-regulated still with the highest levels in the hippocampus (Sharma and Timiras, 1986). In all areas, adrenalectomy increases binding tremendously in the young animal but much less in the adult. Not only is receptor number affected by age, but some physicochemical properties (as reflected in responses to temperature and calcium) are affected as well (Sharma and Timiras, 1987a). Heat activates the nuclear binding of dexamethasone-receptor complexes in the cerebral hemispheres of young and adult rats but more significantly in young, whereas calcium activates nuclear binding to the same extent at both ages (Figure 6).

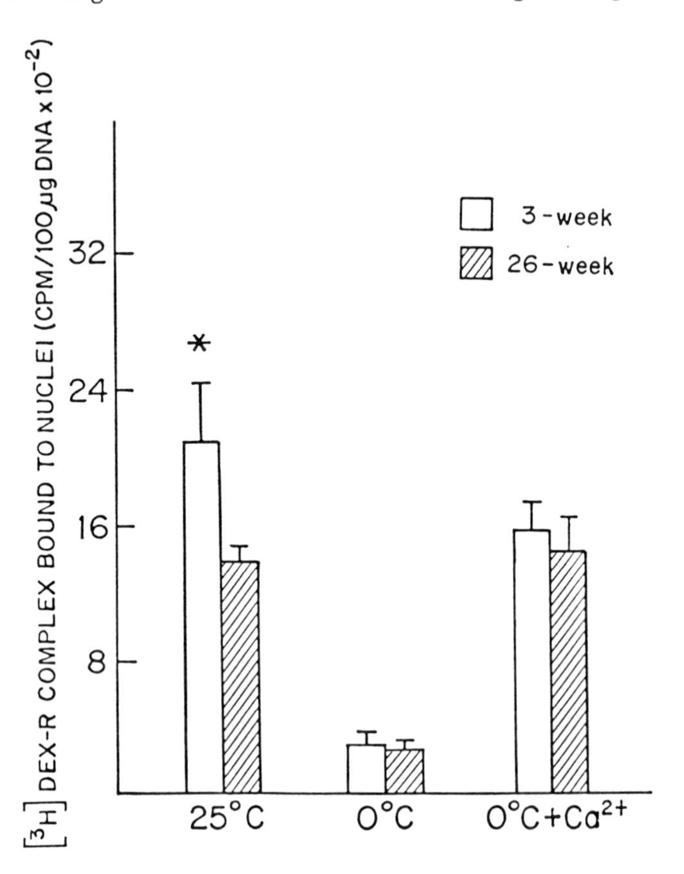

Fig. 6 Age, temperature and calcium influence nuclear binding of corticoid-receptor complexes in the cerebral cortex of rats (Sharma and Timiras, 1987a).

These age-related qualitative differences in binding are specific for different tissues (Sharma and Timiras, 1987b, c). Their exact mechanism is not well understood although conformational changes capable of exposing the DNA- and chromatin binding domains have been proposed. Irrespective of the cause, these changes in glucocorticoid binding with age may be related to the neurotoxic effect of these hormones on hippocampal cells. With aging, loss of neurons in the hippocampus would result in the increased release of hypothalamic neuropeptides capable of stimulating the pituitary-adrenocortical axis. Further secretion of glucocorticoids would induce a feed-forward cascade with degeneration of hippocampal cells and hyperadrenocorticism (Sapolsky *et al.*, 1986). Alterations throughout the lifespan in this hippocampus-hypothalamus-pituitary-adrenocortical circuit have been held responsible for the progressive decline in the ability of the organism, with aging, to adapt to stress.

V. CORTICOSTEROIDS ACT DIRECTLY ON DEVELOPMENT OF ADRENERGIC ENZYMES IN HUMAN NEUROBLASTOMAS.

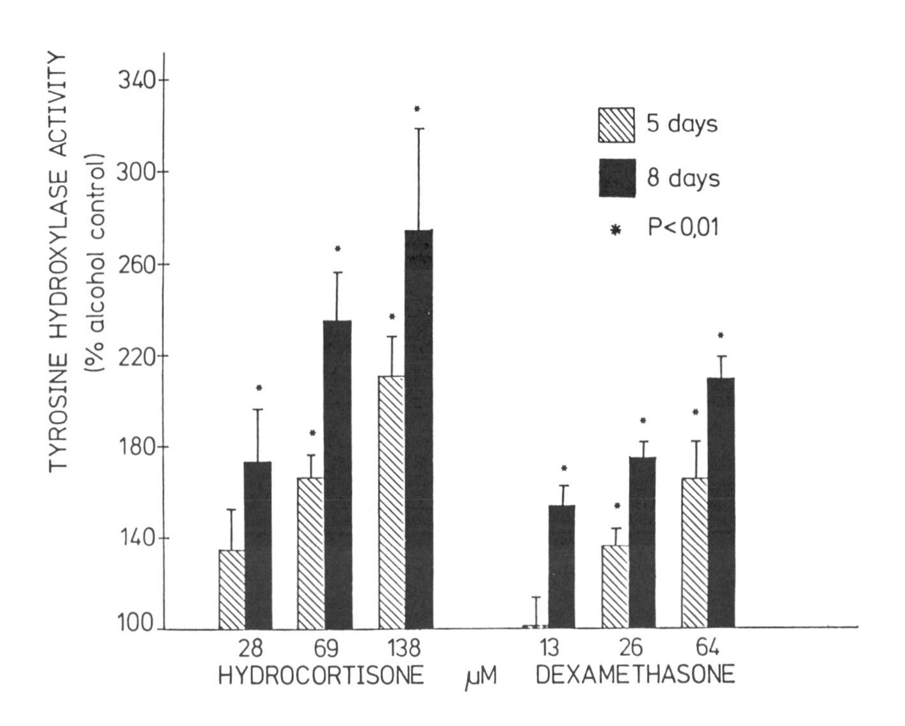

Fig. 7 Glucocorticoids stimulate tyrosine hydroxylase (TH) activity in human neuroblastoma cells (John, 1985).

The importance of corticosteroids in response to stress has long been known and, as shown in recent studies from our and other laboratories, quantitative and qualitative changes occur with age in the binding of these hormones in discrete brain areas, in their putative neurotoxic action and their relation to neurotransmitters. That these actions are direct on neuronal cells and not mediated indirectly through the general metabolic effects of the hormones is demonstrated, in cultured neuroblastoma cells, by the ability of the hormones to stimulate the synthesis of the enzyme tryosine hydroxylase (TH), the rate-limiting enzyme in the catecholamine biosynthetic pathway (John 1985). In human neuroblastoma cells, a line of adrenergic neurons, hydrocortisone and dexamethasone increase TH activity in immature, rapidly dividing cells: this increase is dose-dependent, and is more marked with longer hormone treatment and with the natural steroid (Figure 7). However, following differentiation of the cells (by culture in serum-free medium or addition of retinoic acid), the inducing effect of steroid on TH activity is lost. Binding sites for glucocorticoids are present in both undifferentiated and differentiated neurons, therefore, this loss of responsiveness does not appear to result from the loss of receptor binding. A number of explanations have been suggested such as changes in nuclear availability for binding of steroid-receptor complexes, genomic changes conferring repressive rather than positive transcriptional activity. While these hypotheses need to be substantiated, these data emphasize the direct action of the hormones on neuronal cells and their ability to influence neurotransmitter metabolism.

VI. STRESS MAY INDUCE NORMAL AND ABNORMAL CELLULAR AGING IN CULTURED NEURONAL CELLS.

The potentially neurotoxic actions of corticosteroids *in vivo* and *in vitro* have suggested the possibility of utilizing stress and/or hormones to induce models for human neuronal pathology (Figure 8). *In vivo* studies have utilized comparison with spontaneously-occurring neuropathology in aging animals, or cognitive and behavioral deficits after lesioning of specific brain areas, or the administration of neurotoxic substances. One such approach involves the infusion of ibotenic acid into the nucleus of Meynert in rats (Arendash *et al.*, 1987). In the last few years, our laboratory has been attempting to create an *in vitro* model for Alzheimer disease (AD). Such a model *is based* on reproducing, in

STRESS-INDUCED MODELS OF ALZEHEIMER SENILE DEMENTIA

In vivo :
- Ibotenic acid infused into nucleus of Meynert
- Cerebral cortex damage:
  (Chemical) ACh, CAT, somatostatin, peptide Y
  (Physical) Neurofibrillary tangles, amyloid plaques

In vitro:
- Human neurogenic teratocarcinoma and neuroblastoma cells
- Aluminum, glutamate, aspartate, colchicin, Doxorubicin in medium
- Produce neurofibrillary tangles and amyloid plaques

Fig. 8 Stress-induced models for senile dementia of the Alzheimer type (SDAT). (*In vivo* studies from Arendash *et al.*, 1987; *in vitro* studies from Cole and Timiras, 1987).

cultured neuroblastoma or teratocarcinoma neurons, the characteristic hallmarks of AD pathology (Cole _et al._, 1985; Cole and Timiras, 1987). These characteristics include intracellular neurofibrillary tangles and extracellular amyloid deposited either in plaques or surrounding cerebral vessels. To produce this model, cultured neurons have been caused to differentiate and have then been treated with a variety of toxic substances, including aluminum, excitatory aminoacids such as glutamate and aspartate, and such disrupting agents as cholchicine for microtubules and doxorubicin for DNA. Prolonged treatment with these substances induced cellular aging, as represented by accumulation of age pigments (lipofuscin). It also resulted in the presence of paired-helical filaments, perhaps of microtubular origin, and constituents of the neurofibrillary tangles. Amyloid proteins were also increased; they cross-related with antibodies from human AD and showed immunogel patterns consistent with the recently reported abnormal AD proteins. The diverse mechanisms of action of the pathogenic agents suggest that damage activates a generalized cellular stress response. Such a stress response would incorporate the specific response to each toxic agent and the total would be responsible for cellular aging followed by pathology. This general response is reminiscent of Selye's adaptation syndrome at the organismic level and involves the sympathetic and pituitary-adrenocortical systems. At the cellular level, it may be anlogous to the production of aberrant proteins in response to heat shock (Schlesinger _et al._, 1982). Exposure to high temperature (in bacteria to man) activates the expression of genes previously silent and the end-result of this activation is the production of specific heat-shock proteins. The influence of temperature on the binding of corticosteroid-receptor complexes to nuclear DNA described above indicates a possible site and neuroendocrine mechanism for the action of cellular stress. This opens a new avenue of approach to the study of normal and abnormal neuronal aging.

VII. SUMMARY AND CONCLUSIONS.

The amalgamation of seemingly disparate data provides a scaffold upon which to build a neuroendocrine profile for development and senescence. The salient points are: (1) a neuroendocrine pacemaker orchestrates the timetable for the lifespan; (2) the pacemaker may be localized in specific brain areas, e.g. hypothalamus; (3) the pacemaker action is signalled by neurotransmitters; (4) neuroendocrine receptors respond to the signal; and (5) neuroendocrine-mediated stress mimics aging-related brain pathology.

References

Arendash, G.W., Millard, W.J., Dunn, A.J., and Meyer, E.M. (1987). Science 238, 952-956.

Cole, G.M., Wu, K., and Timiras, P.S. (1985). Int. J. Dev. Neurosci. 3, 23-32.

Cole, G.M. and Timiras, P.S. (1987). In "Model Systems of Development and Aging of the Nervous System" (A. Vernadakis, A. Privat, J.M. Lauder, P.S. Timiras and E. Giacobini, eds.) pp. 453-473, Martinus Nijhoff Publ. Boston.

Everitt, A.V. (1982). In "Biological and Social Aspects of Mortality and Length of Life" (S. Preston, ed.) pp. 279-301, Orinda Edition, Liege.

Finch, C.E. (1976). Quar. Rev. Biol. 51, 49-83.

Finch, C.E. (1977). In "Handbook of the Biology of Aging" (C.E. Finch and L. Hayflick, eds.) pp., 262-280. Van Nostrand Reinhold, New York.

Finch, C.E. and Lancefield, P.W. (1985). In "Handbook of the Biology of Aging" (C.E. Finch and Schneider, E.L. eds.) pp 567-594. Van Nostrand Reinhold, New York.

John, N.J. (1985). Ph.D. Thesis, University of California, Berkeley.

Meites, J., Goya, R. and Takahashi, S. (1987). Exptl. Gerontol. 22, 1-5.

Mizumori, S.J.Y., Patterson, T.A., Sternberg, H., Rosenzweig, M.R., Bennett, E.L., and Timiras, P.S. (1985). Neurobiol. Aging 6, 51-56.

Sapolsky, R.M., Krey, L.C., and McEwen, B.S. (1986). Endocr. Rev. 7, 285-301.

Schlesinger, M.J., Ashburner, M., and Tissieres, A. (1982). Cold Spring Harbor Laboratory, Cold Spring Harbor, New York.

Sharma, R. (1988). In "Physiologic Basis of Aging and Geriatrics" (P.S. Timiras, ed.) pp. 43-58, MacMillan, New York.

Sharma, R. and Timiras, P.S. (1986). Biochem. Int. 13, 609-614.

Sharma, R. and Timiras, P.S. (1987a). Dev. Brain Res. 36, 285-287.

Sharma, R. and Timiras, P.S. (1987b). Mech. Ageing Dev. 37, 249-256.

Sharma, R. and Timiras, P.S. (1987c). Biochem. Biophys. Acta 930, 237-243.

Sternberg, H., Segall, P.E., Bellport, V., and Timiras, P.S. (1987). Dev. Brain Res. 34, 316-317.

Timiras, P.S. (1978). Am. Sci. 66, 605-613.

Timiras, P.S. (1982a). In: "The Development of Attachment and Affiliative Systems" (R.N. Emde and R.J. Harmon, eds.) pp. 47-46. Plenum Press, New York.

Timiras, P.S. (1982b). In "Hormones in Development and Aging" (A. Vernadakis and P.S. Timiras, eds.) pp. 551-586. SP Medical & Scientific Books, New York.

Timiras, P.S. (1985a). In "Relations Between Normal Aging and Disease" (H.A. Johnson, ed.) pp. 151-156. Raven Press, New York.

Timiras, P.S. (1985b). In "Principles and Practice of Geriatric Medicine" (M.S.J. Pathy, ed.) pp. 105-130. John Wiley & Sons, New York.

Timiras, P.S. (1988). MacMillan, New York.

Timiras, P.S., Hudson, D.B., and Oklund, S. (1973a). Prog. Brain Res. 40, 267-275.

Timiras, P.S., Hudson, D.B. and Oklund, S. (1973b). IV Int. Mt. Int. Soc. Neurochem, (Abstracts). Tokyo, Japan.

Timiras, P.S., Choy, V., and Hudson, D.B. (1982a). Age and Ageing 11, 73-88.

Timiras, P.S., Hudson, D.B., and Miller, C. (1982b). In, "The Aging Brain: Cellular and Molecular Mechanisms of Aging in the Nervous System" (E. Giacobini, G. Filogamo, G. Giacobini and A. Vernadakis, eds.), pp. 173-184, Raven Press, New York.

Timiras, P.S., Hudson, D.B. and Segall, P.E. (1984). Neurobiol. Aging 5, 235-242.

Walker, R.F., and Timiras, P.S. (1982). In "Cellular Pacemakers" Volume 2. (D. Carpenter, ed.) pp. 345-365, John Wiley & Sons, New York.

16

THYROID HORMONES IN EARLY DEVELOPMENT, WITH SPECIAL REFERENCE TO TELEOST FISHES[1]

Christopher L. Brown
and
Howard A. Bern[2]

Zoology Department and Cancer Research Laboratory
University of California
Berkeley, California, 94720

I. INTRODUCTION

The question of thyroid involvement in early development in fish has been reviewed recently (Lam, 1985), but ongoing research in a number of laboratories has begun to change our perception of the extent of thyroid hormone intervention. The well-known ability of exogenous thyroid hormones to alter the patterns of growth and development in teleost embryos and larvae (for example, see Lam, 1980 or Lam et al., 1985) may reflect physiological regulatory actions of these hormones, although the evidence in support of this hypothesis is inconclusive. The patterns of thyroid ontogeny in fish are variable and correlations of larval thyroid function with specific developmental events are inconsistent. The issue of thyroid regulation of early development in fish has remained unresolved, in part, because there have been so many contradictory reports on the subject. Differences in experimental species, in technical approach, and in hormone dosages that have extended well into the pharmacological range have contributed to the confusion. With a few exceptions, (most notably, flounder metamorphosis--Miwa and Inui, 1987a) it

---

[1] Aided by the California Sea Grant College Program grant NA80AA-D-00120 Project R/F-101, and California Department of Fish and Game contract 20609.
[2] To whom reprint requests should be sent.

has been difficult to ascribe specific regulatory roles to thyroid hormones in teleost development.

One recurring theme in the most recent research is the likelihood that thyroid and other hormones stored in yolk may be involved in the regulation of early developmental events. It is now known that the yolk of many species of fish eggs contains thyroid hormones (Kobuke et al., 1986; Brown et al., 1987; Tagawa and Hirano, 1987). In view of the sensitivity of fish larvae to thyroid hormones, even before a functional hypothalamus-pituitary-thyroid system has formed, the possibility that maternal thyroid hormones stored in yolk may influence early development requires attention. The availability of hormones from this potential alternate source may aid in the interpretation of the early sensitivity to exogenous thyroid hormones and the absence of a close relationship of the onset of larval thyroid function with events in differentiation, although it raises a variety of other questions. Some of these are addressed in the present paper.

In order to evaluate the evolutionary significance of the proposed maternal contribution to the regulation of embryogenesis and larval differentiation, the case for thyroid involvement in early differentiation in other vertebrate classes is also considered. The information available within this context on amphibians and reptiles is minimal, although some parallels have been drawn between the well-established role of the thyroid in amphibian metamorphosis and metamorphic changes that occur in the transition of fish larvae to fry. The majority of evidence generated by avian developmental endocrinologists suggests that thyroid hormones of endogenous origin are of primary importance in embryonic development, and that there is at present no direct evidence that the hormonal contents of yolk ever reach putative embryonic target organs in birds. In the case of placental mammals, also, the prevailing opinion is that embryos develop in the relative absence of influences of maternal thyroid hormones, although late embryonic or fetal thyroid function is known to be critically important. Thus the possibility that the development of fish embryos may be influenced by maternal thyroid hormones suggests an association between the endocrine function of the female fish and the viability of her offspring, an association which may yet prove significant among other vertebrate groups.

II. THYROID HORMONES IN LARVAL FISHES

Histological studies indicate that some teleost embryos develop active-looking thyroid follicles prior to hatching (such as the Atlantic salmon, *Salmo salar*, Hoar, 1939; the fathead minnow, *Pimephales promelas*, Wabuke-Bunoti and Firling, 1981) and/or active thyrotropic cells in the pituitary (coho salmon, *Oncorhynchus kisutch*, Leatherland and Lin, 1979). In other teleosts, thyroid follicles do not become visible until after the completion of yolksac absorption (such as the tilapia, *Sarotherodon niloticus*, Nacario, 1983; the striped bass, *Morone saxatilis*, Brown et al., 1987). Some radiochemical approaches have also been used in the determination of the timing of the onset of thyroid hormone production. Iodide uptake increases appreciably during prehatching development of the fathead minnow (Wabuke-Bunoti and Firling, 1981) and of two Pacific salmonids (*Oncorhynchus kisutch* and *O. tschawytscha*, Greenblatt, 1987), but remains low until the yolk has been absorbed and thyroid follicles have become evident in the striped bass (around the second week after hatching; Brown et al., 1986). It appears, then, that the onset of independent thyroid function in fish is a developmental event that occurs at different times relative to hatching in different species. Although patterns of relatively early or late thyroid maturation have been associated with particular patterns of early development in birds (see McNabb, 1987), it is unclear whether there are any such relationships in the few fish species that have been studied from this perspective. Thus far, the early life histories (in terms of egg size, incubation time, degree of physical development and behavioral autonomy at the time of hatching, etc.) have been too divergent in the fishes in which thyroid ontogeny has been studied to allow meaningful comparisons or worthwhile generalizations.

A. EXPERIMENTAL HYPO- AND HYPER-THYROIDISM

Another experimental strategy that has been used extensively in the study of thyroid hormone roles in larval fish development has been the classical approach involving hormone supplementation, removal, and replacement. The sensitivity of embryos and larvae to exogenous thyroid hormones has been studied in a variety of teleost species. Experimental hormone supplementation has provided some interesting, although somewhat inconsistent results. Larval responses to thyroid hormone treatment (usually applied by immersion) include accelerated growth, altered body proportions, induction of early fin and scale differentiation, and increased survival rates (discussed by Brown et al., 1987). Newly-hatched fish larvae respond readily to thyroid hormones, even in some species that do not appear to hatch with a functional pituitary-thyroid system (such as the tilapia; Nacario, 1983). There have also been indications of embryonic sensitivity to thyroid treatment in the earlier developmental stages: exposure of fertilized eggs to thyroxin has reportedly accelerated the prehatching development of chum salmon (*Oncorhynchus keta*; Dales and Hoar, 1954) and caused increases in egg viability, hatching rate, and subsequent larval survival rates in the common carp, *Cyprinus carpio* (Lam and Sharma, 1985).

Antithyroid agents such as thiourea and thiouracil have been used to block thyroid hormone utilization by fish larvae; these treatments typically cause attenuated growth and development. Baker-Cohen (1961) tabulated the results of 32 reports in which antithyroid compounds were applied to developing fishes, and the most frequent observations were retarded growth and/or impaired gonadal development (seen in 20 of the 32 studies). Some behavioral effects have also been noted. For example, thiourea treatment has been shown to block the onset of swimming activity, which can be stimulated in brown trout (*Salmo trutta*) larvae by thyroxin administration (Woodhead, 1966). Numerous other specific developmental processes have been impaired by antithyroid drugs, including the differentiation of skin and the formation of a functional swimbladder (Baker, 1964). Although the results of these studies and other studies involving radiothyroidectomy (e.g., Norris, 1969) tend to indicate that the loss of thyroid hormones disrupts growth and development, the interpretation of some of these results has been questioned (see Lam and Loy, 1985; Brown et al., 1987). Impaired growth and differentiation could be side-effects of a non-specific chemical injury, and the toxicity of thiourea to fish has been established (Chambers, 1953). Nevertheless, when investigations have included groups of larvae receiving both antithyroid treatment and concurrent thyroid hormone replacement, accelerated development has been observed (as in the chum salmon; Ali, 1961, and the flounder, Miwa and Inui, 1987b). In these cases, at least, it is evident that induction of the hypothyroid condition produces actions that are the opposite of the growth- and development-promoting effects of thyroid hormone supplements. These results are consistent with the idea that thyroid hormones may be important in the regulation of early developmental processes in teleost fishes.

B. WHOLE-BODY THYROID HORMONE MEASUREMENTS

Some recent observations of whole-body thyroid hormone contents throughout the early developmental stages of fishes have contributed insights into possible hormonal roles in embryonic and larval differentiation. Kobuke et al. (1987)reported that coho salmon (*O. kisutch*) egg yolk contained appreciable amounts of thyroxin, and higher concentrations than those found in larval body tissues. The observed compartmentalization of thyroxin in the yolk (96% of total T_4) and patterns of disappearance of total thyroxin, which closely followed the utilization of the yolk, led the authors to postulate that hormones stored in yolk might be available continually to the developing embryos and larvae. Based on these results and the known stimulatory capacity of thyroid hormones on fish embryos and larvae, Kobuke et al. (1987) suggested that hormones of maternal origin stored in yolk might have some regulatory roles in salmonid development that precede the roles of endogenously-produced hormones. Developmental profiles of whole-body T_3 and T4, extracted from chum salmon eggs and larvae beginning just after fertilization, indicated that the hormonal content of yolk did not change until after hatching (Tagawa and Hirano, 1987). As in the study by Kobuke et al. (1987), the hormonal content of yolk measured by Tagawa and Hirano (1987) accounted for almost all of the thyroid hormone present in the larvae until about the time of the completion of yolk sac absorption. Although the results of both of these studies seemed to indicate that the first surge in larval thyroid hormone synthesis occurred around the time of completion of yolk absorption (which roughly coincides with the onset of active feeding and emergence from gravel), it is probably unwise to draw the conclusion that the pituitary-thyroid system is in a quiescent state based on whole-body extractions alone. This method gives no indication of the source of hormones, and it is possible that an early increase in hormone synthesis might not be detected in whole-body contents, if the hormone clearance rate increased simultaneously. In follow-up studies by Greenblatt (1987), increases in embryonic radioiodide uptake which were sometimes (although not always) accompanied by signs of a sharp increase in extractable thyroid hormones were seen in two species of Pacific salmon prior to hatching. The thyroid hormone concentrations in embryos dissected from the eggs exceeded the concentrations in yolk, which, together with the radioiodide data, suggested that in these progeny, active hormone biosynthesis had begun before the time of hatching. Greenblatt's (1987) study, and to some extent Kobuke et al.'s study (1987), revealed that whole-body thyroid hormone patterns vary among species and among progeny within species. If the hormonal content of yolk is a function of the endocrine status of the maternal fish, as it appears to be, then such variation in the progeny is to be expected.

C. THYROID HORMONES IN LARVAL FISH PLASMA

Until recently, the assessment of thyroid function in fish embryos and larvae has been limited by the small volumes of blood that are available for hormone analysis. It is impractical to obtain sufficient quantities of plasma for thyroid hormone radioimmunoassay except in relatively large larvae, and the pooling of hundreds of samples has been required for a few measurements. Despite these limitations, consistent patterns of change in circulating thyroid hormones have been seen in a variety of larval salmonids (Sullivan et al., 1987; Tagawa and Hirano, unpublished). Plasma thyroxin increases and triiodothyronine decreases to undetectable levels during the time of the second half of yolksac absorption in Atlantic (*Salmo salar*) and several species of Pacific (*Oncorhynchus* spp.) salmon. Both the larval thyroid follicles and the presumably maternal thyroid hormone pool found in yolk were mentioned as possible sources of the hormones detected in the plasma of developing salmonids (Sullivan et al., 1987). The patterns of plasma thyroid hormone changes in salmonid larvae are compatible with the hypothesis that hormones from the yolk enter the larval circulatory system and thereby reach hormone-sensitive peripheral tissues, although other interpretations are possible. The amount of thyroid hormone secreted by the embryonic thyroid follicles during yolk absorption remains to be ascertained.

D. A YOLK-ENRICHMENT STUDY

The developmental pattern of whole-body thyroid hormone levels has been reported in the striped bass (Brown et al., 1987); T_3 levels were undetectable at the end of the second week post-hatching. Fertilized ova hatch within about 48 hours, and the first two weeks of development are particularly significant in striped bass; rapid morphological and behavioral changes that take place within this time include the completion of yolksac absorption, swimbladder formation and inflation, increased swimming ability, and initiation of predatory feeding (Doroshev, 1970). Heavy mortality typically occurs during this critical developmental period (Doroshev et al., 1981). Subsequent increases in both T_3 and T_4, considered together with histological (Brown et al., 1987) and radiochemical data (Brown et al., 1986) suggest that the larval thyroid follicles are relatively inactive until after this decisive series of developmental events has been completed.

Some experimental work conducted by our group over the last two years has been directed at defining a possible influence of maternal hormones in the early differentiation of striped bass larvae. We have found that intramuscular thyroid hormone injections of prespawning female striped bass cause increases in the amount of hormone in egg yolk, confirming the uptake of T_3 by the ovary (Brown et al., manuscript submitted). Furthermore, the supplementation of yolk T_3 has had profound effects on subsequent larval development and survival in both years of study. Larvae from the hormone-injected females consistently display increases in body size, swimbladder inflation, and survival rates. There was a linear relationship of the T_3 content of the yolk in unfertilized eggs and the survival rate within the progeny. All of the observed hormonal effects were evident within the initial two weeks after hatching. We interpret these results as a clear demonstration that hormones transported into oocytes from the maternal circulation can have a vital influence on early development. In light of the ontogenetic pattern of thyroid function in this species and the restriction of the observed changes to the earliest stages of larval development, we suggest that maternal hormones may play an important role in the promotion of some aspects of embryonic or early larval differentiation. At present, we do not know what specific aspects of differentiation were altered by the hormone supplements. It is known that thyroid hormones can promote the development of hypothalamic neurosecretory capacity in shark embryos (*Squalus suckleyi*; Gorbman and Ishii, 1960).

E. OTHER HORMONES IN FISH EGGS

The interest in possible actions of hormones found in the yolk of fish eggs is not limited to the thyroid hormones. Developmental profiles of the steroids testosterone, 11 ketotestosterone, 17α - 20β -dihydroxyprogesterone, and 17β - estradiol have been obtained in coho salmon (G. Feist et al., manuscript submitted). As was the case with the thyroid hormones in this species (Kobuke et al., 1987), these steroids are abundant in yolk, and the total amount of hormone declines precipitously as the yolk is absorbed. Insulin and other hormones are also present in the yolk of salmonid eggs (E. Plisetskaya, personal communication). It appears that the yolk of fish eggs contains a fairly representative distribution of the hormones present in maternal plasma, although we do not know that any attempts have yet been made to measure either growth hormone or prolactin in fish eggs. The biological importance of these and other maternal products that are potentially available to developing embryos (such as other hormones, growth factors, etc.) remains to be determined.

F. IMPLICATIONS OF MATERNAL CONTRIBUTION TO LARVAL DEVELOPMENT

In the traditional approach to fish stock recruitment theory, the size of a population of fish is determined by a combination of factors that include the reproductive output of the parent stock, which places an upper limit on population size, and a variety of environmental variables, which modify the survival rate (Ricker, 1975). There has been a great deal of debate about the relative importance of climate and density-dependent biological factors such as food availability, predation, territoriality, etc., in determining year-class size (see Koslow, 1984; Gutreutter and Anderson, 1985). Most recruitment models make the assumption that all fertile eggs are "created equal", but this is certainly not the case with striped bass. Fish culturists are aware that the viability of striped bass larvae spawned and reared under apparently identical conditions can vary greatly from one progeny to the next, even from two females that are outwardly identical in appearance. In one recent study, the growth rates of larval striped bass were determined over an eight-year period, and the number of days required to reach a length of 14 mm under laboratory conditions varied among the different year-classes by more than 26% (Brown, 1987). Explanations of the pronounced qualitative differences in the early development and survival of striped bass larvae have hinged on a combination of inherent parental variables (e.g., genotypic variation) and environmental stresses, particularly in the form of pollutant burdens, that might eventually impinge on gamete viability (Whipple et al., 1981). Our studies indicating a potentially important influence of maternal hormones in early development in this species raise the possibility that maternal thyroid function may relate to the presently poorly-understood fluctuations in larval growth and survival rates. It is possible that artificial spawning practices, and such variables as broodstock diet and hatchery photoperiods, could alter maternal endocrine function, with potentially adverse effects on the offspring.

III. COMPARATIVE PERSPECTIVES

A. FISHES AND AMPHIBIANS

A series of studies of the endocrine regulation of larval-fry metamorphosis of flounder (*Paralichthys olivaceus*) has indicated that metamorphic changes can be induced by treatment with exogenous thyroid hormones, and can be obstructed by antithyroid treatments (Inui and Miwa, 1985; Miwa and Inui,

1987b). Further support for the argument that thyroid hormones play a physiological role in the regulation of metamorphosis in this species has been derived from the observation of profound histological changes in the immunoreactive TSH- and T_4-producing cells during metamorphosis (Miwa and Inui, 1987a). The increasing activity of the pituitary-thyroid axis during flounder metamorphosis resembles a typical amphibian metamorphic climax, and suggests some degree of evolutionary conservation of this extreme sort of morphogenic action of thyroid hormones. A dose-dependent induction of metamorphosis by thyroxin in eels (*Anguilla anguilla*; Tesch, 1977) would appear to rule out the possibility that the promotion of metamorphosis by thyroid hormones in fish is unique to the flounder, although the physiological regulatory mechanisms active during eel metamorphosis are by no means completely understood.

The specialized morphogenic actions of thyroid hormones in teleost and amphibian metamorphosis may have corollaries in other, less dramatic developmental changes. Most fishes do not undergo metamorphosis *per se*, although early teleost development has been described as a series of step-wise intervals, bordered by "thresholds or leaps" of morphological transformation (Balon, 1975). The endocrine regulation of specific events in embryogenesis and larval development in the lower vertebrates has received little direct attention, although some inferences can be made from existing data. The general acceleration of growth, development, and survival rates in response to thyroid-hormone supplementation (e.g., Lam et al., 1985) may be a consequence of improved efficiency in the transition through critical developmental intervals, possibly reflecting direct or indirect morphogenic actions of thyroid hormones. Whether thyroid hormones may be involved similarly in early amphibian development remains open to question. To the best of our knowledge, neither the possible presence of thyroid hormones in amphibian eggs, nor the sensitivity of embryos to exogenous thyroid hormones, has been tested. It is well established that the capacity of amphibian larvae to respond to thyroxin is established well before metamorphosis normally occurs (Gudernatsch, 1912; Etkin, 1950), and there are some indications that thyroid hormones are present before changes associated with metamorphosis have begun (Hanaoka et al., 1973), even in early larval life (Kaltenbach, 1982).

B. BIRDS

The ontogeny of thyroid function in the developing chick embryo has been the subject of extensive study (reviewed by Thommes, 1987). Some synthesis of thyroid hormones may begin as early as day 4.5 of incubation, but a pituitary-thyroid regulatory system starts to function at about day 10.5-11.5, and hypothalamic dependence is established at about the same time (Thommes, 1987). It has been suggested that thyroxin deposited in yolk from maternal circulation may reach the embryonic plasma during yolk absorption (Hilfer and Searls, 1980), but later experimental results indicate that the majority of the thyroxin found in the plasma of 13.5- to 16.5-day-old chick embryos comes from the embryonic thyroid, as thyroidectomy causes a significant reduction in plasma T_4 (Thommes et al., 1984). The decrease in plasma thyroxin is rapid: two hours after thyroidectomy, plasma T_4 concentrations were reduced to 0.87 ng/ml in 13.5 day-old embryos (about an order of magnitude lower than controls) and 9.38 ng/ml on day 16.5 (a 50% decrease). Decapitated embryos aged 13.5 to 16.5 days had plasma T_4 concentrations in the 1 ng/ml range (Thommes and Jameson, 1980). These data indicate that the removal of the thyroid or the disruption of its regulatory system cause a substantial decline in circulating T_4 in chicks approximately midway through incubation or later, and provide convincing evidence that the primary source of T_4 during these developmental stages is endogenous hormone biosynthesis. However, none of the experimental treatments of chick embryos caused plasma T_4 to decline to undetectable levels. The effects of thyroidectomy on plasma T_3, and perhaps more importantly, tissue T_3 contents, have not been reported. T_3 has been found at concentrations of more than 6 ng/ml in the brain and liver of chick embryos as early as day 6 of incubation, with peak levels that precede the onset of regulated thyroid secretion (Mazumber and Banerjee, 1985). T_3 receptors have been found in the embryonic chick brain from the earliest sample date at which they have been sought (age 9 days; Bellabarba et al., 1983). Thus, T_3 receptors are present and tissue T_3 levels are elevated considerably before the thyroid system has become fully functional, and during a time when plasma T_4 levels are relatively low, but plasma T_3 is curiously elevated (data from several sources compiled by Scanes et al., 1987). The source or sources of T_3 in chick plasma and tissues and the cause of the high T_3/T_4 ratio in plasma prior to the maturation of the hypothalamic/pituitary/thyroid axis remain to be identified. The two most likely possibilities are 1) that the embryonic thyroid initially secretes hormones independently of higher control, and/or 2) that some quantity of thyroid hormones from the yolk reaches the ostensibly hormone-sensitive peripheral tissues. We have found that the concentration of T_3 in chicken egg yolk is almost twofold that of T_4 (21.2 ± 1.1 ng T_3 per gram yolk and 12.9 ± 0.7 ng/g T_4; n = one dozen; Brown and Weber, unpublished). It follows that any uptake of yolk and consequent uptake of thyroid hormones by the embryo might tend to increase the plasma T_3/T_4 ratio.

C. MAMMALS

As in avian development, a large body of evidence has been gathered in support of the hypothesis that embryos develop in the virtual absence of influence of maternal thyroid hormones. The cornerstone of this argument is the observation that, in most mammals, the placenta is relatively impermeable to thyroid hormones (see reviews by Fisher et al., 1977; Bachrach and Burrow, 1985). Maternal thyroid hormone treatment does not reverse fetal hypothyroidism in a variety of species (reviewed by Bachrach and Burrow, 1985), except in some cases involving large hormone doses (Peterson and Young, 1952; Knobil and Josimovich, 1959). Transplanted fetal rat paws grow and differentiate regardless of the thyroid status of the host (Cooke et al., 1984), and bone development does not appear to become thyroid-dependent until the postnatal period in rats (Liu and Nicoll, 1986).

Relatively few studies contradict the notion of maternal thyroid-independence in the development of mammalian embryos. Some results have suggested that although placental transfer is limited, potentially biologically significant amounts of thyroid hormones may reach fetal rat serum (Geloso and Bernard, 1967). The degree of transport of thyroid hormones across the placenta appears to be species-dependent (Fisher et al., 1977). The progeny of hypothyroid rats have reduced levels of plasma T_3 and T_4, and a pattern of increased TSH production that persists through adulthood (Porterfield and Hendrich, 1981). Thyroid hormones are present in rat embryo-trophoblasts long before the onset of embryonic thyroid function (Obregon et al., 1984), and the thyroid hormone contents in the rat conceptus are reduced in the hypothyroid pregnant rat (Morreale de Escobar et al., 1985). Although few in number, these results raise some doubt about the complete independence of mammalian embryos from maternal thyroid hormones, particularly in the early phases of differentiation, when such hormones could reach the developing embryo before the placenta begins to present a barrier to diffusion.

IV. CONCLUSIONS

In view of the very early effects of thyroid hormones from yolk on fish embryos, we suggest that possible actions of thyroid hormones during the early stages of avian and mammalian embryogenesis receive further consideration. It remains possible that hormones from the maternal circulation could influence development before the activation of the embryonic thyroid gland, without necessarily causing pronounced elevations in embryonic plasma hormone levels. Tonic low levels of maternal thyroid hormones in the circulatory system (from either the yolk, the uterus, or the placenta) may meet the requirements of hormone-sensitive embryonic tissues, such as the brain and liver.

Further investigation of the deposition of thyroid hormones in fish eggs and the consequences of these compounds to the larvae could contribute to a better understanding of the basic mechanisms of embryogenesis. Since thyroid hormones seem to be crucially important in the differentiation of the central nervous system in mammals (Nicholson and Altman, 1972a, b), a role of maternal thyroid hormones in the initial formation of the central nervous system in fish embryos might be a logical avenue for further research. It is plausible that many of the effects of experimental thyroid and antithyroid treatments on young fish could be explained as consequences of accelerated or impaired CNS differentiation. At present, however, only minimal attention has been paid to direct actions of thyroid hormones on the early differentiation of neuroendocrine systems.

Insofar as yolk-stored thyroid hormones are concerned, they are present in ovarian (Greenblatt, 1987) and in fertilized eggs (Kobuke et al., 1987; Brown et al., 1987); thus, a role in oocyte maturation should also be considered. This statement obviously applies to other hormones as well.

V. ACKNOWLEDGMENTS

We wish to thank the California Department of Fish and Game for their support and cooperation throughout our studies of larval fish development. Mr. M. Cochran and his staff at the Central Valley Hatchery have generously donated time and materials. Thanks are also expressed to Professor S. Doroshov, for his many important contributions to these studies, and to Professor T. Hirano, Mr. M. Greenblatt, Mr. M. Tagawa, Mr. G. Feist, Dr. J. Specker, Dr. E. Plisetskaya, Dr. C. Sullivan, and Mr. J. Daniels for their discussions and/or provision of unpublished data.

REFERENCES

Ali, M. A. (1961). Effect of thyroxine plus thiourea on the early development of the chum salmon (*Oncorhynchus keta*). Nature 191: 1214-1215.

Bachrach, L. K. and Burrow, G. N. (1985). Thyroid function in pregnancy. *In*: Pediatric and adolescent endocrinology (Laron, Z., ed.), Volume 14; Pediatric thyroidology (Delange, F., Fisher, D. A., and Malvaux, P., eds.). Karger, Basel.

Baker, B. I. (1964). Pituitary-thyroid relationship during development in the teleost *Herichthys cyanoguttatus*: a histophysiologic study. Gen. Comp. Endocrinol. 4:164-175.

Baker-Cohen, D. F. (1961). The role of the thyroid in the development of platyfish. Zoologica 46: 181-223.

Balon, E. K. (1975). Terminology of intervals in fish development. J. Fish. Res. Board Can. 32: 1663-1670.

Bellabarba, D., Bedard, S., Fortier, S., and LeHoux, J.-G. (1983). 3,5,3' Triiodothyronine nuclear receptor in chick embryo. Properties and ontogeny of brain and lung receptor. Endocrinology 112: 353-359.

Brown, C. L., Doroshov, S. I., Nunez, J., Hadley, C., Nishioka, R. S. and Bern, H. A. (submitted). Maternal triiodothyronine injections cause increases in larval growth, development, and survival rates in striped bass, *Morone saxatilis*.

Brown, C. L., Sullivan, C. V., Bern, H. A., and Dickhoff, W. W. (1987). Occurrence of thyroid hormones in early developmental stages of teleost fish. Am. Fish. Soc. Sympos. 2: 144-150.

Brown, C. L., Greenblatt, M. and Bern, H. A. (1986). Growth-related changes in thyroid function in larval striped bass. (abstract) Amer. Zool. 26(4): 24A.

Brown, R. L. (1987). 1985-1986 Report of the interagency ecological studies program for the Sacramento-San Joaquin estuary. Technical Report 11. Striped bass egg and larvae survey. California Department of Water Resources, Sacramento, CA

Chambers, H. A. (1953). Toxic effects of thiourea on the liver of the adult male killifish, Fundulus heteroclitus. Biol. Bull. 14: 69-93.

Cooke, P. S., Yonemura, C. U., and Nicoll, C. S. (1984). Development of thyroid hormone dependence for growth in the rat: a study involving transplanted fetal, neonatal, and juvenile tissues. Endocrinology 115: 2059-2064.

Doroshev, S. I., Cornacchia, J. W., and Hogan, K. (1981). Initial swimbladder inflation in the larvae of physoclistous fishes and its importance for larval culture. Rapp. P.-v. Reun. Cons. Int. Explor. Mer, 178: 495-500.

Doroshev, S. I. (1970). Biological features of the eggs, larvae, and young of the striped bass [Roccus saxatilis (Walbaum)] in connection with the problem of its acclimatization in the USSR. J. Ichthyol. 10: 235-248.

Dales, S. and Hoar, W. S. (1954). Effects of thyroxine and thiourea on the early development of chum salmon (Oncorhynchus keta). Can. J. Zool. 32: 244-251.

Etkin, W. (1950). The acquisition of thyroxine-sensitivity by tadpole tissues. (Abstract). Anat. Rec. 108:541.

Feist, G., Schreck, C. B., Fitzpatrick, M. S., and Redding, J. M. (submitted). Sex steroid profiles of coho salmon, Oncorhynchus kisutch, during early development and sexual differentiation.

Fisher, D. A., Dussault, J. H., Sack, J., and Chopra, I. J. (1977). Ontogenesis of hypothalamic-pituitary-thyroid function and metabolism in man, sheep, and rat. Rec. Prog. Horm. Res. 33: 59-116.

Geloso, J. P. and Bernard, G. (1967). Effets de l'ablation de la thyroide maternelle ou foetale sur le taux des hormones circulantes ches le foetus de rat. Acta Endocrinol. 56: 561-566.

Gorbman, A. and Ishii, S. (1960). Stimulation of neurosecretion in shark embryos by thyroid hormones. Proc. Soc. Exp. Biol. Med. 103: 865-867.

Greenblatt, M. N. (1987). Changes in thyroid hormone content and iodide uptake during early development of coho and chinook salmon. M. A. Thesis in Endocrinology, University of California, Berkeley, CA 94720.

Gudernatsch, J. F. (1912). Feeding experiments on tadpoles. Arch. Entwicklungsmech. Organ. 35: 457-483.

Gutreutter, S. J. and Anderson, R. O. (1985). Importance of body size to the recruitment process in largemeouth bass populations. Trans. Am. Fish. Soc. 114: 317-327.

Hanaoka, Y., Koya, S. M., Kondo, Y., and Yamamoto, K. (1973). Morphological and functional maturation of the thyroid during early development of anuran larvae. Gen. Comp. Endocrinol. 21: 410-423.

Hilfer, S. R. and Searls, R. L. (1980). Differentiation of the thyroid in the hypophysectomized chick embryo. Devel. Biol. 79: 107-118.

Hoar, W. S. (1939). The thyroid gland of the Atlantic salmon. J. Morphol. 65: 257-295.

Inui, Y. and Miwa, S. (1985). Thyroid hormone induces metamorphosis of flounder larvae. Gen. Comp. Endocrinol. 60: 450-454.

Kaltenbach, J. C. (1982). Circulating thyroid hormone levels in amphibia. In "Gunma Symposium on Endocrinology", Vol. 19, "Phylogenetic Aspects of Thyroid Hormone Actions". Pp. 63-74. Center for Academic Publications, Tokyo, Japan.

Knobil, E. and Josimovich, J. B. (1959). Placental transfer of thyrotropic hormone, thyroxine, triiodothyronine, and insulin in the rat. Ann. N. Y. Acad. Sci. 75: 895-904.

Kobuke, L., Specker, J. L. and Bern, H. A. (1987). Thyroxine content in eggs and larvae of coho salmon, *Oncorhynchus kisutch*. J. Exp. Zool. 242(1): 89-94.

Koslow, J. A. (1984). Recruitment patterns in northwest Atlantic fish stocks. Can. J. Fish. Aquat. Sci.41(12): 1722-1729.

Lam, T. J. (1985). Role of thyroid hormone on larval growth and development in fish. In: Lofts, B. and Holmes, W. N. (editors) Current trends in comparative endocrinology, pp.481-485. Hong Kong University Press, Hong Kong.

Lam, T. J. (1980). Thyroxine enhances larval development and survival in Sarotherodon (Tilapia) mossambicus. Aquaculture 21: 287-291.

Lam, T. J., Juario, J. V. and Banno, J. (1985). Effect of thyroxine on growth and development in post-yolk-sac larvae of milkfish, Chanos chanos. Aquaculture 46: 179-184.

Lam, T. J. and Loy, G. L. (1985). Effect of L-thyroxine on ovarian development and gestation in the viviparous guppy, Poecilia reticulata. Gen. Comp. Endocrinol. 60: 324-330.

Lam, T. J. and Sharma, R. (1985). Effects of salinity and thyroxine on larval survival, growth, and development in the carp, Cyprinus carpio. Aquaculture 44: 201-212.

Leatherland, J. F. and Lin, L. (1975). Activity of the pituitary gland in embryo and larval stages of coho salmon, Oncorhynchus kisutch. Can. J. Zool. 53: 297-310.

Liu, L. and Nicoll, C. S. (1986). Postnatal development of dependence on thyroid hormones for growth and differentiation of rat skeletal structures. Growth 50: 472-484.

Mazumber, A. and Banerjee, S.K. (1985). Triiodothyronine levels of embryonic and adult chick brain. I.R.C.S. Med. Sci. 13: 529-530.

McNabb, F. M. A. (1987). Comparative thyroid development in precocial japanese quail and altricial ring doves. J. Exp. Zool. Suppl. 1: 281-290.

Miwa, S. and Inui, Y. (1987a). Histological changes in the pituitary-thyroid axis during spontaneous and artificially-induced metamorphosis of larvae of the flounder Paralichthys olivaceus. Cell Tissue Res. 249: 117-123.

Miwa, S. and Inui, Y. (1987b). Effects of various doses of thyroxine and triiodothyronine on the metamorphosis of the flounder (Paralichthys olivaceus). Gen. Comp. Endocrinol. 67: 356-363.

Morreale de Escobar, G., Pastor, R., Obregon, M. J., and Escobar del Rey, F. (1985). Effects of maternal hypothyroidism on the weight and thyroid hormone content of rat embryonic tissues, before and after the onset of fetal thyroid function. Endocrinology 117: 1890-1900.

Nacario, J. F. (1983). The effect of thyroxine on the larvae and fry of Sarotherodon niloticus L. (Tilapia nilotica). Aquaculture 34: 78-83.

Nicholson, J. A. and Altman, J. (1972a). The effects of early hypo- and hyper-thyroidism on the development of rat cerebellar cortex. 1: cell proliferation and differentiation. Brain Res. 44(13): 13-23.

Nicholson, J. A. and Altman, J. (1972b). The effects of early hypo- and hyper-thyroidism on the development of rat cerebellar cortex. 2: synaptogenesis in the molecular layer. Brain Res. 44(13): 25-36.

Norris, D. O. (1969). Depression of growth following radiothyroidectomy of larval chinook salmon and steelhead trout. Trans. Am. Fish. Soc. 98: 104-106.

Obregon, M. J., Mallol, J., Pastor, R., Morrele de Escobar, G., and Escobar del Rey, F. (1984). L-Thyroxine and 3,5,3'-triiodo-L-thyronine in rat embryos before onset of fetal thyroid function. Endocrinology 114: 305-307.

Peterson, R. R. and Young, W. C. (1952). The problem of placental permeability for thyrotrophin, propylthiouracil, and thyroxine in the guinea pig. Endocrinology 50: 218-225.

Porterfield, S. P., and Hendrich, C. E. (1981). Alterations of serum thyroxine, triiodothyronine, and thyrotropin in the progeny of hypothyroid rats. Endocrinology 108: 1060-1063.

Ricker, W. E. 1975. Stock and Recruitment. J. Fish. Res. Board Can. 11: 559-623.

Scanes, C. G., Hart, L.E., Decuypere, E., and Kuhn, E. R. (1987). Endocrinology of the avian embryo: an overview. J. Exp. Zool. Suppl. 1: 253-264.

Sullivan, C. V., Iwamoto, R. N. and Dickhoff, W. W. (1987). Thyroid hormones in blood plasma of developing salmon embryos. Gen. Comp. Endocrinol. 65: 337-345.

Tagawa, M. and Hirano, T. (1987). Presence of thyroxine in eggs and changes in its content during early development of chum salmon, Oncorhynchus keta. Gen. Comp. Endocrinol. 68: 129-135.

Tesch, F. W. (1977). The Eel: Biology and Management of Anguilla anguilla Eels. (English Edition edited by Greenwood, P. H.). John Wiley, N.Y.

Thommes, R. C. (1987). Ontogenesis of thyroid function and regulation in the developing chick embryo. J. Exp. Zool. Suppl. 1: 273-279.

Thommes, R. C. and Jameson, K. M. (1980). Hypothalamo-adenohypophyseal-thyroid interrelationships in the chick embryo. III. Total T_4 levels in the plasma of decapitated chick embryos with adenohypophyseal transplants. Gen. Comp. Endocrinol. 42: 267-269.

Thommes, R. C., Clarke, N. B., Mok, L. L. S., and Malone, S. (1984). Hypothalamo-adenohypophyseal-thyroid interrelationships in the chick embryo. V. the effects of thyroidectomy on T_4 levels in blood plasma. Gen. Comp. Endocrinol. 54: 324-327.

Wabuke-Bunoti, M. A. N., and Firling, C. E. (1983). The pre hatching development of the thyroid gland of the fathead minnow, Pimephales promelas (Rafinesque). Gen. Comp. Endocrinol. 49: 320-331.

Whipple J. A., Eldridge, M. B., Benville, P., Bowers, M. J., Jarvis, B., and Stapp, N. (1981). The effect of inherent parental factors on gamete condition and viability in striped bass (Morone saxatilis). (abstract) Rapp. P.-v. Reun. Cons. Int. Explor. Mer 178: 93-94.

Woodhead, A. D. (1966). Effects of thyroid drugs on the larvae of the brown trout, Salmo trutta. J. Zool. Lond. 149: 394-413.

17

THE HYPOTHALAMO-PITUITARY (GROWTH HORMONE)- SOMATOMEDIN AXIS

Colin G. Scanes

Department of Animal Sciences
Rutgers University
New Brunswick, NJ 08903

I. INTRODUCTION

Growth hormone (GH) release from the anterior pituitary gland is controlled by stimulatory and inhibitory hypothalamic releasing factors. Growth hormone releasing factor (GRF) and somatostatin (SRIF) are the major stimulatory and inhibitory factors respectively. In addition, thyrotropin releasing hormone (TRH) is also involved in the control of GH secretion in many species. GH is involved in the control of growth but has other roles including regulation of lipid metabolism. The effects of GH on growth (particularly bone growth) are mediated via somatomedin(s); specifically-somatomedin C or insulin-like growth factor I (IGFI). This is produced by the liver and, also, peripherally for instance by the cartilage, where it has a local (paracine or autocrine) effect. Of the major somatomedins, IGFI is thought to be predominant ly GH dependent while IGFII is GH independent. The hypothalamo-pituitary (GH) -somatomedin axis includes the hypothalamic components, GH and the somatomedins.

This chapter will briefly consider the evolution of the hypothalamo-pituitary GH-somatomedin axis, the role of GH in growth, the ontogeny of GH secretion, GH and aging and the ontogeny of somatomedins (release and role).

Development, Maturation, and Senescence of
Neuroendocrine Systems: A Comparative Approach

II. EVOLUTION OF THE HYPOTHALAMO-GH-SOMATOMEDIN AXIS.

The hypothalamo-pituitary GH-somatomedin axis has been found to exist in mammals and birds. In lower vertebrates, there is strong evidence that the hypothalamo and pituitary components of the axis exist. For instance in fish, GH release is affected by mammalian releasing factors being stimulated by TRH (Cook and Peter, 1984) and inhibited by SRIF (Fryer et al., 1979; Cook and Peter, 1984). Moreover, GH stimulates growth in various fish (discussed in detail below in Section B). However, there is little data on somatomedins in poikilotherms.

Why is vertebrate growth controlled by the hypothalamo-GH-somatomedin axis? The existence of the hypothalamo-pituitary GH axis must be of major importance to the survival of vertebrate species. Support, for this contention comes from several distinct arguments. Firstly, GH is found in all vertebrate species studied. If GH were of little importance, even in a group of species, it is reasonable to assume that GH and/or somatomedins would be lost in some species. This is indeed possible as some individual humans, mice and cattle without GH have been observed. However, the absence of GH or somatomedins is not found in wild living animals. Applied with the argument (existence implies importance) is the observed complexity for control of GH release. There are two (or three) releasing factors together with feedback control by IGFI. Thus not only must GH be present but it is advantageous to have optimal secretion rates.

The hypothalamo-GH somatomedin axis regulates animal growth, and hence size, for optimal survival. This involves central nervous integration of data on environmental factors (including temperature and nutrition) which might influence growth. If growth were simply substrate driven, an animal could not regulate its growth to reach a determined size optimal for its environments (or niche). Substrate driven growth would also be haphazard and obviously detrimental to species survival. Many invertebrates have solved this problem by the existence of an exoskeleton limiting growth/size together with hormonal control of moulting which acts to synchronize the system. This is not practical in vertebrates. Thus hormonal control of growth (in part by GH axis) allows growth up to a maximum size for its environment. During selection (by natural selection or by man), the existence of the axis constrains the rate of evolutionary change but would then increase the likelihood of successful adaptation a population to a new environment.

Hypothalamo-Pituitary GH-Somatomedin Axis in the Invertebrates

The axis does not appear to exist outside the vertebrates; there being no pituitary gland in any invertebrate. However, some of the polypeptide components of the system exist. The hypothalamic peptides, TRH and SRIF, have been detected by immunoreactivity in invertebrates (TRH eg. in molluscs-Grimm-Jorgensen *et al.*, 1975; SRIF in planaria-Bautz and Schilt, 1986; molluscs-Grimm-Jorgensen 1983; insects-Doerr-Schott *et al.*, 1978; tunicates-Falkmer *et al.*, 1978). Furthermore there is some evidence that these peptides are involved in the control of growth with SRIF playing a stimulatory role in shell growth in a gastropod (Grimm-Jorgensen, 1983) and an inhibitory role in regeneration in planaria (Bautz and Schilt, 1986).

While GRF, GH and IGFI have not been definitively identified in invertebrates, this may not reflect their absence. The techniques required to detect the presence of vertebrate hormones in invertebrates (immunoassay, immunocytochemistry) depends on conservation of amino-acid residue sequence during evolution. While this is the case for the simple peptides-TRH and SRIF, it is not true for GRF, GH and IGFI even within mammals. It is, therefore, probable that some, or even all the hormones of the hypothalamus-pituitary (GH) - somatomedin axis pre-dated the emergence of the vertebrates. With the evolution of the vertebrates it may be speculated that there was greater integration of this important control system.

III GROWTH HORMONE AND GROWTH

Role of GH in Mammalian Fetus

There are essentially two views on the role of the high plasma concentrations of GH in the fetus. The high fetal GH secretion may reflect an obligatory but not-functional stage in the development of the axis. Alternatively the fetal GH may have a definite role in growth and/or development. The first possibility (high fetal GH secretion being obligatory but non-functional) involves two separate contentions. Evidence that high fetal GH secretion is obligatory is supported by its consistent observation in individual animals and different species (see below). It has been argued that the high rate of GH release reflects immaturity of the inhibitory mechanisms (including SRIF) which reduce plasma GH concentrations (Gluckman *et al.*, 1981; Macdonald *et al.*, 1985). Moreover, fetal GH does not appear to be required for growth. Hypophysectomy in fetal lambs does not influence body weight (Liggins and Kennedy, 1968). Similarly removal of fetal pituitary hormones by decapitation has surprisingly no effect on body growth in rats (Wells, 1947), rabbits (Jost, 1947), and pigs (Martin *et al.*, 1984). In addition, removal of fetal pituitary hormones has little effect on the circulating concentration of somatomedins (Brinsmead and Liggins, 1979). Liver GH receptors are not found until following birth in sheep (Gluckman *et al.*, 1983). However, these observations do not preclude GH having some role in the fetus or of fetal GH influencing post-natal growth.

There are several arguments supporting the tentative conclusion that GH has a physiological role in the fetus. These include the simplist view that, "if its there, it must be doing something". Indeed Macdonald and colleagues (1985) suggested that the presence of high concentrations of GH which change with physiological state support a role for the hormone. Altering the environment of fetus by maternal diabetes is associated by reduced circulating concentrations of GH in the pig fetus (Kasser *et al.*, 1982). Changes in fetal metabolism are associated with differences in the plasma concentrations of GH; circulating concentrations of GH being lower in obese than in lean fetal pigs (Martin *et al.*, 1985).

Studies with hypophysectomy support a role for GH in fetal development. Hypophysectomized fetal lambs have excess adipose tissue and delayed bone development (Liggins and Kennedy, 1968). Similarly, decapitated fetal pigs have decreased ash and increased carcass lipid (Kasser *et al.*, 1983); and circulating concentrations of triglyceride and free fatty acids (Martin *et al.*, 1984). Definitive evidence of a role of GH in the fetus would come from ablation/replacement therapy studies with GH. This has not been done. There is also a need for *in vitro* studies to characterize the putative effects of GH on fetal tissue.

Role of GH in the Avian Embryo

It is reasonable to assume that GH is not involved in the control of the growth of the chick embryo. Plasma concentrations of GH are not detectable until very late in the development of the chick embryo. There is a question as to whether chick embryo tissues are responsive to GH (reviewed Scanes, 1987). However, in the absence of appreciable circulating concentrations of GH, a role for GH in embryonic growth is unlikely. However, the hypothalamo-pituitary-thyroid axis is involved in *in ovo* growth (reviewed Scanes, 1987).

Role of GH in Post Natal Growth of Homeotherms

Growth hormone is required for growth, or its full expression. In mammals, post-natal hypophysectomy reduces or abolishes growth while GH replacement therapy restores full growth (rat, eg. Glasscock and Nicoll, 1981). There are three phases of growth in rats and probably other mammals each with different hormonal requirements: - a) fetal growth (largely independent of GH and thyroid hormones), b) perinatal growth (partially dependent on GH and thyroid hormones), and c) post natal growth (requiring GH) (Glasscock and Nicoll, 1981). During the latter phase, exogenous GH will stimulate further growth in intact animals. In birds, there is a requirement for GH in post-hatching growth (reviewed Scanes, 1987).

Role of GH in Poikilotherms During Growth/Development

In reptiles, exogenous GH can stimulate growth as can prolactin (eg. Licht and Hoyer, 1968). There is little information on the physiological role of reptile GH. In amphibians, GH appears to be the stimulator of growth following metamorphorsis (Zipser *et al*, 1972). In boney fish, the physiology of GH appears to resemble that in post-natal mammals; hypophysectomy reducing growth (Pickford, 1954) while exogenous GH stimulating growth even in intact animals.

GH influences other aspects of development. Some fish undergo a major environmental change during development; migrating from fresh water to sea water. In salmonid fish, this is preceded by preadaption (moultification). There is evidence that GH may have a role in this: - reducing liver lipid and increasing triglyceride lipase (Sheridan, 1986) and gill Na^+/K^+ -ATPase activity (Rickman and Zaugg, 1987).

IV ONTOGENY OF GH SECRETION

The ontogenic pattern of GH secretion has been examined in some detail in a few mammals and birds. Consistently in mammals maximal plasma concentrations of GH occurs during fetal development. In birds, the maximum is found during post hatching growth.

(a) Rats Plasma concentrations of GH are non-detectable in fetal 18 day rat pups (Rieutort, 1974) but rise rapidly to a maximum at 2 days of age (figure 1). Plasma concentrations of GH then decrease to a nadir at about 20 days of age but subsequently rise during sexual maturation (figure 1). Changes in pituitary GH content are shown in figure 2. GH is not detectible in the 18 day fetus. Thereafter, pituitary GH content increases rapidly; doubling every 0.6 days between 19 and 21 days of fetal life. A slower but still logarithmic increase in pituitary GH content occurs between 5 and 50 days old with the content doubling every 9.6 days. A plateau for the pituitary GH content is attained at about 60 days of age.

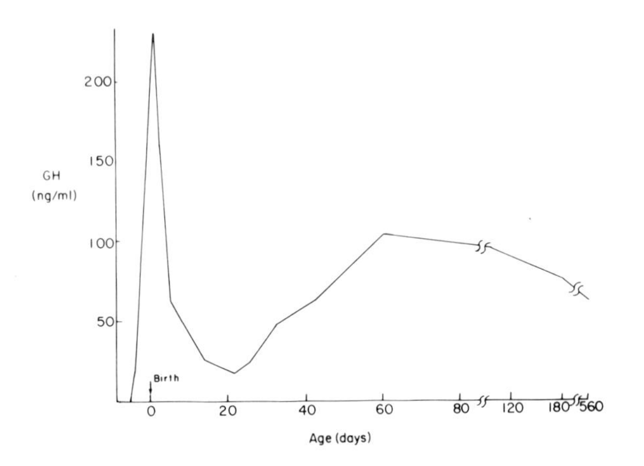

Figure 1 Plasma Concentrations of GH During Development and Aging in the Rat. (Pooled data calculated from Blazquez et al., 1974; Dickerman et al., 1972; Rieutort, 1974; Ojeda and Jameson, 1977; Drazner *et* al., 1979; Eden *et al.*, 1979)

The increase in plasma concentrations of GH (and pituitary content) between 18 days of fetal life and 2 day old (post-birth) reflects development and maturation of the hypothalamo-pituitary-GH axis. The rapid increase in plasma concentrations of GH in the peri-natal period is due to increases in: (a) the releasable GH stores in the pituitary gland, (b) the sensitivity of pituitary cells to TRH and GRF and (c) the availability of TRH and GRF to provoke GH release. A role for SRIF is unlikely in view of its inability to effect GH release from fetal rat pituitary tissue (Rieutort, 1981; Klorram *et al.*, 1983). There is strong evidence from *in vitro* studies for an increase in pituitary sensitivity to GRF/TRH during the peri-natal period (Rieutort, 1981; Khorram *et al.*, 1983; Baird *et al.*, 1984).

While the fetal pituitary is responsive to exogenous

GRF and TRH that does not *per se* indicate that endogenous secretagogues are responsible for the perinatal increase in circulating concentrations of GH in the rat. Evidence that the hypothalamus controls of GH secretion in the fetus comes from the reduction on the high plasma concentration of GH, if fetal hypothalamic functioning is disrupted surgically (Jest *et al.*, 1974) or pharmacologically (Stuart *et al.*, 1976). Secretion of GH in the fetus is partially independent on hypothalamic GRF/TRH control. Autonomous GH secretion increases as the pituitary content of GH increases. The hypothalamic dependent component of GH release increases at a greater rate. This is due to the maturation of the hypothalamus and the portal blood vessels as well as increases in pituitary sensitivity/responsiveness to the TRH and GRF. The hypothalamus develops between 18 days from conception. Hypothalamic TRH content increases immediately/prior to birth (Oliver *et al.*, 1980). Similarly, GRF release does not occur until late in fetal development. Neural fibers containing GRF are not observed in the median eminence until 19.5 days of gestation and thereafter increase in number (Daikoku *et al.*, 1985). The portal blood vessel develop at approximately twenty days following conception (Glydon, 1957).

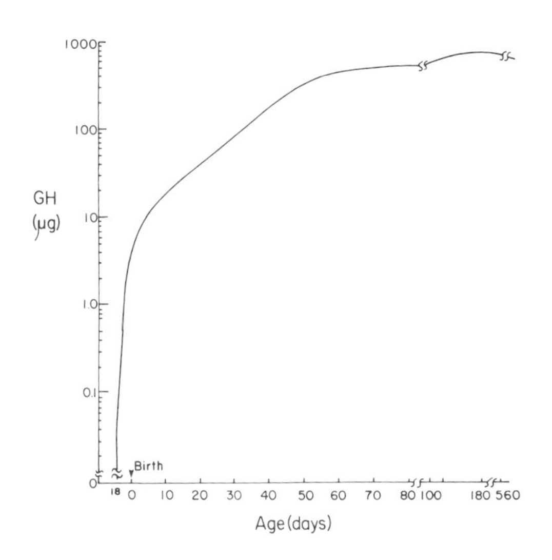

Figure 2 Changes in Pituitary GH Content With Age in the Data. (Data from Dickerman et al., 1972; Rieutort, 1974; Ojeda and Jameson, 1977).

Despite the rise in pituitary GH content (figure 2). Plasma concentrations of GH decrease between days 4 and 20 post-natally (figure 1). This precipitous decline in plasma concentrations of GH is concomitant with the onset of responsiveness to SRIF. *In vivo* SRIF does not affect plasma concentration of GH in the rat fetus or in 2 or 3 day old rat pups but SRIF is partially effective at 4 days of age and fully active by 7 days of age (Rieutort, 1981). Similarly rat pituitary *in vitro* is not sensitive to SRIF until 5 days of age (Rieutort, 1981; Khorram *et al*., 1983; Ailtter *et al*., 1986). Not only is the sensitivity to SRIF increasing but so is SRIF availability; hypothalamic SRIF concentrations rising between 2 and 20 days of age (Walker *et al*., 1977). The role of SRIF as a major cause in the post-natal decline in plasma concentration of GH is also supported by the high but negative correlation (r=-0.74) between plasma concentration of GH and the hypothalamic SRIF concentration (Walker *et al*., 1977). During post-natal developmental, (supplying the portal blood vessels and hence the pituitary) which begins at about 5 days following birth and is complete by sexual maturity (Glydon, 1957).

It is possible that changes in pituitary sensitivity to GRF and/or TRH also play some role in the post-natal decline in plasma concentration of GH. The evidence for differences in the response to GRF are contradictory with a decrease in GRF stimulated GH released *in vitro* observed between 2 and 12 day old (Szabo and Cuttler, 1986) but no change between 4 and 16 days of age (Welsh *et al*., 1986). The GH response to TRH decreases during post-natal development; the response to TRH declining from a maximal at 12 days of age to non-detectable at 30 days of age (Welsh *et al*., 1986). A similar age related decrease in the synergistic GH release with TRH and GRF occurs (Welsh *et al*., 1986).

The pubertal increase in plasma concentrations of GH is well established. This reflects both gonadal and hypothalamic influences on GH secretion.

(b) *Sheep* The sheep has been extensively employed as a model for fetal development (reviewed by Gluckman *et al*., 1981).

The plasma concentrations of GH are high in the fetal lambs with peak at approximately 65 days of gestation and immediately prior to parturition (days 131-140) but are low following birth (see figure 3).

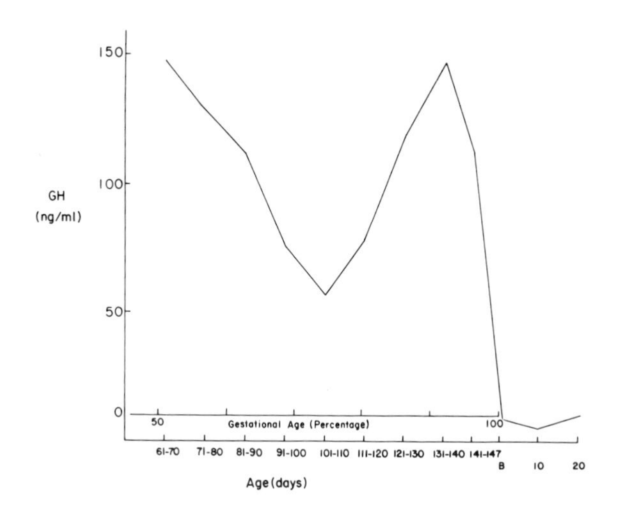

Figure 3 Ontogeny of Plasma Concentrations of GH in Sheep (Data from Bassett *et al.*, 1970; Gluckman *et al.*, 1979).

Fetal plasma concentrations of GH reflect high secretion rates and low clearance rates (Bassett *et al.*, 1970; Wallace *et al.*, 1973) (see table 1). Secretion of GH in the fetus depends on: - (a) number of somatotroph and their ability to synthesize and release GH, (b) somatotrophs sensitivity to releasing factors and (c) the concentration of these secretagogues in the portal blood.

TABLE 1

Comparison of GH Secretory Dynamics in Fetal and Neonatal Sheep (Data from Wallace *et al* 1973)

	Fetus (132 d.)	Lamb (6d)
Plasma GH (ng/ml)	120	75
Metabolic Clearance Rate (MCR) (ml/min)	7.7	11.3
Secretion Rate (SR) (ng/ml)	924	85
Half Life (min)	25	15

Comparison of GH Secretory Dynamics in Young and Adult Chickens (Data from Lauterio and Scanes, 1987)

	Young	Adult
Plasma GH (ng/ml)	76	14
Metabolic Clearance Rate (nl/min)	2.8	4.2
Secretion Rate (SR) (ng/ml)	209	59

The importance of somatotroph number on GH stores in the pattern of GH secretion in the fetal lamb is not clear. There is little apparent relationship between the profile of plasma concentrations of GH and the changes in pituitary GH content. Pituitary immunoreactive GH content increases in a linear manner between 60 and 110 days of gestation (Gluckman *et al.*, 1981). Pituitary bioassayable GH content rises in linear manner between 115 days of gestation (fetus) and 100 day old (lambs) (Charrier *et al.*, 1973).

Although the hypothalamus is involved in the control of GH secretion, there is insufficient data to explain the profile of plasma concentration of GH by changes in hypothalamic control. The ovine fetal pituitary gland is responsive to GRF (Ohmura *et al.*, 1984) and SRIF (McMillen *et al.*, 1978; Gluckman *et al.*, 1979) but not to TRH (Thomsett *et al.*, 1980). The response to GRF declines from 70 days of gestation through to the time of birth (Ohmura *et al.*, 1984). The pituitary has been observed to be sensitive to SRIF in fetuses older than 100 day (McMillen *et al.*, 1978; Gluckman *et al.*, 1979) but there is no evidence of a change in responsiveness. Stimulation of the fetal hypothalamus by pharmacological agents affects GH release by increasing GRF or SRIF release. Plasma concentrations of GH are elevated in the fetus by the administration of α-endorphin (Gluckman *et al.*, 1980) or the serotonin precursor, 5 hydroxy tryptophan (Marti-Henneberg *et al.*, 1980) with the

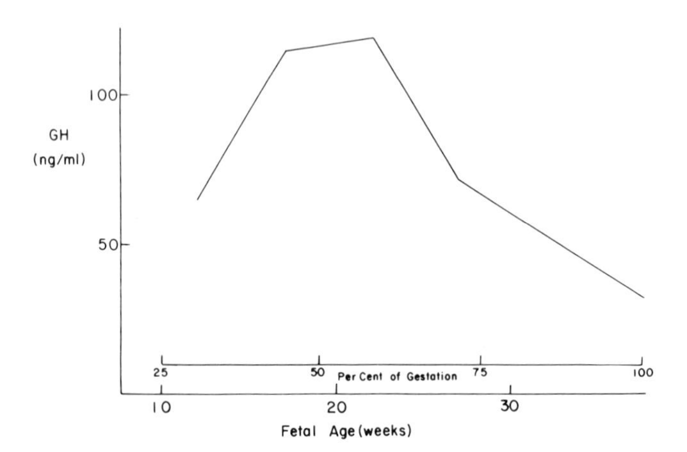

Figure 4 Ontogeny of Plasma Concentrations of GH in the Human (Re-drawn from Kaplan *et al.*, 1972).

(c) Human Plasma concentration of GH have been followed through fetal development (Kaplan *et al.*, 1972) (figure 4).
response to both being greater in young (less than 110 days) than older fetuses (greater than 110 days of gestation). These may act by increasing GRF release. In addition, plasma concentrations of GH in the fetus are depressed by a β adrenergic agonist (isoproterenol) (Gluckman *et al.*, 1981) and a GABA agonist (muscimol) (Gluckman *et al.*, 1981). Fetal plasma concentration of GH are not increased by clonidine (α_2 adrenergic agonist) which stimulates GH secretion in post-natal lambs (Marti-Henniberg *et al.*, 1980).

The maximal plasma concentrations of GH in the fetus in the second trimester but decline in the third trimester. No correlation has been observed between the increasing pituitary GH content (figure 5) and plasma concentrations of GH (Kaplan *et al.*, 1972). In early fetal development (13 to 17 weeks) the pituitary GH content increases rapidly (the

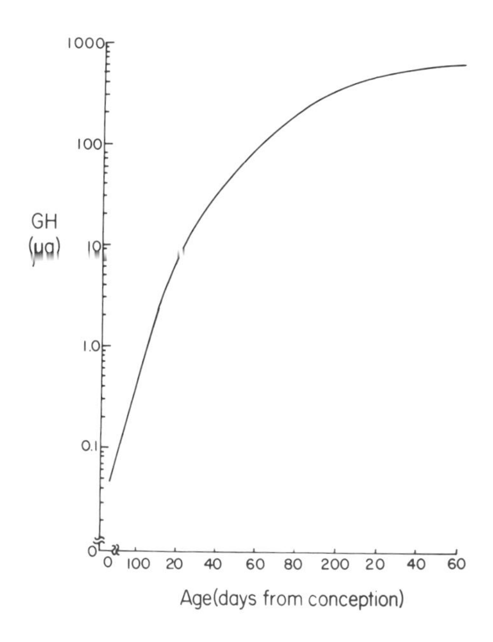

Figure 5 Changes in Pituitary GH Content During Human Development (data from Kaplan *et al.*, 1972).

content calculated to be doubling every 4-5 days). Pituitary GH content continues to increase, albeit at a slower rate (calculated to double every 18 days between 22 weeks and 28 weeks of gestation).

The mechanism accounting for the high plasma concentrations of GH during the second trimester and the subsequent decline are not well understood. Studies with fetal pituitary tissue _in vitro_ provide some evidence to explain this. The initial rise in plasma concentrations of GH may reflect increase in pituitary GH content and hence releasable GH. _In vitro_ studies indicate that basal GH release increase logarithmically from 5 week to term (from 6 to 150,000 ng GH released _in vitro_ per pituitary) (Silver-Khodr _et al._, 1974). The role of the hypothalamic in mediating changes in the secretion is not well understood. There is little evidence for changes in pituitary sensitivity to either GRF or SRIF during fetal development. _In vitro_ by 10 weeks of development, pituitary cells are responsive to GRF (Goodyear _et al._, 1987), to cAMP, (the presumed second messanger for GRF) (Goodyear _et al._, 1977) and SRIF (Goodyear _et al._, 1977; 1987). However TRH appears to have little effect on GH release from fetal human pituitary cells _in vitro_ (Goodyear _et al._, 1977). Both GRF and SRIF are present in the fetal median eminence as early as the 19th week of fetal life (Bresson _et al._, 1983).

(d) _In cattle_, plasma concentrations of GH rise through fetal development (figure 6) (Oxender _et al._, 1972a).

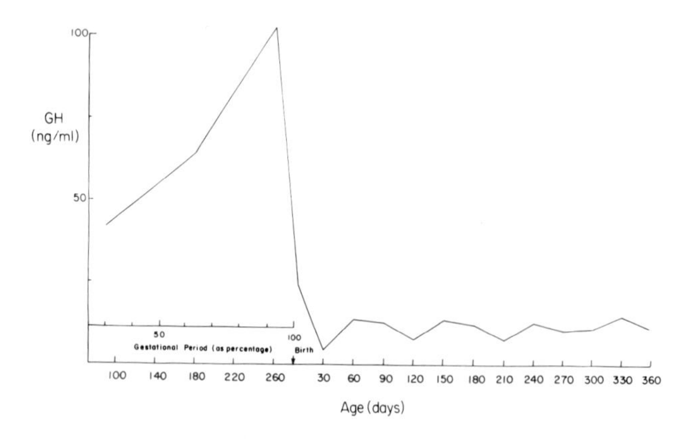

Figure 6 Changes in Plasma Concentrations of GH During Development in Cattle (Data from Purchas _et al._, 1970; Oxender _et al._, 1972a).

At approximately the time of birth, there is a precipitous decline in the plasma concentrations with low levels being maintained thereafter (Purchas *et al.*, 1970). During fetal development, there is a parallel increase in the concentrations of GH in the pituitary gland (90 day - 4.2 μg/mg; 180 day 8.9 μg/mg; 260 day 18.1 μg/mg) (Oxender *et al.*, 1972b). The pituitary concentration of GH increases in the first three months of post-natal growth but subsequently declines (Purchas *et al.*, 1970). There is little information on the physiological basis for these changes.

(e) In the pig, plasma concentrations of GH are high during the last 40% of fetal life but decline post-natally (figure 7) (Atimmo *et al.*, 1976) (figure 7). These changes is not well understood. The high fetal plasma concentrations of GH reflects a combination of basal and secretagogue induced GH release. There is evidence of hypothalamic control of GH secretion. TRH increases GH release *in utero* (Macdonald *et al.*, 1985) indicating both that pituitary sensitivity to the tripeptide and that GH secretion is not maximal.

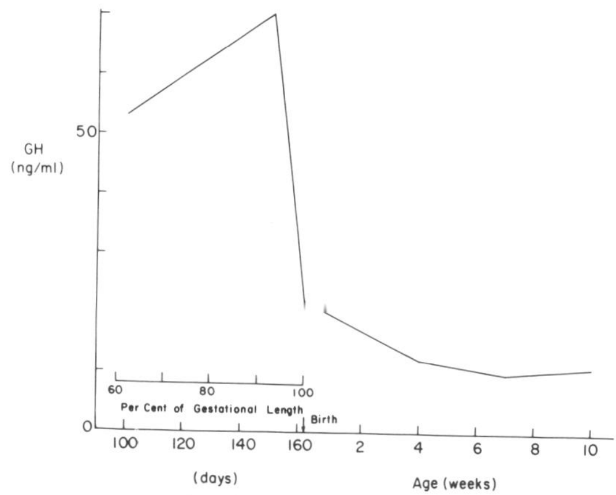

Figure 7 Ontogeny of Plasma Concentrations of GH in the Pig (Data from Altimo *et al.*, 1976).

(f) In rabbits, plasma concentrations of GH are non-detectable in mid-gestation fetuses (eg. day 22). Plasma concentrations of GH have been found to rise between day 23 and day 26 (day 23 13.8 ng/ml; day 24 45.3 ng/ml, day 25 118 ng/ml; and day 26 228 ng/ml (Jost *et al.*, 1979).

(g) Birds Plasma concentrations of GH are high during post-natal growth but low in adults and in embryos (figure 8a).

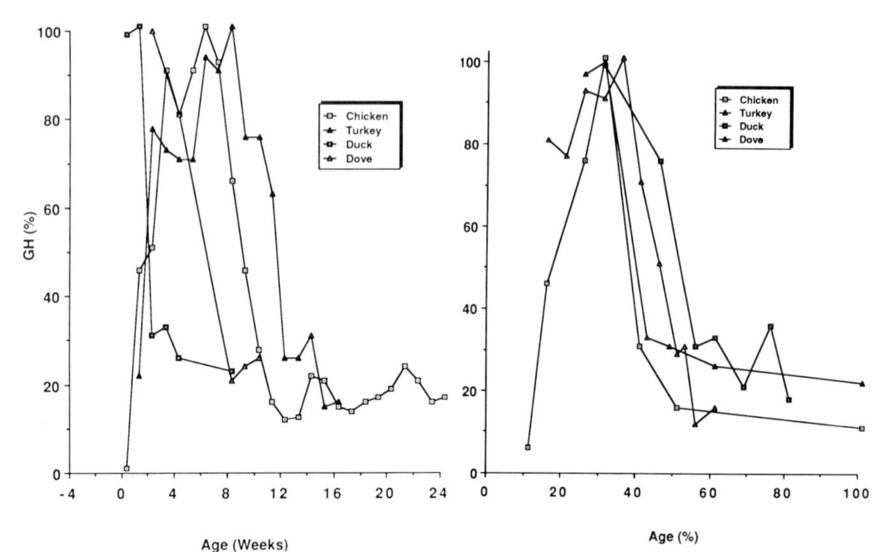

Figure 8 Developmental Profile for Plasma Concentrations of GH in Birds (Left 8a Age in Weeks; Right 8b Age as a Percentage of Time from Conception to Maturity) (Re-drawn from Scanes *et al.*, 1984)

There are marked species differences in the time of the maximal plasma concentration of GH and the rapid decline. However if the age of the birds is expressed as a percentage of the time from conception to maximal body weight, the patterns are synchronized (figure 8b). This suggests a tight coupling of stage of development and plasma concentration of GH.

In chick embryos, GH is not detected in the plasma until late (day 17) in embryonic development (Harvey *et al.*, 1979) with the GH producing somatotrophs first observed on day 12 of development (Josza *et al.*, 1979). The peri-natal rise in plasma concentrations of GH reflects increases in somatotroph number/GH content and sensitivity to both TRH and GRF (reviewed Scanes, 1987).

Pituitary GH content increases during the post-natal period (figure 9), while plasma concentrations of GH plateau (figure 8a).

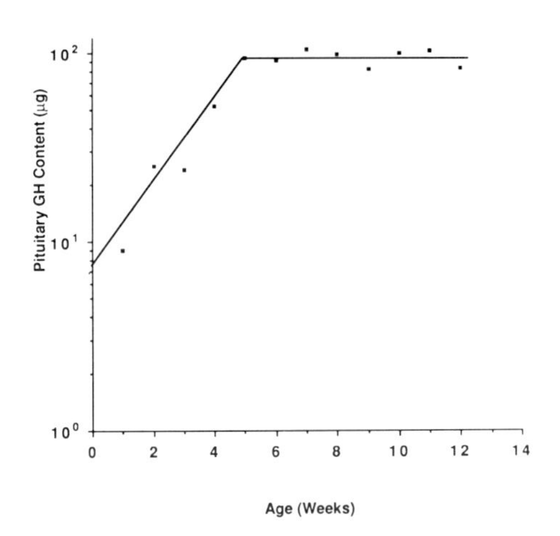

Figure 9 Changes in Pituitary GH Content with Age in Chickens (Data from Vasilatos-Younken, 1986).

Plasma concentrations of GH decline precipitously between weeks 6 and 12 in chickens (figure 8a) (reviewed Scanes, 1987). This decrease in plasma concentrations of GH precedes sexual maturation and is not related to changes in sex steroids as gonadectomy fails to influence the decrease. Differences between plasma concentrations of GH in young and adult chickens are due to decreased secretion and elevated clearance (table 1). The change in secretion rate is occurring despite the increase in somatotroph number with age (Malamed and Scanes, unpublished observations). Explanations for the reduction in GH release with age include: - (a) changes in somatotroph structure, (b) decreased pituitary sensitivity to TRH and/or GRF, (c) increased sensitivity to SRIF and (d) greater negative "feedback"-like effects of IGFI and/or triiodothyronine (T_3). There is evidence for all of these. The secretory granules of the somatotroph became smaller with age which may decrease their ability to be released (Malamed et al., 1985). The maximal GH secretory responses to either TRH or GRF are lower in adults than young anesthetized chickens (Harvey and Scanes, 1984). Increases in IGFI feedback may occur as plasma concentrations of IGFI rise with age (see below and figure 4). A role for thyroid hormones in the decline in GH release is likely as hypothyroidism prevents the decline in plasma concentration of GH with age (reviewed Scanes, 1987).

Evidence for a shift in the concentration of T_3 or sensitivity to T_3 is more circumspect. There is little difference in the plasma concentration of T_3 with age but it is not possible to preclude changes in free T_3. There is no data on changes in sensitivity to T_3.

(g) Poikilotherms There is little information of changes in GH secretion during growth, development and aging in poikilotherms. At present full longitudinal studies have not been reported in any lower vertebrate. However, changes in plasma concentration of GH have been measured during a portion of growth being estimated weekly for 6 weeks coho salmon (Wagner and McKeown, 1986). The pattern of GH was suggestive of a decline with age. Release of GH from fish pituitary tissue does appear to change during growth. *In vitro* basal release of GH from (large) tilapia was greater than in small fish. However GH release is more labile pituitary tissue from younger fish; being more sensitive to stimulation by cortisol or high osmotic pressure and to inhibition by SRIF (Helma *et al.*, 1987).

V. AGING AND GROWTH HORMONE

Compared to the pattern of GH secretion during ontogeny and growth, changes during aging have received less attention. In mammals, studies on GH secretion during aging are confined to rats and humans. Senscence related changes of GH release have yet to be examined in non-mammalian vertebrates.

GH release in humans

Plasma concentration of GH have been reported to either decrease and not to change with increasing age in humans, depending on physiological circumstances. For instance, middle aged humans (47-52) showed less pulsatile GH release during either wakefullness or sleep than young adults (23-42) (Finkelstein *et al* 1972). Similarly, sleep (slow wave) associated increases in plasma concentrations of GH are observed in young adults but not in over 60% of individuals over 50 years old (Carlson *et al.*, 1972). No relationship was observed between age and circulating concentrations of GH following either an overnight fast or arginine infusion (Dudl *et al.*, 1973). The decline in plasma concentrations of GH with age may contribute to the negative nitrogen balance in older people. There is little evidence, however, that GH therapy can alleviate this problem.

Rats Plasma concentrations of GH are decreased in aged (>180 days old) rats (Dickerman _et al._, 1972) (figure 1). The situation is complicated in the female. Plasma concentrations of GH are elevated during estrus in cycling (young) rats but old female rats cease to cycle and show continuous estrus. Plasma concentrations of GH are reduced to a level intermediate to that observed at proestrus and diestrous in young rats (Dickerman _et al._, 1972). This decline in plasma concentrations of GH in female rats might be attributed to decreased levels of estradiol and hence estradiol stimulating GH release (Dickerman _et al._,1972).

In male rats there is definite senescence of the hypothalamo-pituitary-GH axis. The decline in mean plasma concentrations of GH (Dickerman _et al._, 1972) reflect decrease in the amplitude of GH pulsatile releases (Sonntag _et al._, 1980). Possible explanations for this include decreases in: (1) pituitary releasable GH stores, (2) GRF responsiveness, or (3) GRF release from the hypothalamus; or increases in: (4) blood volume (ie dilution or clearance), (5) sensitivity to SRIF or (6) SRIF released from hypothalamus. There is evidence for some of these being involved in the decrease in plasma concentrations of GH with age. Pituitary content of GH in old male rats (18 months old) is only 56.4% that of young male rats 40.3% (Sonntag _et al._, 1980). Furthermore, increased blood volume to be expected with greater body weight (young adults - 380 g; old 557 g (Sonntag _et al._, 1980) and would lead to reduced plasma concentrations of GH. There is also evidence for a decrease in pituitary responsiveness to GRF in old rats. This being observed _in vivo_ by Sonntag _et al._, (1983) and Ceda _et al._, (1986) but not by Wehrenberg and Ling (1983). Age related differences in responsiveness to GRF have been observed _in vitro_. Cells for old rats were less responsive to GRF and to cAMP. In addition, GRF induces a smaller accumulation of cAMP in pituitary cells from old than young rats (Ceda _et al._, 1980). The possibility that changes in sensitivity to SRIF may be involved in the age related decline in GH has received little attention but hypothalamic SRIF content is reduced in old rats (Sonntag _et al._, 1980). This may reflect differences in release rates.

VI ONTOGENY OF SOMATOMEDINS

Ontogeny of Circulating Concentrations of Somatomedins

The developmental patterns for both IGFI and IGFII are similar in different species (figure 10). Plasma concentrations of IGFI rise through fetal growth (human - Bennett

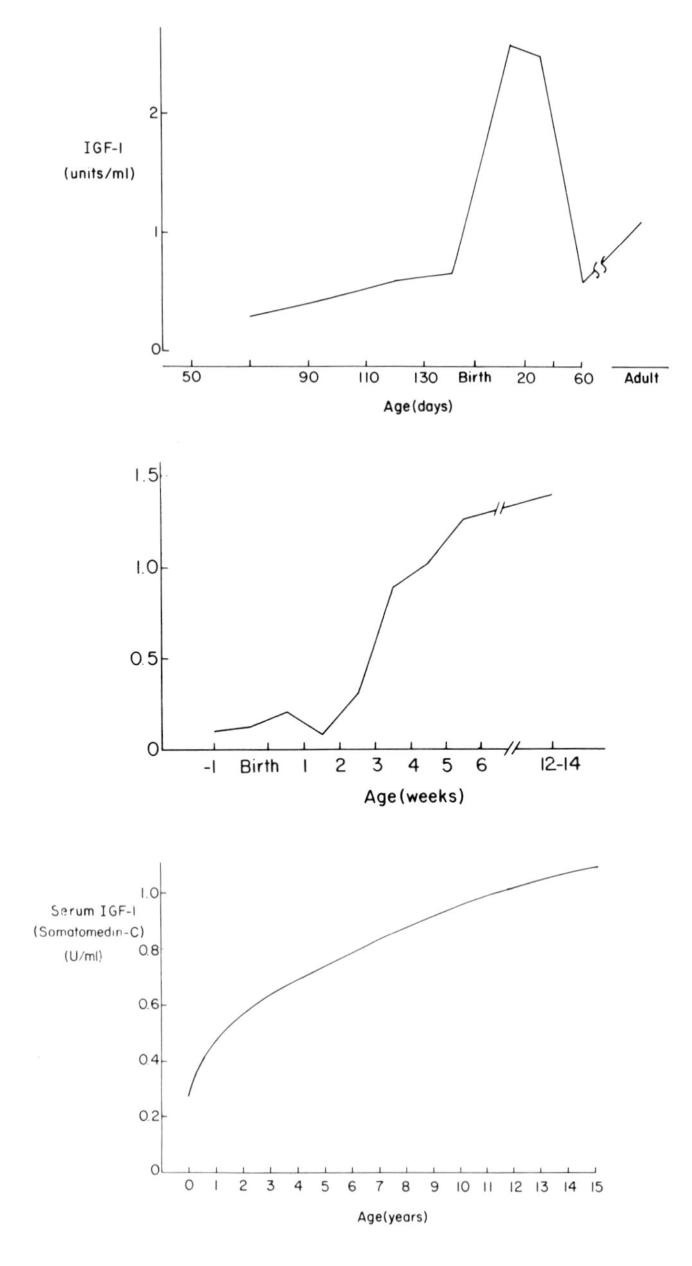

Figure 10 Ontogeny of Circulating Concentrations of IGFI (somatomedin C) in Sheep (top, re-drawn from Gluckman and Butler, 1983); mouse (middle, data from D'Ercole and Underwood, 1980); human (lower, re-drawn from D'Ercole *et al.*, 1977);

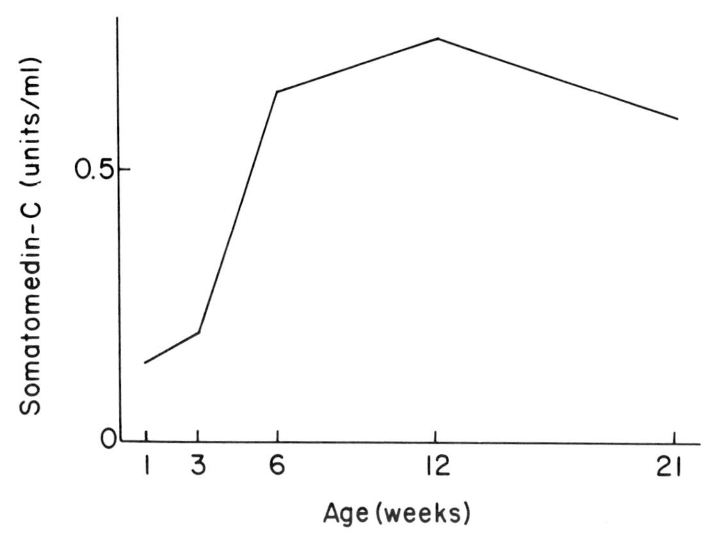

Figure 11 Ontogeny of plasma concentrations of immunoreactive IGFI in the chicken (re-drawn from Huybreachts *et al.*, 1985)

et al., 1983; sheep - Gluckman and Butler, 1983; mouse - D'Ercole, 1980; and chicken - Huybrechts *et al.*, 1985) (figure 10 and 11). In most species examined, plasma concentrations of IGFI are reduced in adults (eg human - Luna *et al.*, 1983; chickens - Huybrechts, *et al.*, 1985). Plasma concentrations of IGFI increase during puberty in the human. This may account for the growth spurt and may be partially related to estrogens directly (or indirectly via GH) increasing IGFI production (eg. Cutler *et al.*, 1985). Growth appears to be dependent on IGFI. Plasma concentrations of IGFI correlate with growth rate, are reduced in hypopituitary (slow growing) children, and are increased by GH therapy which also increases growth (D'Ercole *et al.*, 1977).

Plasma concentrations of IGFII are high in the fetus, decline precipitously at about the time of birth and show little post natal variation (rat - Moses *et al.*, 1980, Daughaday *et al.*, 1982; human - Bennett *et al.*, 1983; sheep: - Gluckman and Butler, 1983) (figure 12).

Control of Somatomedin Release During Growth.

In the fetus, plasma concentrations of somatomedins (IGFI and IGFII) do not appear to be under pituitary control. For instance in the fetal lamb, hypophysectomy has been reported to have little effect on circulating somatomedin activity (Parkes and Hill, 1985) or plasma

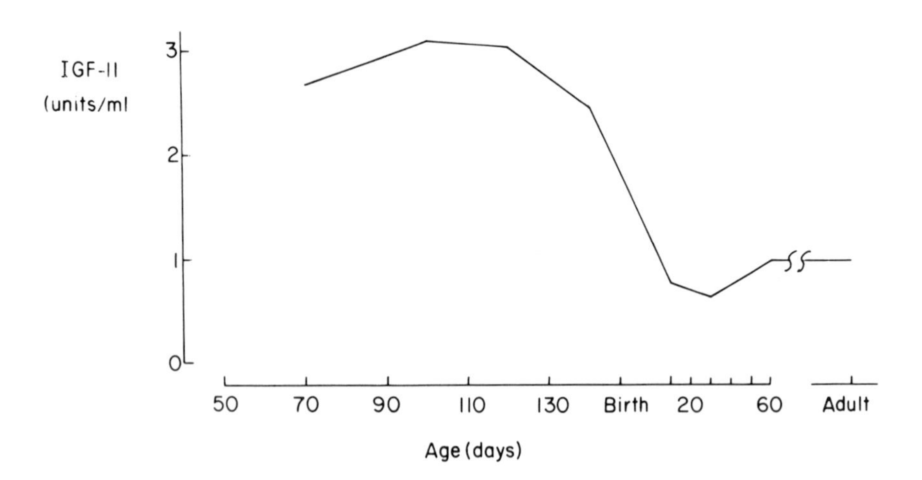

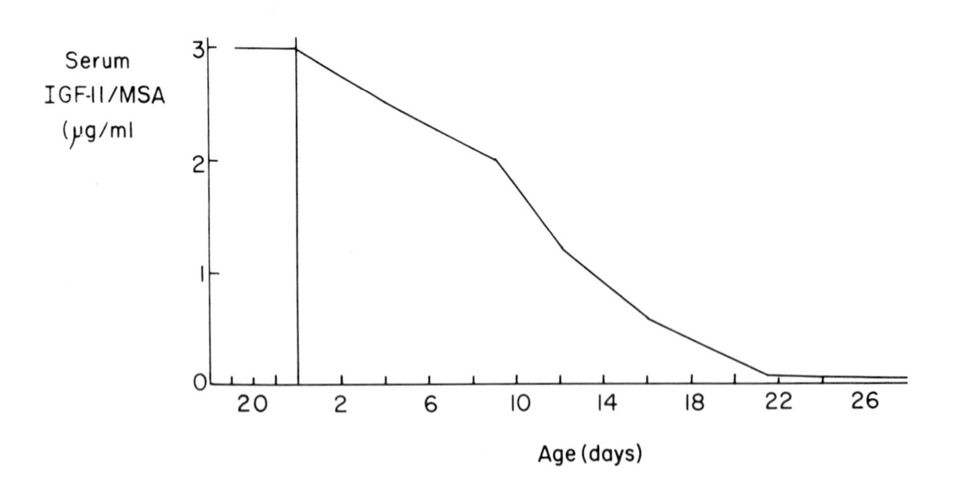

Figure 12 Ontogeny of circulating concentrations of IGFII in sheep (upper, re-drawn from Gluckman and Butler, 1983) and rat (lower; re-drawn from Moses et al., 1980).

concentrations of either IGFI or IGFII (Mesiano et al., 1987). Somatomedin release may be autonomous, and/or influenced by hormones other than GH. For instance, insulin increases plasma somatomedin activity in the fetal pig (Spencer et al., 1983a). Release of IGFII and IGFI may be stimulated by placental lactogen; placental lactogen increasing IGFII release from rat fibroblasts (Adams et al.,

1983). Moreover fetal liver cells have receptors which are specific for placental lactogen (Freeman *et al.*, 1987).

The switch from IGFII to IGFI production during the perinatal period is intriguing. The physiological basis for this is not known but may involve a developmental clock and/or parturation related hormones. In the post natal period, plasma concentrations of IGFI are partially dependent on the presence of GH (eg. human: - D'Ercole *et al.*, 1977; chicken - Huybrechts *et al.*, 1985). In addition, plasma concentrations of IGFI are depressed with poor nutrition; this effect being at least partially independent of changes in GH secretion.

VII CONCLUSIONS

The hypothalamo-pituitary GH-somatomedin axis is found in all vertebrates examined. The ontogeny of the axis shows some common features and some species differences. In mammals, plasma concentrations of GH are high during fetal or perinatal life. In birds, plasma concentrations of GH are maximal at a later stage being high during the early post-natally. These maximal concentrations of GH reflect both high secretion rate and low clearance rate. The mechanism by which plasma concentrations of GH decline following the maxima involves increased SRIF sensitivity in some but probably not all species. The period of maximal plasma concentrations of GH coincides paradoxically with relatively poor growth responses to GH and few liver GH receptors. In all vertebrates, GH is required for post natal (or post-hatching or post-metamorphic) growth. GH has other roles during development including involvement in the control of energy balance (lipid mobilization) and perhaps in sexual maturation. During senescence, there tends to be some decline in plasma concentrations of GH and GH secretion. The developmental profiles for IGFI and IGFII show consistent patterns with plasma concentrations of IGFII being high in the fetus while IGFI is high in post-natal growth. Reflecting their roles as pre-natal (IGFII) and post-natal (IGFI) growth factors.

REFERENCES

Adams, S.O., Nissley, S.P., Handwerger, S., and Rechler, M.M. (1983). Nature 302, 150.

Atinmo, T., Baldijao, C., Pond, W.G., and Barnes, R.H. (1976). J. Nutr. 106, 940.

Baird, A., Wehrenberg, W.B., and Ling, N. (1984). Regulatory Peptides 10, 23.

Bassett, J.M., Thorburn, G.D., and Wallace, A.L.C. (1970). J. Endocrinol. 48, 251.
Bautz, A., and Schilt, J. (1986). Gen. Comp. Endocrinol 64, 267.
Bennett, A., Wilson, D.M., Liu, F., Nagashima R., Rosenfeld, R.G., and Hintz, R.L. (1983). J. Clin. Endocrinol. Metab. 57, 609.
Blazquez, E., Simon, F.A., Balzquez, M., and Foa, P.P. (1974). Proc. Soc. Exp. Biol. Med. 147, 780.
Bresson, J.-L., Clavequin, M.-C., Fellmann, D., and Bugnon, C. (1983). C. R. Seases Biol. 177, 45.
Brinsmead, M.W., and Liggins, G.C. (1979). Endocrinology 105, 297.
Carlson, H.E., Gillin, J.C., Gorden, P., and Synder, F. (1972). J. Clin. Endocrinol. Metab. 34, 1102.
Ceda, G.P., Valenti, G., Butturini, V., and Hoffman, A.R. (1986). Endocrinology 118, 2109.
Charrier, J. (1973). Ann. Biol. Anim. Biochem. Biophys. 13, 155.
Cook, A.J., and Peter, R.E. (1984). Gen. Comp. Endocrinol. 54, 109.
Cuttler, L., Van Vliet, G., Conte, F.A., Kaplan, S.L., and Grumbach, M.M. (1985). J. Clin. Endocrinol. Met. 60, 1087.
Cuttler, L., Welsh, J.B. and Szabo, M. (1986). Endocrinology 119, 152.
Daikoku, S., Kawano, H., Noguchi, M., Tokuzen, M., Chihara, K., Saito, H., and Shibasaki, T. (1985). Cell Tiss. Res. 242, 511.
Daughaday, W.H., Parker, K.A., Borowsky, S., Trivedi, B., and Kapadia, M. (1982). Endocrinology 110, 575.
D'Ercole, A.J., and Underwood, L.E. (1980). Dev. Biol. 79, 33.
D'Ercole, A.J., Underwood, L.E., and Van Wyk, J.J. (1977). J. Pediatrics 90, 375.
Dickerman, E., Dickerman, S., and Meites, J. (1972). In:Growth and Growth Hormone (Eds. A. Pecile and E.E. Muller) pp. 252-260. Excerpta Medica, Amsterdam.
Doerr-Schott, J., Joly, L., and Dubois, M.P. (1978). C. R. Acad. Sci. (Paris) 286, 93.
Draznin, B., Morris, H.G., Burstein, P.J., and Schalch, D.S. (1979). Proc. Soc. Exp. Biol. Med. 162, 131.
Dudl, R.J., Ensinck, J.W., Palmer, H.E., and Williams, R.H. (1973). J. Clin. Endocrinol. Metab. 37, 11.
Eden, S., Albertsson-Wikland, K., and Isaksson, O. (1978). Acta Endocrinol. 88, 676.
Faulkmer, S., Elde, R.P., Hellerstrom, C., and Peterson, B. (1978). Metabolism 27 (Suppl. 1), 1193.

Finkelstein, J.W., Rolfwarg, H.P., Boyar, R.M., Kream, J., and Hellman, L. (1972). J. Clin. Endocrinol. Metab. 35, 665.

Freemark, M., Comer, M., Korner, G., and Handwerger, S. (1987). Endocrinology 120, 1865.

Fryer, J.N., Nishioka, R.S., and Bern, H.A. (1979). Gen. Comp. Endocrinol. 39, 244.

Glasscock, G.F., and Nicoll, C.S. (1981). Endocrinology 109, 176.

Glydon, R. St. J. (1957). J. Anat. 91, 237.

Goodyear, C.G., Hall, C.S.G., Guyda, H., Robert, F., and Giroud, C.J.P. (1977). J. Clin. Endocrinol. Metab. 45, 73.

Goodyer, C.G., Sellen, J.M., Fuks, M., Branchaud, C.L., and Lefebvre, Y. (1987). Nutri. Develop. 27, 461.

Gluckman, P.D., and Butler, J.H. (1983). J. Endocrinol. 99, 223.

Gluckman, P.D., Mueller, P.L., Kaplan, S.L., Rudolph, A.M. and Grumbach, M.M. (1979). Endocrinology 104, 162.

Gluckman, P.D., Mueller, P.L., Kaplan, S.L., Rudolph, A.M., and Grumbach, M.M. (1979). Endocrinology 104, 974.

Gluckman, P.D., Marti-Henneberg, Kaplan, S.L., Li, C.H., and Grumbach, M.M. (1980). Endocrinology 107, 76.

Gluckman, P.D., Grumbach, M.M., and Kaplan, S.L. (1981). Endocrine Rev. 2, 363.

Gluckman, P.D., Butler, J.H., and Elliot, T.B. (1983). Endocrinology 112, 1607.

Grimm-Jorgensen, Y. (1983). Gen. Comp. Endocrinol. 49, 108.

Grimm-Jorgensen, Y., McKelvy, J.F., and Jackson, I.M.D. (1975). Nature (London) 254, 620.

Harvey, S. and Scanes, C.G. (1984). Neuroendocrinol. 39, 314.

Harvey, S., Davison, T.F. and Chadwick, A. (1979). Gen. Comp. Endocrinol. 39, 270.

Harvey, S., Scanes, C.G., and Marsh, J.A. (1984). Gen. Comp. Endocrinol. 55, 493.

Helms, L.M.H., Grau, E.G., Shimoda, S.K., Nishioka, R.S., and Bern, H.A. (1987). Gen. Comp. Endocrinol. 65, 48.

Huybrechts, L.M., King, D.B., Lauterio, T.J., Marsh, J., and Scanes, C.G. (1985). J. Endocrinol. 104, 233.

Jost, A. (1947). C.r. Lebd. Seance. Acad. Sci, Paris 225, 322.

Jost, A., Dupouy, J.P., and Rieutort, M. (1974). Proc. Brain Res. 41, 209.

Jost, A., Rieutort, M., and Bourbon, J. (1979). C. R. Acad. Sci, 288, 347.

Kaplan, S.L., Grumbach, M.M., and Shapard, T.H. (1972). J. Clin. Invest. 51, 3080.
Kasser, T.R., Gahagan, J.H., and Martin, R.J. (1982). J. Anim. Sci. 55, 1351.
Kasser, T.R., Hausman, G.J., Campion, D.R., and Martin, R.J. (1983). J. Anim. Sci. 56, 579.
Khorram, O., DePalakis, L.R., and McCann, S.M. (1983). Endocrinology 113, 720.
Licht, P., and Hoyer, H. (1968). Gen. Comp. Endocrinol. 11, 338.
Liggins, G.C., and Kennedy, P.C. (1968). J. Endocrinol. 40, 371.
Luna, A.M., Wilson, D.M., Wibbelsman, C.J., Brown, R.C., Nagashima, R.J., Hintz, R.L., and Rosenfeld, R.G. (1983). J. Clin. Endocrinol. Metab. 57, 268.
MacDonald, A.A., Spencer, G.S.G., and Hallett, K.G. (1985). Acta Endocrinol. 109, 126.
Malamed, S., Gibney, J.A., Loesser, K.E., and Scanes, C.G. (1985). Cell Tissue Res. 239, 87.
McMillen, I.C., Jenkin, G., Thorburn, G.D., and Robinson, J.S. (1978). J. Endocrinol. 78, 453.
Marti-Henneberg, C., Gluckman, P.D., Kaplan, S.L., and Grumbach, M.M. (1980). Endocrinology 107, 1273.
Martin, R.J., Campion, D.R., Hausman, G.J., and Gahagan, J.H. (1984). Growth 48, 158.
Martin, R.J., Ramsay, T.G., Campion, D.R., and Hausman, G.J. (1985). Growth 49, 400.
Mesiano, S., Young, I.R., Baxter, R.C., Hintz, R.L., Browne, C.A., and Thorburn, G.D. (1987). Endocrinology 120, 1821.
Moses, A.C., Nissley, S.P., Short, P.A., Rechler, M.M., White, R.M., Knight, A.B., and Higa, O.Z. (1980). Proc. Natl. Acad. Sci. USA 77, 3649.
Ohmura, E., Jansen, A., Chernik, V., Winter, J., Friesen, H.G., Rivier, J., and Vale, W. (1984). Endocrinology 114, 299.
Ojeda, S.R., and Jameson, H.E. (1977). Endocrinology 100, 811.
Oliver, C., Eskay, R.L., and Porter, J.C. (1980). Biol. Neonate 37, 145.
Oxender, W.D., Hafs, H.D., and Ingalls, W.G. (1972a). J. Ani. Sci. 35, 56.
Oxender, W.D., Convey, E.M., and Hafs, H.D. (1972b). Proc. Soc. Exp. Biol. 139, 1017.
Parkes, M.J., and Hill, D.J. (1985). J. Endocrinol. 104, 193.
Purchas, R.W., Macmillan, K.L., and Hafs, H.D. (1970). J. Anim. Sci. 31, 358.

Richman, N.H., and Zaugg, W.S. (1987). Gen. Comp. Endocrinol. 65, 189.
Rieutort, M. (1974). J. Endocrinol. 60, 261.
Rieutort, M. (1981). J. Endocrinol. 89, 355.
Scanes, C.G. (1987). Critical Reviews on Poultry Biology 1, 51.
Scanes, C.G., Lauterio, T.J., and Buonomo, F.C. (1983). In: Avian Endocrinology (ed. Mikami, S.) pp. 307-326. Springer Verlag.
Sheridan, M.A. (1986). Gen. Comp. Endocrinol. 64, 220.
Siler-Khodr, T.M., Morgenstern, L.L., and Greenwood, F.C. (1974). J. Clin. Endocrinol. Metab. 39, 891.
Sonntag, W.E., Steger, R.W., Forman, L.J., and Meites, J. (1980). Endocrinology 107, 1875.
Sonntag, W.E., Hylka, V.W., and Meites, J. (1983). Endocrinology 113, 2305.
Stuart, M., Lazarus, L., Smythe, G.A., Moore, S., and Sara, V. (1976). Neuroendocrinol. 22, 337.
Szabo, M., and Cuttler, L. (1986). Endocrinology 118, 69.
Thomsett, M.J., Marti-Henneberg, C., Gluckman, P.D., Kaplan, S.L., Rudolph, A.M., and Grumbach, M.M. Endocrinology 106, 1074.
Vasilatos-Younken, R. (1986). Gen. Comp. Endocrinol. 64, 99.
Wagner, G.F., and McKeown, B.A. (1986). Gen. Comp. Endocrinol. 62, 452.
Walker, P., Dussault, J.H., Alvarado-Urbina, G., and Dupont, A. (1977). Endocrinology 101, 787.
Wallace, A.L.C., Stacy, B.D., and Thorburn, G.D. (1973). J. Endocrinol. 58, 89.
Wehrenberg, W.B., and Ling, N. (1983). Neuroendocrinol. 37, 463.
Wells, L.J. (1947). Anat. Rec. 97, 409.
Welsh, J.B., Cuttler, L., and Szabo, M. (1986). Endocrinology 119, 2368.
Zipser, R.D., Licht, P., and Bern, H.A. (1969). Gen. Comp. Endocrinol. 13, 382.

18

DEVELOPMENT AND AGING OF THE THYROID IN HOMEOTHERMS

F. M. Anne McNabb

Department of Biology
Virginia Polytechnic Institute and State University
Blacksburg, Virginia 24061, USA

I. INTRODUCTION

This review compares the patterns of precocial development in birds and mammals, with emphasis on the ways in which the sequence of developmental events is similar in the two classes of homeotherms. Birds are discussed first, and in more detail, because there are no reviews focusing on the progression of thyroid function throughout the avian lifespan. In contrast, there have been regular reviews of both thyroid development and of function during aging in mammals.

II. DEVELOPMENTAL PATTERNS IN BIRDS AND MAMMALS

Development of the thyroid gland (TG) and its control by the hypothalamic-pituitary axis (HPT) in birds and mammals proceeds by different patterns in species with precocial vs. altricial modes of development. Precocial young are hatched or born at a relatively advanced stage of development and have long gestation periods. They show thermoregulatory responses to cooling at hatching or birth; completion of thermoregulatory ability occurs during early juvenile life. Altricial young are hatched or born at a less developed stage, generally after a shorter gestation period, and are completely dependent on parental care and brooding for some time. They remain ectothermic for some period after hatching/birth and then show relatively rapid development of thermoregulatory ability. These patterns represent two ends of a continuum. Thyroid development is correlated with the pattern and timing of thermoregulatory development, an association that is not surprising considering that thyroid hormones (TH) are important in the control of metabolism. Thus, thyroid function develops much earlier in precocial than in altricial birds and mammals.

In birds the hen has one-time maternal input into the egg and no continuing physiological/biochemical influence on the developing embryo. In contrast, mammals developing *in utero* have the potential for being influenced by alterations in the mother's physiology including any maternal hormones to which the placenta may be permeable. However, in both cases the young appear to develop equally autonomously; the avian embryo appears to depend on only TH produced by its own TG (Thommes et al., 1984), and the mammalian placenta permits almost no transfer in either direction of iodothyronines or thyrotropin (TSH) when these materials are present at physiological concentrations (Fisher et al., 1977).

A. The Thyroid Gland

1. Birds

a. **Precocials.** Avian TG development has been studied from the beginning of embryonic life only in chicken embryos (precocial); later embryonic stages have been studied in Japanese quail (also precocial). The pattern of TG growth and development is essentially identical for embryos, hatchlings and juveniles of these species when adjustments are made for the differences in the incubation times (21 days in chickens vs. 16.5 days in quail; McNabb, 1987, 1988). In embryonic chickens the TG primordium appears early in incubation (day 2/21), followed by iodine accumulation shortly thereafter, and stepwise acquisition of the production of precursors and TH before follicular organization (Romanoff, 1960). Histologically, the TG progresses through the same three stages that have been referred to as precolloid, beginning colloid and follicular-growth phases in mammals (Fisher et al., 1977). The TG is organized into follicles by mid-incubation (day 11/21 in chicken embryos: Romanoff, 1960; day 9/16.5 in quail: McNabb and McNabb, 1977). TG-T4 has been detected by immunocytochemistry as early as day 4.5/21 and serum T4 has been detected by RIA on day 6.5/21 in chicken embryos. These indications of early, but very limited TG function, suggest autonomous TG development prior to HPT axis maturation (Thommes, 1987).

During the latter half of incubation, TG function develops much more rapidly than the rate of TG growth; TG-TH stores increase markedly in late incubation (chickens: Daugeras et al., 1976; quail: McNichols and McNabb, 1988). In quail, the embryonic T3/T4 ratio is relatively high at days 11-12/16.5 then falls inversely with the increasing iodine accumulation in the TG (McNichols and McNabb, 1988). The TG-T3/T4 ratios in late embryos are similar to those in adult quail and other birds (Astier, 1980), with an extreme preponderance of T4 over T3 in TG-TH stores. Studies in quail indicate a shift from TG-TH production and storage toward TH release as the embryo enters the perinatal period. Circumstantial evidence in chickens and quail (Thommes, 1987; McNabb, 1987) indicates that TH release is high during the perinatal period. Thus, the TG of precocial birds achieves considerable maturity prior to hatching. Changes in TG size after hatching in quail parallel changes in body size; there is an early postnatal decrease in TG-TH content followed by gradual increases in TG-TH content (McNichols and McNabb, 1988). In chickens, the T4 secretion rate decreases during the first two weeks of juvenile life (Tanabe, 1965).

b. **Altricials.** The Ring dove is the only altricial bird in which thyroid development has been described. In this species the TG shows follicular organization, with colloid in some follicles by day 10/16.5 of incubation. Comparison of embryonic doves and Japanese quail (same length incubation period) shows the TGs are of similar size and development in 10 day embryos but that the dove TG grows much more slowly and appears much less active throughout embryonic life (McNabb and McNabb, 1977). Thus, the follicular growth phase appears to occur primarily posthatch in this species.

Dove TG-TH are extremely low throughout embryonic life, remain relatively low until 8 days after hatching, then increase rapidly in 8-16 day nestlings The TGs of doves do not contain as much TH as those of perinatal quail until the doves are 10 day nestlings. The T3/T4 ratio in the embryonic dove TG is high until 2 days prior to hatching then drops to low values characteristic of adults (McNabb and Cheng, 1985; McNichols and McNabb, 1988).

2. Mammals

a. **Precocials.** As in birds, development of the mammalian TG differs in the timing of development in precocial vs. altricial species. This review focuses on the most studied species, sheep and humans (both relatively precocial with respect to thyroid development and thermoregulation) and rats (which are altricial in these regards). Several reviews have provided a comrehensive treatment of the development of HPT function in these species (Nathanielsz, 1975; Fisher et al., 1977).

The human TG primordium appears very early in embryonic development and has reached the final (follicular-growth) stage by 1/4 to 1/3 of gestation. In embryonic sheep, follicles are first visible after 1/3 of gestation (50/150 days) and the gland is fully organized into colloid-filled follicles by 1/2 of gestation. Thus, in both these species, as in precocial birds, the TG shows considerable histological evidence of maturation midway through embryonic/fetal development (Nathanielsz, 1975; Fisher et al., 1977).

b. **Altricials.** In contrast to precocial mammals, in the altricial rat, TG development is very slow. The TG primordium only appears on day 11/23 of gestation (Kawaoi and Tsuneda, 1985). By 3/4 of gestation the rat TG is still in the precolloid or early colloid stages and iodine concentrating ability is beginning. TH synthesis begins just before birth and coincides with the follicular organization. Most TG maturation occurs after birth in the rat, so it is considerably delayed compared to precocial mammals (Fisher et al., 1977).

B. Control of the Thyroid

1. Birds

a. **Precocials.** Thommes (1987) has reviewed the evidence that the individual glands of the HPT axis in birds initially develop autonomously followed by later maturation of the interactive control system. The embryonic chicken TG produces, stores and releases TH at very low levels prior to stimulation by TSH from the pituitary.

Serum TSH has not been measured in avian embryos because specific avian TSH antibodies are not available and heterologous TSH antibodies do not show sufficient cross-reactivity for RIAs. However, heterologous antibodies reveal a marked increase in the numbers and immunoreactivity of putative TSH-producing cells in the pituitary at 10.5-11.5/21 days in chicken embryos (Thommes et al., 1983). The TG comes under pituitary control and the feedback of TH on the pituitary also matures at about 11 days of embryonic age (Thommes et al., 1977). Hypothalamic control of the pituitary also has matured by this stage of embryonic life and the hypothalamus as well as the pituitary is necessary for "normal" thyroid function during the latter part of incubation (Thommes, 1987). Our studies in quail indicate an equivalent pattern of pituitary maturation to that in the chicken (McNabb et al., 1984b).

b. **Altricials.** In embryonic Ring doves three lines of evidence suggest that pituitary control of the TG does not mature until after hatching. First, serum TH are very low in embryos. Second, embryos do not show a serum T4 response to injected bovine TSH, although this response does occur from 2 days post-hatching and onward. Third, estimates of pituitary TSH by bioassay found no detectable TSH in embryos and nestlings up to 4 days post-hatching, then increasing amounts of TSH thereafter (McNichols and McNabb, 1988).

2. Mammals

a. **Precocials.** In sheep and humans HPT maturation occurs during the latter half of gestation. Each gland first attains autonomous function at low activity levels, then the linking of the glands by "forward" control and negative feedback occurs. During the latter half of embryonic/fetal life, under HPT axis control, there are steady increases in function of the TG and in the levels of serum T4 (Fisher et al., 1977; Klein and Fisher, 1980).

b. **Altricials.** In the rat, the stages of TG development that occur very late in embryonic life correspond to those in the first 1/2 of embryonic life in precocial species. The capability for negative feedback of TH on the pituitary matures at day 18/23 of gestation (80% through gestation; Pic and Bouquin, 1985). However, fetal serum TSH declines between 20-21 days of gestation and reaches a low at birth and for several days after birth (Fukiishi and Hasegawa, 1985). These low TSH levels appear to result from inhibition of TSH secretion by somatostatin (Theodoropoulos, 1985). The pituitary does not respond to thyroid releasing hormone (TRH) for the first 5 postnatal days, but the response is present in 7-14 day young (Oliver et al., 1981). Most HPT axis maturation occurs after birth, with steady increases in both pituitary TSH content and serum TSH in the first few postnatal weeks (Fisher et al., 1977).

C. Peripheral Thyroid Function

1. Birds

a. **Precocials.** During the latter 1/2 or 1/3 of incubation in precocial embryos, serum T4 increases steadily but T3 remains very low (chickens: Thommes 1987; quail: McNabb, 1987). This occurs because the key deiodination pathway at this time is 5-deiodination (5D) which converts T4 to reverse-T3 (rT3); 5′deiodination (5′D) of T4 to T3 is extremely low (chickens: (Borges et al., 1980; Galton and Hiebert, 1987; quail: Hughes and McNabb, 1986). Serum TH increase markedly during the perinatal period, with increases in T4 reflecting HPT stimulation of the TG and increases in T3 reflecting increases in the 5′D at the beginning of the perinatal period (chickens: Decuypere et al., 1982; quail: Hughes and McNabb, 1986). Thus, there is a shift toward an increasingly important role of peripheral tissues in the control of serum T3. The significance of this shift turns the assumption that T3 is the metabolically active TH in birds as it is in mammals (see II D below).

After hatching, serum TH concentrations decrease and then increase during early juvenile life to stabilize at adult levels by a few weeks of age (chickens: Wentworth and Ringer, 1986; quail: McNabb et al., 1984a). Serum T3/T4 ratios in juvenile to adult birds are usually at least 10-fold higher than the TG T3/T4 ratios, reflecting the importance of the peripheral production of T3 by 5′D.

b. **Altricials.** In Ring doves, serum TH are very low in embryos and there is no evidence of a perinatal peak of serum TH like that in precocial species. After hatching, serum TH increase steadily until about day 8 with subsequent increases to adult levels occurring more gradually (McNabb et al., 1984a; McNabb and Cheng, 1985). The attainment of relatively stable serum TH concentrations after about 6-8 days coincides with the beginning of nestling thermoregulatory responses and a decrease in parental brooding (McNabb et al., 1984a).

2. Mammals

a. **Precocials.** In humans and sheep, the increase in serum TSH release at mid-gestation stimulates the TG, with resultant increases in serum T4 and other indications of TG activity such as increased radioiodine uptake. Serum T3 remains very low for 3/4 of gestation, then increases slightly. In the last few hours before parturition there is a sharp rise in serum T3 that is independent of HPT axis control (Klein et al., 1978; Nwosu et al., 1978). In contrast to the serum T3 pattern, rT3 tends to parallel serum T4 because the major deiodination pathway is 5D which converts T4 to rT3 (Kaplan, 1986). In the first few hours after birth there are marked increases in serum TSH and T3 and a slower increase in serum T4. The neonatal surge in TSH, which has been attributed to a HPT response to cooling during birth, stimulates TG-TH release and the increase in serum T4 (Fisher et al., 1977). The serum T3 increase results primarily from an increase in peripheral 5′D of T4 to T3 and appears to be mediated by the sympathetic nervous system plus the rise in serum cortisol that occurs at this time (Kaplan, 1986). At least part of the difference between fetal and later levels of 5′D activity is due to lower concentrations of cofactor in tissues such as liver. Fetal and neonatal tissues have been studied *in vitro,* but the relative roles of different tissues in the neonatal shifts in peripheral TH availability have not been elucidated (Kaplan, 1986; Wu et al., 1986b). After birth serum T3 and T4 fall and then stabilize at levels characteristic of those in infants and adults (Nathanielsz, 1975; Fisher et al., 1977).

b. **Altricials.** Associated with the low activity of the embryonic TG, serum TH are low in newborn and early postnatal rats. Serum T4 increases between 2 and 15-20 days postnatally (due to TSH stimulation), then gradually falls to adult levels. Serum T3, which is extremely low at birth increases gradually during the first month then falls slowly to adult levels. TH turnover and production rates of juveniles exceed those in adults, a situation that corresponds to late embryonic life in precocial species (Fisher et al., 1977).

In rats, the postnatal increase in serum T3 is associated with a shift from early predominance of 5D (rT3 production) to 5′D (T3 production). The increase in 5′D activity results from an increase in enzyme as well as increases in the tissue availability of cofactor (Wu et al., 1986a). T4 to T3 conversion during late gestation and early postnatal development has been studied in a number of tissues *in vitro* to attempt to build a picture of the interacting developmental events that determine TH availability (McCann et al., 1984; Iglesias et al. 1987; Kaplan, 1986). However, kinetic studies that account for TH distribution and disposal also are needed to complete our understanding of these events (Jang and DiStefano, 1985).

D. Effects of Thyroid Hormones During Development

1. T3 Receptors

The physiological actions of TH in mammals are presumed to be mediated by the binding of T3 to chromatin-associated nuclear T3 receptors. T4 is assumed to be a prohormone that is activated by 5′D to T3 (Oppenheimer, 1987). Nuclear T3 receptors, equivalent to those in mammals in their binding properties and numbers, have been demonstrated in a number of tissues in chickens (e.g. Bellabarba and Lehoux, 1985; Dainat et al., 1986) and in liver of quail (Weirich and McNabb, 1984). Interest in receptors during development has focused on when receptors appear in relation to the developing pattern of TH availability, whether receptor numbers and properties change with development and how receptor development correlates with tissue growth and functional development.

In embryonic chicken liver both the binding capacity and the binding affinity of the nuclear T3 receptors increase progressively between 9 and 19 days of incubation (Bellabarba and Lehoux, 1981). Receptor numbers in three types of fast and slow muscle fibers are correlated with the contractile characteristics of the fiber types and their relative metabolic activity during development (Dainat et al., 1986). Thus, the development of receptors may provide for more precise tissue regulation of TH action than do changes in serum TH availability. Studies of TH receptors in birds have been conducted exclusively on precocial species.

In mammals, T3 receptors have been studied in a variety of tissues and at a wide range of developmental ages. Different tissues show distinctly different patterns of receptor development that correspond in general to the pattern of tissue/organ development characteristic of each species. Receptor numbers and binding affinity increase with developmental age and in association with key developmental stages although features such as relative binding affinities remain constant (e.g. rats: Valcana and Timiras, 1978; Bellabarba et al., 1984; sheep: Ferreiro et al., 1987). Maturation of the receptors continues for a considerable period after birth ($>$ 30 days) in the altricial rat. However, some receptors can be detected coincident with the first appearance of the TG, i.e. long before thyroid function is fully established (Perez-Castillo et al., 1985).

2. Effects on Growth and Differentiation

Thyroid hormones are necessary, in addition to growth hormone and tissue growth factors, for normal postnatal growth and development in birds (Scanes et al., 1984) and mammals (Legrand, 1986). In birds, studies on precocial species also have demonstrated a TH requirement during embryonic development (King and May, 1984). In altricial rats embryonic development appears to be TH independent (Schwartz, 1983; Cooke et al., 1984), and most work has focused on the postnatal role of TH. However, recent work in altricials suggests that some embryonic tissues may require TH prenatally (e.g. lung, Devaskar et al., 1987; brain, Legrand, 1986). Studies of precocial mammals (e.g. sheep, in which the pattern of HPT axis development is similar to that in humans) indicate that they, like precocial birds, require TH for normal embryonic growth and development (Legrand, 1986).

Despite documentation of the necessity for TH in growth of birds and mammals, the roles of TH in these processes are not well understood and little is known of tissue TH availability. Few studies have attempted to distinguish between effects on growth, which may be defined as increases in mass and length, and maturation/differentiation events, which result in functional and morphological changes. In these developmental contexts, TH may act in a number of ways, among them production, release and interactions of growth hormone and tissue growth factor. In mammals, TH, like growth hormone, may function through their influence on the production of somatomedins which in turn have direct effects on cell growth (Schwartz, 1983; Legrand, 1986). However, this idea does not account for the picture of TH and somatomedins in developing birds (Scanes et al., 1986).

In a number of tissues, T3 triggers specific differentiation and maturation events; e.g. the maturation of lung tissue just prior to the initiation of pulmonary respiration (chicken embryos; Wittmann et al. 1983) and the initiation of pulmonary surfactant production (human tissue; Gonzales et al., 1986). TH also trigger specific biochemical changes and appear to be critical to the maturation of intestinal function just prior to hatching (Black and Moog, 1978). Many other examples of differentiation, maturation, and growth effects of TH have been studied in both birds and mammals, but no unifying theme has been revealed (Schwartz, 1983). The induction of specific enzymes by T3 has been studied in a number of systems and some, such as malic enzyme, seem promising as model systems for understanding developmental events (rats: Barton and Bailey, 1987; chicken embryos: Goodridge, 1983).

3. Effects on Metabolism

Circumstantial evidence that TH are important in the development of thermoregulatory ability is provided by the general correlations between the patterns of serum TH and the patterns of thermoregulatory development in precocial and altricial birds and mammals (Fisher et al., 1977; McNabb et al., 1984a). Studies of the effects of TH and thyroid inhibitors on neonates have provided more direct evidence of the roles of TH in the initiation of thermogenic responses to cooling (e.g. Freeman, 1970). In precocial species the perinatal surge of TH is closely associated with the initiation of thermoregulatory responses that occur shortly after hatching or birth. In altricial species, TH remain low in early postnatal life then increase approximately coincident with beginning thermoregulatory responses (Fisher et al., 1977; McNabb et al., 1984a). However, these correlations are general and have not been addressed by specific, detailed studies of the role of TH in stimulating thermogenic responses over a range of different developmental ages. Current studies of 5′D in thermogenic tissues such as brown fat (Wu et al., 1986a) during thermoregulatory development should help to develop this area.

III. AGING AND THYROID FUNCTION IN BIRDS AND MAMMALS

Studies of thyroid function and aging in birds and mammals have been dominated by work on humans and rats with little focus on birds. Much of the work suffers from a problem common in studies of aging, namely that only "old" vs "young" groups have been studied, so maturation cannot be separated from aging. In this review, the information on maturation is included with development.

The literature on the thyroid and aging appears to provide general answers to several questions. I will present those first, then review the data under the same categories used above for development. (1) Is aging simply the reversal of development? No; the patterns of thyroid function during aging do not suggest that aging is simply a regression or reversal of earlier developmental events. (2) Do alterations in thyroid function play an important causal role in aging? Probably not; changes in thyroid function with age are relatively subtle and it is questionable whether they should be considered outside the range of "normal" thyroid function. There is no evidence that either hypo- or hyperthyroid conditions as such are associated with acceleration of aging or with decreases in longevity. (3) Is there evidence that a breakdown of the regulatory (HPT) axis control of thyroid function occurs and is a contributor to aging? Probably not; there are subtle changes in the function of the HPT axis with aging but there is no evidence that these constitute an important breakdown of regulatory function that plays a causal role in aging. (4) Does the aging process produce changes in thyroid function? Perhaps, but there is no adequate way, with the data available, to judge this vs. the question asked in (2) above.

A. The Thyroid Gland

Studies in humans and rats are in agreement that there are alterations in TG size (either larger or smaller), increases in TG nodularity, increases in TG connective tissue and decreased TG radioiodine uptake in the aged (Sartin, 1983). In rats non-functional "cold" follicles probably result from aging effects on the cell cytoskeleton and resultant failure of endocytotic processes involved in TH release (Gerber et al., 1987). Decreased TG-TH production in humans is suggested by decreases in T4 clearance (with no change in serum T4), but studies of TG responsiveness indicate this is probably a response to the peripheral TH dynamics rather than an inadequacy of the TG (Minaker, et al., 1985). Overall, these changes in TG structure and function may not be physiologically important, because they are not associated with general changes in thyroid status or with marked alterations in serum TH (Florini and Regan, 1985; Robuschi et al., 1987).

There do not appear to be any studies of TG structure with aging in birds. A number of references suggest that there is a decrease in the thyroid secretion rate with "advancing age" in chickens. However, all the references cited in association with these statements are to studies of juvenile birds up to 100 days of age or less (i.e. about half the age at which chickens reach sexual maturity), some with comparisons to adults of undefined ages. Such results as these bear on development and maturation, not aging.

B. Control of the Thyroid

In humans, studies of the elderly have often shown elevated serum TSH although this has not been a consistent finding. Individuals with markedly elevated TSH often have autoimmune thyroid disease (Robuschi et al., 1987). Harman et al. (1984) studied young, middle and aged adult men who were medically well-characterized from their participation in a longevity study and were free of non-thyroidal illness. They found significantly elevated serum TSH concentrations with aging although there were no individuals with primary hypothyroidism (decreased total and free serum TH). Sensitive evaluations HPT axis function (serum TSH and TH changes after TRH administration), have shown decreased responses of the HPT axis in some studies including Harman's where hypothyroidism and non-thyroidal illness were not factors. However, the daily turnover rate of TSH in the elderly does not differ from that in younger subjects (Florini and Regan, 1985; Minaker et al., 1985). The general consensus seems to be that HPT axis functions are intact in the elderly, except perhaps in the context of specific disease conditions.

In rats, the pituitary content of TSH does not change with age but the picture of serum TSH concentrations is inconsistent from study to study (no change or decrease with aging). The pituitary response to TRH is unchanged with age but there is evidence that the TG response to TSH is decreased with age (Minaker, et al., 1985). Studies of cold exposure add even more complexity. The TSH response to cold is decreased with aging in rats (Tang et al., 1986) but Gambert and Barboriak (1982) found that aged rats were better able to survive cold exposure than were the young.

One study that compared immature chickens (13-17 wks) with retired breeders (20-25 mon) found pituitary responses to TRH injection (with increases in cAMP-dependent protein kinase activity) in the young birds but not in the old birds (Carr and Chiasson, 1983). However, the TG in the old birds showed elevated cAMP-dependent protein kinase content in the same experiment, a paradoxical result considering the apparent lack of pituitary response.

C. Peripheral Thyroid Function

Although some aspects of aging seem similar to the symptoms of hypothyroidism, the ideas that either thyroid deficiency plays an important causal role in aging or alternatively that aging produces a hypothyroid state, have not been fruitful. Early clinical studies argued for a general tendency toward hypothyroidism in the elderly, but these data often were skewed by the inclusion of individuals with illnesses that suppress thyroid function. Recent studies have attempted to compare the "healthy" elderly with younger subjects. In humans, recent studies indicate that serum total T4 and free T4 concentrations are not altered with aging (Robuschi et al., 1987). Serum total T3 and free T3 concentrations are decreased significantly, but when well characterized older groups are compared with younger groups the values for the older groups are still within the range usually considered normal (Harman et al., 1984). However, depressions of mean serum T3 concentrations by 10-20%, as have been reported in a number of studies, could be important functionally if the condition existed for years. Decreases in T4 clearance in the aged in combination with other features of the TH profile, have been used to infer that there are decreases in both TG-T4 production and peripheral T4 to T3 conversion. However, the serum TH picture indicates peripheral hormone availability within the normal range, so hypothyroidism clearly is not a general characteristic of the elderly. Age associated increases in the incidence of primary hypothyroidism in humans do occur and vary considerably with the populations of different countries (Florini and Regan, 1985; Minaker et al., 1985; Robuschi et al., 1987).

Most studies in rats indicate that serum total T4, free T4, total T3 and free T3 all decrease in old age in this species (Sartin, 1983; Minaker et al., 1985). However, when young adult and middle aged adult rats were compared in a study of the effects of diet and age on hormone dynamics, it was found that the decrease in serum TH were correlated with body weight but not with age (Jang and DiStefano, 1985). Body weight does not seem to have been considered in the interpretation of most thyroid/aging studies. There are conflicting reports of normal, increased and decreased conversion of T4 to T3 with aging in rats. Most of these studies measured deiodinase activity *in vitro* in tissues such as liver and kidney (Minaker et al., 1985). Kinetic analyses indicate that whole body T4 to T3 conversion increases almost 3-fold in middle-aged compared to young adult rats although indices of T4 metabolism do not change (Jang and DiStefano, 1985). These authors suggest that erroneous conclusions can result from studies of deiodinase in only fast-exchanging tissues like liver and kidney without studying slow-exchanging tissues such as muscle, which kinetic studies suggest make much greater contributions to the whole body T3 pool. This study does not change the general conclusion that thyroid function is "normal" or adequate to meet the needs of older animals. However, it puts additional emphasis on the importance of TH regulation by peripheral T3 production in determining thyroid status in older animals (Jang and DiStefano, 1985). Analyses of this type do not appear to have been done on truly aged individuals (e.g. 2 yrs of age).

No studies have compared thyroid function in young, middle-aged and old birds. However, an examination of serum TH data in cases where the actual ages of adult birds are given does not suggest that there are consistent or conspicuous changes in thyroid function with aging. We followed serum TH concentrations in Japanese quail hens for 56 weeks after the onset of laying in a study of the effects of dietary iodine on thyroid function (McNabb et al., 1985). There was no evidence of a consistent pattern of changes in serum TH with increasing age in any of the treatment groups.

D. Effects of Thyroid Hormones During Aging

1. T3 Receptors

There may be alterations in the numbers or characteristics of T3 receptors with aging and such changes could have important ramifications for the physiological effects of TH in target tissues. The studies cited below are illustrative. TH binding sites were investigated in mononuclear blood cells from healthy young (16-30 years), middle-aged (31-60) and old (61-90) adults free of non-thyroidal illness. There were no differences between the age groups in serum TSH but serum free T3 was decreased with each age increment. The maximal specific binding capacity of T3 receptors decreased between the young and middle group (maturational change) but did not differ between the middle and old groups (i.e. no "age" effect). The association constant of the T3 receptor did not differ between the young and middle groups but decreased significantly between the middle and older groups (i.e. an "age" effect; Kvetny, 1985). The complexity of the differences that may be attributed to maturation vs. those that may be attributed to aging are well illustrated by the inclusion of three adult groups in this study. In rats, Margarity et al. (1985) have found that young (2 month old; newly sexually mature) vs. old (24 months old; median lifespan) adults the in vivo nuclear binding of T3 (in liver and brain) did not differ with age (serum T3 also did not differ), but the proportion of bound T3 arising from peripheral conversion of T4 to T3 was depressed. The authors conclude that these and other results from their studies indicate a suboptimal thyroid state with aging due to alterations in production, binding and processing of hormones. Latham and Tseng (1985) found that although the affinity constant of the T3 receptor was conserved with age, some tissues (heart and cerebellum) showed decreases in receptor numbers during post-pubertal aging in rats.

2. Effects on Metabolism

Evaluations of metabolism in elderly humans indicate that although the basal metabolic rate (BMR) decreases with age, when age-associated decreases in muscle mass are accounted for BMR does not change (Robuschi et al., 1987). Thus, the minor alterations in serum TH (no change in serum T4, minor decreases within the normal range of T3) do not appear to play an important role physiologically. Studies of rats suggest that the pituitary TSH response to cold may be impaired in the aged (Tang et al., 1986). However, the physiological significance of this is not clear because there were no differences in serum TH response to cold between any of the age groups in that study. This apparent lack of an aging effect in the peripheral TH response to cold and the work of Gambert and Barboriak (1982) which showed better survival of cold exposure in old rats than young rats, suggest that thermogenic responses are not decreased with aging.

A variety of cellular systems have been investigated with respect to possible thyroid/aging interactions. For example, human fibroblasts cultered from aged vs. young patients showed aging associated alterations in T3 induction of cell membrane Na-K ATPase activity (Guernsey et al., 1986). Likewise, rat thymocytes show age related decreases in cell responsivity (of sugar uptake) to TH (Segal and Troen, 1986). To date, unifying themes about the nature of TH effects on cell responsivity are lacking.

IV. CONCLUSIONS

Thyroid development occurs in different patterns in precocial vs. altricial modes of development in both birds and mammals. The glands of the HPT axis develop autonomously, then establish interactive axis control, followed by an increase in general thyroid activity. In precocial species this maturation and increased activity occurs during embryonic/fetal life, in altricial species it occurs after hatching/birth. Thyroid hormones are important in normal growth and differentiation.

Aging in mammals is associated with only very subtle changes (tendency toward decreases in some indices) in thyroid status or in the function of the HPT axis. (In birds the information is extremely limited and inconclusive). Peripheral TH availability appears to be within the "normal" range in aged animals. Some studies indicate higher incidence of individuals with hypothyroidism in aged populations. Changes in thyroid function do not seem to play an important role in aging, nor does aging seem to play a key role in altering thyroid function.

REFERENCES

Astier, H. (1980). In *Avian Endocrinology.* (Epple, A. and Stetson, M. H., eds.), pp. 167-189. Academic Press, New York.

Barton, C. H., and Bailey, E. (1987). *J. Dev. Physiol.* **9**, 215-224.

Bellabarba, D., and Lehoux, J-G. (1981). *Endocrinology* **109**, 1017-1025.

Bellabarba, D., and Lehoux, J-G. (1985). *Mech. Aging Dev.* **30**, 325-331.

Bellabarba, D., Fortier, S., Belisle, S., and Lehoux, J-G. (1984). *Bio. Neonate* **45**, 41-48.

Black, B. L., and Moog, F. (1978). *Dev. Biol.* **66**, 232-249.

Borges, M., LaBourene, J., and Ingbar, S. H. (1980). *Endocrinology* **107**, 1751-1761.

Carr, B. L., and Chaisson, R. B. (1983). *Gen. Comp. Endocrinol.* **50**, 18-23.

Cooke, P. S., Yonemura, C. V., and Nicoll, C. S. (1984). *Endocrinology* **115**, 2059-2064.

Dainat, J., Bressot, C., Rebiere, A., and Vigneron, P. (1986). *Gen. Comp. Endocrinol.* **62**, 479-484.

Daugeras, N., Brisson, A., Lapointe-Boulu, F., and Lachiver, F. L. (1976). *Endocrinol.* **98**, 1321-1331.

Decuypere, E., Kuhn, E. R., Clijmans, B, Nouwen, E. J., and. Michels, H. (1982). *Gen. Comp. Endocrinol.* **47**, 15-17.

Devaskar, U., Church, J. C., Chechani, V., and Sadiq, F. (1987). *Biochem. Biophys. Res. Commun.* **146**, 524-529.

Fisher, D. A., Dussault, J. H.. Sack, J., and Chopra, I. J. (1977). *Recent Progr. Horm. Res.* **33**, 59-116.

Ferreiro, B., Bernal, J., and Brian, J. (1987). *Acta Endocrinol.* **116**, 205-210.

Florini, J. R., and Regan, J. F. (1985). *Rev. Biol. Res. Aging* **2**, 227-250.

Freeman, B. M. (1970). *Comp. Biochem. Physiol.* **33**, 219-230.

Fukiishi, Y., and Hasegawa, Y. (1985). *Acta Endocrinol.* **110**, 95-100.

Galton, V. A., and Hiebert, A. (1987). *Endocrinology* **120**, 2604-2610.

Gambert, S. R., and Barboriak, J. J. (1982). *J. Gerontol.* **37**, 684-687.

Gerber, H., Peter, H. J., and Studer, H. (1987). *Endocrinology* **120**, 1758-1764.

Gonzales, L. W., Ballard, P. L., Ertsey, R., and Williams, M. C. (1986). *J. Clin. Endocrinol. Metab.* **62**, 678-691.

Goodridge, A. G. (1983). In *Molecular Basis of Thyroid Hormone Action*. (J. H. Oppenheimer and H. H. Samuels, eds.), pp. 246-263. Academic Press, New York.
Guernsey, D. L., Koebbe, M., Thomas, J. E., Myerly, T. K., and Zmolek, D. (1986). *Mech. Ageing Dev.* **33**, 283-293.
Harman, S. M., Wehmann, R. E., and Blackman, M. R. (1984). *J. Clin. Endocrinol.* **58**, 320-326.
Hughes, T. E., and McNabb, F. M. A. (1986). *J. Exp. Zool.* **238**, 393-399.
Iglesias, R., Fernandez, J. A., Mampel, T., Obregon, M. J., and Villarroya, F. (1987). *Biochim. Biophys. Acta.* **923**, 233-240.
Jang, M., and DiStefano, J. J. III. (1985). *Endocrinology* **116**, 457-468.
Kaplan, M. M. (1986). In *Thyroid Hormone Metabolism*. (G. Henneman, ed.), pp. 231-253. Marcel Dekker Inc., New York.
Kawaoi, A., and Tsuneda, M. (1985). *Acta Endocrinol.* **108**, 518-524.
King, D. B., and May, J. D. (1984). *J. Exp. Zool.* **232**, 453-460.
Klein, A. H., and Fisher, D. A. (1980). *Endocrinology* **106**, 697-701.
Klein, A. H., Oddie, T. H., and Fisher, D. A. (1978). *Endocrinology* **103**, 1453-1457.
Kvetny, J. (1985). *Horm. Metab. Res.* **17**, 35-38.
Latham, K. R., and Tseng, Y. C. L. (1985). *Age* **8**, 48-54.
Legrand, J. (1986). In *Thyroid Hormone Metabolism*. (G. Henneman, ed.), pp. 503-534. Marcell Dekker Inc., New York.
Margarity, M., Valcana, T., and Timiras, P. (1985). *Mech. Ageing Dev.* **29**, 181-189.
McCann, U. D., Shaw, E. A., and Kaplan, M. M. (1984). *Endocrinology* **114**, 1513-1521.
McNabb, F. M. A. (1987). *J. Exp. Zool. Suppl.* **1**, 281-290.
McNabb, F. M. A. (1988). In Press, *Amer. Zool.*
McNabb, F. M. A., and Cheng, M-F. (1985). *Gen. Comp. Endocrinol.* **58**, 243-251.
McNabb, F. M. A., and McNabb, R. A. (1977). *The Auk* **94**, 736-742.
McNabb, F. M. A., Stanton, F. W., and Dicken, S. G. (1984a). *Comp. Biochem. Physiol.* **78A**, 629-635.
McNabb, F. M. A., Stanton, F. W., Weirich, R. T., and Hughes, T. E. (1948b). *Endocrinology* **114**, 1283-1244.
McNabb, F. M. A., Blackman, J. R., and Cherry, J. A. (1985). *Domestic Anim. Endocrinol.* **2**, 25-34.

McNichols, M. J., and McNabb, F. M. A. (1988). In Press, *Gen. Comp. Endocrinol.*
Minaker, K. L., Meneilly, G. S., and Rowe, J. W. (1985). In *Handbook of the Biology of Aging*. (C. Finch, and L. Hayflick, eds.), Second ed., pp. 433-456. Van Nostrand Reinhold, New York.
Nathanielsz, P. W. (1976). In *Monographs in Fetal Physiology*, **Vol. 1**, North-Holland Publ. Co., New York.
Nwosu, U. C., Kaplan, M. M., Utiger, R. D., and Delivoria-Papadopulos, M. (1978). *Am. J. Obstet. Gynecol.* **132**, 489-494.
Oliver, C., Giraud, P., Lissitzky, J. C., Conte-Devolx, B. and Gillioz, P. (1981). *Endocrinology* **108**, 179-182.
Oppenheimer, J. H., Schwartz, H. L., Mariash, C. N., Kinlaw, W. B., Wong, N. C. W., and Freake, H. C. (1987). *Endocrine Rev.* **8**, 288-308.
Perez-Castillo, A., Bernal, J., Ferreiro, B., and Pans, T. (1985). *Endocrinology* **117**, 2457-2461.
Pic, P., and Bouquin, J. P. (1985). *J. Dev. Physiol.* **7**, 207-214.
Robuschi, G., Safran, M., Braverman, L. E., Gnudi, A., Roti, E. (1987). *Endocrine Rev.* **8**, 142-153.
Romanoff, A. L. (1960). *The Avian Embryo.* Macmillan, New York.
Sartin, J. L. (1983). *Rev. Biol. Res. Aging* **1**, 181-193.
Scanes, C. G., Harvey, S., Marsh, J. A., and King, D. B. (1984). *Poult. Sci.* **63**, 2062-2074.
Scanes, C. G., Lauterio, T. J., and Perez, F. M. (1986). *IRCS Med. Sci.* **14**, 515-518.
Schwartz, H. L. (1983). In *Molecular Basis of Thyroid Hormone Action.* (J. H. Oppenheimer and H. H. Samuels, eds.), pp. 413-444, Academic Press, New York.
Segal, J., and Troen, B. R. (1986). *J. Endocrinol.* **110**, 511-515.
Tanabe, Y. (1965). *Poult. Sci.* **44**, 591-596.
Tang, F., Fung, K. B., Lo, Y. M., and Man, S. Y. (1986). *Horm. Metab. Res.* **18**, 70-71.
Theodoropoulos, T. J. (1985). *Endocrinology* **117**, 1683-1686.
Thommes, R. C. (1987). *J. Exp. Zool. Suppl.* **1**, 273-279.
Thommes, R. C., Vieth, R. L., and Levasseur, S. (1977). *Gen. Comp. Endocrinol.* **31**, 29-36.
Thommes, R. C., Martens, J. B., Hopkins, W. E., Caliendo, J., Sorrentino, M. J., and Woods, J. E. (1983). *Gen. Comp. Endocrinol.* **51**, 434-443.
Thommes, R. C., Clark, N. B., Mok, L. L. S., and Malone, S. (1984). *Gen. Comp. Endocrinol.* **54**, 324-327.

Valcana, T., and Timiras, P. S. (1978). *Molec. Cell. Endocrinol.* **11**, 31-41.
Weirich, R. T., and McNabb, F. M. A. (1984). *Gen. Comp. Endocrinol.* **53**, 90-99.
Wentworth, B. C., and Ringer, R. K. (1986). In *Avian Physiology.* (P. D. Sturkie, ed.), 4th ed., pp. 452-465 Springer-Verlag, New York.
Wittmann, J., Kluger, W., and Petry, H. (1983). *Comp. Biochem. Physiol.* **75A**, 379-384.
Wu, S. Y., Polk, D. H., and Fisher, D. A. (1986a). *Endocrinology* **118**, 1334-1339.
Wu, S. Y., Polk, D. H., Klein, A. H., and Fisher, D. A. (1986b). *J. Dev. Physiol.* **8**, 43-47.

19

THE ADRENALS[1]

Rocco V. Carsia

Department of Animal Sciences
Rutgers - The State University
New Brunswick, New Jersey 08903

Sasha Malamed

Department of Anatomy
University of Medicine and Dentistry of New Jersey -
Robert Wood Johnson Medical School
Piscataway, New Jersey 08854

I. INTRODUCTION

Steroid-secreting (adrenocortical) cells of the interrenal gland and adrenal cortex, and catecholamine-secreting cells of some paraganglia (adrenomedullary chromaffin tissue) demonstrate an increasing anatomical and functional association with the evolutionary ascent of the vertebrate classes

---

[1] Supported by United States Department of Agriculture Grant No. 85-CRCR-1-1846, National Institutes of Health Grant Nos. RR-5576, AG00468, New Jersey American Heart Association Chapters, Foundation of the College of Medicine and Dentistry of New Jersey, Rutgers University Research Council Grants, 1982, 1983, 1985 and the New Jersey Agricultural Experiment Station Project Nos. NJ06109, NJ06514.

(Chester Jones, 1957, 1976; Chester Jones and Phillips, 1986; Gorbman, 1986). Thus, in the elasmobranchs and a few bony fishes, steroid secreting cells and catecholamine-secreting cells are separated, whereas in mammals, these two cell types are within a single structure. However, the selective advantages underlying the evolutionary survival of this association are poorly understood. Among the important reasons for studying the structural and functional relationships of these disparate tissues during ontogeny and senescence is that the functional advantages of their increasingly close anatomic relationship might be revealed. However, although there is a growing body of information on the cytological and functional features of these tissues in most vertebrate classes, information on structural and functional changes accompanying ontogeny, maturation (embryonic and postembryonic development) and senescence is very spotty.

The scope of this chapter is to discuss recent representative work (over approximately the past 15-20 years) on the ontogeny, maturation and sensescence of the vertebrate adrenal gland with emphasis on events occurring at the cellular level. As expected, some vertebrate classes are not represented under various topics, whereas others, most notably Aves and Mammalia, are heavily represented. In the DISCUSSION AND CONCLUSIONS (Section IV), conserved themes of functional change with age are discussed.

II. ADRENOMEDULLARY CHROMAFFIN TISSUE

A. Ontogeny and Maturation

1. Structural Features

Detailed structural and cytological studies on the ontogeny and maturation (embryonic and postembryonic development) of adrenomedullary chromaffin tissue are scant. However, these investigations suggest a conserved pattern.

During the embryonic period of the amphibian, *Bufo bufo*, there is a cephalo-caudal gradient of differentiation of catecholamine-containing cells (Accordi *et al*., 1975; Accordi and Milano, 1977). During this period, the cells stain faintly, suggesting low functional activity, and two cell types [norepinephrine (NE)- and epinephrine (E)- containing cells] are not distinguishable. However, histochemical and ultrastructural evidence indicates that during the postembryonic period, NE- and E- containing cells are distinguishable at early metamorphosis, that differentiation of cell types is completed at metamorphic climax, and that NE- containing cells predominate; this NE predominance in the gland carries over into adult life (Ostlund, 1954; Piezzi, 1965).

In avian and mammalian species, caudal thoracic populations of neural crest cells are destined to become adrenomedullary chromaffin cells (Coupland, 1965). Elegant experiments with quail-chick chimeras (LeDouarin and Teillet, 1971; Teillet and LeDouarin, 1974) have shown that a specific section of neural crest, from somites 18-24, are destined for the adrenal gland. Along their course of migration, these primitive sympathetic cells acquire increasing transmitter specificity and lose much of their mutability (Section II. A,2). The primitive sympathetic cells and their progeny are thought to interact with somitic and ventral neural tubular cells which direct them to the cortical primordia (Norr, 1973; LeDouarin and Teillet, 1974; Black, 1982). Once intermingled or centralized within cortical elements, the cells undergo close contact, form clumps (Hall and Hughes, 1970), and tranform into definitive chromaffin cells (Coupland, 1965), probably through induction by unknown factors in the intercellular fluid of the adrenal gland (LeDouarin and Teillet, 1974). In the domestic fowl (*Galus gallus domesticus*), unmyelinated nerve fibers become associated with differentiating chromaffin cells at day 12 of embryonic life (Hall and Hughes, 1970). Changing ultrastructural features of avian neural crest cells (Mastrolia and Manelli, 1969) along the course of migration and during differentiation to definitive chromaffin cells, share many characteristics with those of amphibian cells. Thus, primitive sympathetic cells (lacking chromaffin granules) presumably transform into pheochromoblasts (initiation of chromaffin granule formation) which in turn, transform into pheochromocytes (definitive chromaffin cells having abundant rough endoplasmic reticulum, voluminous Golgi membranes with terminal dilatations containing electron-dense material, and abundant secretory granules). There is some suggestion that initially NE granules form by the coalescence of small granules with a unit membrane-bound vacuole (Hall and Hughes, 1970); presumably this initial mode of granule formation is replaced by budding of Golgi terminal dilatations.

The aforementioned pattern of cytological differentiation changes very little with mammalian species. However, there are some interesting features related to the increased centralization of the adrenomedullary chromaffin cells within the adrenal primordia in mammalian species that deserve consideration.

Within the Class Mammalia, there is considerable variation in the age of gestation at which the precursor chromaffin cells invade the cortical primordia. What is quite consistent however is the characteristic size of the fetus when this event occurs, about 14-20 mm crown-rump length (Carmichael et al., 1987). An exception is the marsupial, the opossum (Didelphus virginiana), in which the invasion and subsequent differentiation of catecholamine-secreting cells occurs postnatally, with chromaffin cell maturation occurring by day eight (Carmichael et al., 1987; Spagnoli et al., 1987). Invasion and centralization do not necessarily confer mature functional status (Section II. A,2). Unified cellular compaction of the chromaffin tissue may be necessary for normal secretory function that will carry over to adult life (Yeasting, 1986).

Coherent with the chromaffin cell centralization and adrenal gland zonation in mammals is the dependency of chromaffin cell maturation on location. For example, in several mammals during development, ostensibly differentiated adrenomedullary chromaffin cells contain both NE and E granules [human fetus ((Hervonen, 1971); rabbit (Coupland and Weakley, 1968); rat (El-Maghraby and Lever, 1980)]. However, in several mammals after final centralization, two cell types are apparent, each with its own specific type of secretory granule; the dark granules contain NE and the light granules contain E (Hopwood, 1971). NE-containing cells are located adjacent to cortical cells and E-containing cells are located more centrally (Pelto-Huikko et al., 1985). This localization is paradoxical because blood flow in the adrenal gland is centripetal (Harrison and Hoey, 1960), and the E-forming enzyme, phenylethanolamine N-methyltransferase (PNMT) is induced by glucocorticoids (Wurtman and Axelrod, 1966). However, a recent study suggests 1) a lack of direct communications between cortical and medullary sinusoids, 2) that the medullary arteries, which arborize at the cortico-medullary junction, are the predominant blood supply to the medullary sinusoids, and 3) that the cortical sinusoids drain directly into radicles of the central vein (Kikuta and Murakami, 1982). Accordingly, chromaffin cells located more centrally would receive a greater concentration of glucocorticoids than cells at the periphery, and this postulate is consistent with the localization of adrenomedullary cell types. In this connection it is interesting to note that although glucocorticoids can regulate E-containing cell function (Wurtman and Axelrod, 1966; Pohorecky and Wurtman, 1971; Ciaranello et al., 1976), there is no developmental evidence that a chromaffin cell-cortical cell intercellular relationship affects E-containing cell differentiation from pheochromocytes (El-Maghraby and Lever, 1980).

Preceding chromaffin cell differentiation into NE- and E-containing cell types is the development of synapses with preganglionic sympathetic terminals (Coupland and Weakley, 1968; Hervonen, 1971; El-Maghraby and Lever, 1980); axonal terminal dilation followed by the appearance of agranular or clear (cholinergic) vesicles precede the development of pre- and postsynaptic membranes (Hervonen, 1971). The mature pattern of synaptic connection may be different for NE- and E- containing cells. For example, in the hamster adrenal gland a NE-containing cell receives several terminals, whereas an E-containing cell receives a single terminal such that the synaptic area on E-containing cells is about one half that on NE-containing cells (Grynszpan-Winograd, 1974). Clearly, the ubiquity of this morphologic dichotonomy in synaptic form among mammals (and other vertebrates as well) awaits investigation. There is some evidence that the rat, rabbit and cat may also have this feature. One can speculate that this dichotomy in synaptic form may play a role in the maturation of pheochromocytes into NE- and E-containing cells, and underlies the differential release of NE and E observed in several mammalian species subjected to various stressors (Grynszpan-Winograd, 1974).

2. Functional Features

The function of adrenomedullary chromaffin tissue during ontogeny and maturation of vertebrates below Aves is poorly understood. However, there is information on the functional changes that destined progenitor cells undergo from embryogenesis to definitive expression and modulation of catecholamine secretion (reviewed by Black, 1982). Early expression of catecholamines (histofluorescence) by migrating progenitor cells is strongly influenced by, but not totally dependent upon, critical interactions of the progenitor cells with ventral neural tube and somitic mesenchyme. New characters of catecholaminergic expression are added to the secretory repertoire of the differentiating progenitor cells as they infiltrate the cortical primordia, such that a definitive expression develops, that is, the addition of an adrenergic phenotype to the existing noradrenergic phenotype. Here again, initial expression of PNMT, which converts NE to E, is not dependent on glucocorticoid secretion by the cortical primordia. For example, at day 17 of gestation in the rat, the expression of PNMT appears to occur independent of glucocorticoid action. However, subsequent increases in PNMT during neonatal and postnatal life, to provide sufficient quantities of epinephrine for proper function, are dependent upon glucocorticoid action and are regulated by the pituitary-adrenal axis. Thus, although morphogenetic processes, such as migration, cellular interaction, mitosis, adrenal centralization and cellular compaction, strongly influence the molecular species of catecholamine secreted, the relationship between morphogenetic processes provides enough developmental plasticity to prevent catastrophic derangement of adrenomedullary chromaffin cell function under abnormal circumstances.

Although in mammals morphologic synapses are formed between preganglionic sympathetic axonal terminals and chromaffin cells prior to gestational differentiation of cells into NE- and E-containing cells (Coupland and Weakley, 1968; Hervonen, 1971; El-Maghraby and Lever, 1980), a functional innervation is not demonstrable until late gestation, neonatal or postnatal periods. Prior to the establishment of a functional innervation, stress-related circulating factors directly regulate catecholamine (predominantly NE) secretion (Lagercrantz and Slotkin, 1986). For example in the human, complete maturation of adrenomedullary chromaffin function occurs during the first three years of life (Stanton and Woo, 1978), and during fetal and neonatal life, catecholamines are secreted in direct response to stress (hypoxia, hypoglycemia) apparently without much participation from the medullary innervation (Phillipe, 1983; Lagercrantz and Slotkin, 1986). In general, in altricial mammals (rat) this functional innervation-independent mechanism of secretion is maximal late in gestation and declines thereafter, but extends into neonatal and postnatal periods, whereas in precocial mammals (cow, horse, sheep), it is maximal earlier in gestation and declines thereafter to functional insignificance by birth. In fact, maturation of a functional innervation induces the demise of the functional innervation-independent mechanism (Lagercrantz and Slotkin, 1986).

B. Senescence

1. Structural Features

Information on the structural alterations in adrenomedullary chromaffin tissue with senescence is limited to mammals, namely the rat. Nodular hyperplasia (Tischler et al., 1985) and pleomorphic hyperplastic medullary cells with increased incidence of pheochromocytomas (Wexler, 1981) have been reported in aged rats. Ultrastructural studies reveal hypertrophy of the Golgi apparatus and rough endoplasmic reticulum, formation of giant mitochondria, and an abundance of secretory vesicles in the cytoplasm (Shaposhnikov, 1985). In addition, there is ultrastructural evidence for an enhanced state of secretory function that has a decreased dependence on hypothalamic control (Shaposhnikov and Bezrukov, 1985). Thus overall, the structural features suggest an enhanced secretory function with senescence.

2. Functional Features

The general view that aging increases adrenomedullary chromaffin tissue function in vertebrates, is based on studies of a limited number of vertebrate species. In the reptile, the soft-shelled turtle (*Lissemys punctata punctata*), aging does not affect diurnal variation of adrenal NE and E content but does increase the peak diurnal values of catecholamine content (Mahapatra *et al*., 1987). In addition, an age-related increase in sympathoadrenomedullary activity has been demonstrated in a large number of avian species (Mahata and Ghosh, 1986). Furthermore, with few exceptions, senescence in mammals has been reported to increase levels of circulating catecholamines and adrenomedullary chromaffin tissue dopamine β-hydroxylase activity [rat, Chinese hamster, Mongolian gerbil (Banerji *et al*., 1984)], to increase preganglionic sympathetic neuronal activity and adrenal secretion of NE and E [rat (Ito *et al*., 1986)], to increase stress-induced (cellular glucoprivation, intermittent foot shock, cold) plasma NE and E concentrations [rat (Chiueh *et al*., 1980; McCarty, 1984, 1986)]. In addition, delayed post-stress normalization of urinary NE and E following psychosocial stress [man (Faucheux *et al*., 1981)], suggests prolonged adrenal secretion of catecholamines. These reports of increased function with senescence in mammals support the interpretations of microscopic and ultrastructural observations (Section II. B,1).

III. ADRENOCORTICAL (INTERRENAL) TISSUE

A. Ontogeny and Maturation

1. Structural Features of Adrenocortical Tissue of Nonmammalian Vertebrates

Cells that secrete steroids have mitochondria whose inner membranes are invaginated into their matrices as tubes rather than as the familiar lamellae (cristae) characteristic of non-steroidogenic cells (Malamed, 1975; Fawcett, 1981). This is indicated by the presence of circular cross sections of these membranous tubes. Only in the absence of these circular profiles would it be proper to interpret the parallel pairs of membranes that are also present in the matrix as longitudinal sections of lamellae rather than tubes. Unfortunately, this misinterpretation has been made for most of the electron micrographs of embryonic adrenocortical tissue published. Circular profiles appear in the mitochondira of the youngest adrenocortical cells in these studies (for example, Hillman et al., 1975; Stark et al., 1975; Albano et al., 1976) and thus, because available evidence suggests that embryonic cells of even earlier developmental stages do have lamellar cristae (Enders et al., 1967; Stern et al., 1971; Nagele et al., 1987), it appears that the cells shown have already begun to differentiate. Accordingly, the relationship of at least the mitochondrial aspect of morphological differentiation to functional differentiation cannot be evaluated confidently from previous work. Nor can the abundance of lipid droplets or the smooth endoplasmic reticulum (SER) be used for this purpose. The presence of these two characteristic features of adrenocortical cells (albeit, not unique to them) has been reported in developmental ultrastructural studies but seldom quantitatively on the basis of standard sampling procedures. The following review of reports on the ultrastructural aspects of adrenocortical development is presented with these caveats in mind.

Although the reports are few, more is known about the development of vertebrate adrenocortical tissue than about adrenomedullary chromaffin tissue. Again the birds and the mammals have been studied more than the other classes.

Among the cyclostomes, adrenocortical ultrastructure of the lamprey (*Lampetra planeri*) has been studied (Seiler and Seiler, 1973; Seiler et al., 1973). Presumptive adrenocortical cells are reported to undergo ultrastructural changes that suggest steroidogenic function. Few active cell types showing hydroxysteroid dehydrogenase activity are present in the larval stage, but these increase through metamorphosis to adulthood (Seiler et al., 1981) with parallel changes in steroidogenic function (Seiler et al., 1983).

Ultrastructural studies of the salamander (*Hynobius nebulosus*) are reported to reveal striking changes with development (Setoguti et al., 1985). As larval salamanders progress toward metamorphosis, adrenocortical cells become filled with SER and lipid droplets increase. In the young adult there are less SER and mitochondria and the cells fill with lipid droplets. The mitochondria contain crystalloid structures thought to be cholesterol. These features are interpreted as indicating a decrease in steroidogenesis.

Similar alterations are seen in _Xenopus_ _laevis_ during metamorphosis (Hanke and Pehlemann, 1969).

Presumptive adrenocortical cells of the embryonic chick (_Gallus_ _gallus_ _domesticus_) develop interdigitating microvilli and desmosomes; SER and lipid droplets increase (Hall and Hughes, 1970). The presence of cholesterol in the 10-day adrenal cortex (Sivaram, 1969) suggests that steroidogenesis may begin at that time (Sivaram, 1969). As in amphibians these developmental changes are thought to be regulated by ACTH (Kalliecharan, 1981).

2. Structural Features of Mammalian Adrenocortical Tissue

The adrenocortical primordia organize early in gestation. Initially all cells of the primordia have similar ultrastructural characteristics in the golden hamster. Later two zones are established. Cells of the superficial zone (zona glomerulosa) acquire lipid droplets and small Golgi apparatus. Those of the deep zone (presumably the zona fasciculata and zona reticularis) acquire well developed areas of Golgi apparatus, vesicular SER and lysosomes (Hillman _et_ _al_., 1975). Cocomitant with these ultrastructural changes there is an increase in adrenal weight of the fetal rat apparently under the control of ACTH (Dupouy and Dubois, 1975). Albano _et_ _al_. (1976) have presented evidence that in the fetal rabbit SER proliferates before adenylyl cyclase activity is observed. However, Stark _et_ _al_. (1975) report that in the fetal cat the steroidogenic function and responsiveness to ACTH can be detected at a stage when fine structural characteristics of adrenocortical cells are as yet poorly defined.

Postnatally and through maturation there are changes in various parameters of adrenocortical cell organelles (Magalhães _et_ _al_., 1981) some of which are sex-related (Majchrzak and Malendowicz, 1983; Chuvilina and Kirillov, 1984; Nikicicz _et_ _al_., 1984). In the neonatal rat, there is a stress non-responsive period which persists for up to 7 days post partum (Schapiro _et_ _al_., 1962; Milković and Milković, 1963); this state may be related to certain ultrastructural features of adrenocortical cells. Thus, during this and other periods of adrenocortical inactivity the suppression of steroidogenesis might provide the metabolic basis for an accumulation of precursor cholesterol thought to be arranged into paracrystalline structures within lysosomes (Szabó _et_ _al_., 1982) and mitochondria (Murakoshi _et_ _al_., 1985; Setoguti _et_ _al_., 1985). With gradual recrudescence of adrenocortical function after the stress non-responsive period, adrenocortical cells acquire adult expression of glutathione-peroxidase in the cytosol and mitochondria

(Murakoshi *et al.*, 1987). It is thought that this metabolic expression protects cell organelles against lipid peroxidation.

3. Adrenocortical Zonation

Through development of avian species, there is an intermingling of adrenocortical and adrenomedullary tissues (Chester Jones and Phillips, 1986). However, ultrastructural examination reveals a rudimentary zonation of steroidogenic tissues that is similar to that in mammals. Two regions of adrenocortical tissue are apparent: a subcapsular zone containing cells which have irregularly shaped nuclei, relatively little SER and mitochondria resembling those of mammalian zona glomerulosa cells, and an inner zone, containing relatively smaller cells which have rounded nuclei, more abundant SER and mitochondria resembling those of mammalian zona fasciculata and zona reticularis cells (Pearce *et al.* 1978; Holmes and Cronshaw, 1984).

Although the evolutionary basis for mammalian adrenocortical zonation is unclear, evidence from studies *in vivo* and *in vitro* permits some speculation on how the zones may be formed and maintained in the presence of the continual centripetal migration of subcapsular cells (Zajicek *et al.*, 1986). According to this scheme, early steroidogenic response to ACTH is suppressed, possibly by factors recently described [growth factors operating through a protein kinase C-dependent pathway (Mason *et al.*, 1986; McAllister and Hornsby, 1987) and transforming growth factor-ß (Hotta and Baird, 1987)]. At the same time local concentrations of glucocorticoids within the adrenal cortex influence zone formation and differential steroidogenic expression (mineralocorticoids and glucocorticoids) along a corticosteroid concentration gradient established by the centripetal blood flow from cortex to medulla (Hornsby, 1982; Crivello *et al.*, 1983; Hornsby and Crivello, 1983; Dickerman *et al.*, 1984). Thus zone formation depends on the removal of the suppression of steroidogenesis. In the absence of suppression, steroids are provided for the gradient which controls zone formation.

The cells composing the separate zones are continually renewed by a centripetal migration of new cells from the subcapsular region where progenitor cells lie (Bertholet, 1980; Zajicek *et al.*, 1986). Their migration ends in the zona reticularis where they ultimately degenerate and die. During this migration, the cells are exposed to the gradient and are transformed progressively into the different cell types of the adrenocortical zones. The supply of progenitor cells may be neurally controlled; there is mounting evidence for adrenocortical innervation, for its concentration in the subcapsular region, and for its control of hypertrophy after unilateral adrenalectomy (Unsicker, 1969; Engeland and Dallman, 1975; Dallman et al., 1977; Migally, 1979, Malamed, 1984; Kleitman and Holzwarth, 1985a, 1985b). In the dog, however, the progenitor region is thought to lie between the zona glomerulosa (arcuata) and the zona fasciculata with cells migrating both centripetally and centrifugally (Hullinger, 1978).

4. Special Features of Adrenocortical Zonation in Mammals

Mammalian adrenocortical tissue is displayed generally as three concentric zones: zona glomerulosa (nearest the capsule), zona fasciculata (deep to the zona glomerulosa) and zona reticularis (adjacent to the medulla). The variations on this basic structural theme have been reviewed in detail (Bourne, 1949; Chester Jones, 1957, 1976; Chester Jones and Phillips, 1986). In many mammals, the cells of these zones have distinct ultrastructural features, especially their mitochondria (Malamed, 1975). It is postulated that as the actions of mineralocorticoids and glucocorticoids became more specific with the evolution of mammals, the concomitant evolution of adrenocortical zonation provided an increased precision of regulation of the different corticosteroids (Vinson et al., 1979).

Furthermore, some distinct additional zones of the adrenal cortex are found in mammals. These zones are the X-zone of mice (Mus), an intermediate zone between the zona fasciculata and zona reticularis in the Mongolian gerbil (Meriones unguiculatus), the "special" zone of the brush-tailed possum (Trichosurus vulpecula), and the fetal zone of primates. The juxtamedullary X-zone of mice degenerates at puberty in males and in the female after the first pregnancy (Holmes and Dickson, 1971). The only enzyme system detected in the X-zone is 20α-hydroxysteroid dehydrogenase (Ungar and Stabler, 1980). The intermediate zone of the Mongolian gerbil contains cells that have concentric whorls of rough endoplasmic reticulum. These cell are thought to be recruited as actively secreting steroid cells in times of stress (Nickerson, 1977). The "special" zone of the brush-tailed possum is formed only in the female at the time of puberty (Weiss and Ford, 1977), is under FSH control (Weiss, 1984), as is capable of 17α- hydroxylation, 5β-reduction, and 3α- and 20α-hydroxysteroid dehydrogenation (Weiss and Ford, 1982). The fetal zone of primates is much larger than the outer definitive adrenal cortex, and is the source of androgens (dehydroepiandrosterone, dehydroepiandrosterone sulfate) that are precursors for proper estrogen levels during pregnancy (Cameron *et al*., 1969; Yeasting, 1986). In man, after birth, there is a rapid activation of 3β-hydroxysteroid dehydrogenase in the regressing fetal zone such that it is the major source of androstenedione and testosterone in males during the first year of life (Bidlingmaier *et al*., 1986). After birth, the fetal zone undergoes rapid involution. The outer definitive cortex collapses inward compressing the medullary chromaffin tissue, and the remaining fetal zone stromal elements form a zone of connective tissue between the zona reticularis and medulla (Yeasting, 1986).

Apart from these distinct additional zones, modifications of the zona reticularis should be noted. In the guinea pig, the zona reticularis has low corticosteroidogenic capacity (Nishakawa and Strott, 1984), and high Δ^4-hydrogenase activity, possibly for the degradation of cortisol (Martin and Black, 1982, 1983). In humans, several years prior to puberty, the secretion of androgens and sulfated androgens by the zona reticularis increases, an event called adrenarche (Cutler, Jr. and Loriaux, 1980). Adrenarche is closely linked with gonadal maturation. In nonhuman primates, the adrenarchal process is related more to adrenal growth than to sexual maturation (Smail *et al*., 1982). It is interesting to note that in several marmoset monkey species, an androgen-secreting zona reticularis fails to develop (Levine *et al*., 1982).

5. Functional Features

Little information is available on the function of corticosteroid-secreting tissue of piscine species. In lampreys (*Lampetra planeri*, *Petromyzon marinus*), ontogeny and maturation of function is delayed until metamorphosis. Thus, both hydroxysteroid dehydrogenase (HSD) activity (Seiler *et al*., 1983) and *in vitro* corticosteroid production (Weisbart, 1975; Weisbart and Youson, 1975) of presumptive adrenocortical tissue increase with metamorphosis of larval stages to filter feeding stage to final adult parasitic form. However, in advanced piscine species which do not undergo metamorphosis, an increase in adrenocortical function occurs during the transition from embryonic life *in ovo* to post-embryonic life. For example in several trout species having different durations of development (*Salmo gairdneri*, *Salmo trutta*, *Salvelinus fontinalis*), there is a consistent increase in corticosteroid synthesis by embryos approaching hatch (Pillai *et al*., 1974).

Detailed studies on amphibian adrenocortical function

during development are limited to anuran species. Here again, as seen with metamorphic cyclostomes, adrenocortical function is relatively quiescent until metamorphosis, except for a slight increase in embryonic aldosterone production with the posthatching completion of operculum development [*Bufo arenarum* (Castañe *et al.*, 1987)]. Circulating aldosterone levels rise during premetamorphosis, whereas corticosterone and cortisol levels rise during prometamorphosis [*Rana catesbeiana* (Krug *et al.*, 1983)]. These circulating corticosteroid levels peak at metamorphic climax and then decline in the young frog (Jaffe, 1981; Krug *et al.*, 1983). Interestingly, the potential for corticosteroid production (HSD activity) appears and peaks prior to metamorphosis during the posthatching embryonic period (Hsu *et al.*, 1980). The rise in circulating corticosteroids presumably augments the metamorphic actions of thyroid hormones. Aldosterone and corticosterone increase 3,5,3'-triiodothyronine binding to tadpole tail fin tissue, aldosterone being more potent and active than corticosterone (Suzuki and Kikuyama, 1983).

Definitive studies on the ontogeny and maturation of reptilian adrenocortical function are lacking. In contrast, there are many detailed investigations on avian adrenocortical function. However, these involve precocial species, mainly the domestic fowl (*Gallus gallus domesticus*), rather than altricial species.

In the domestic fowl, glucocorticoid concentrations in adrenal tissue (Kalliecharan and Hall, 1976; Marie, 1981; Tanabe *et al.*, 1986), in allantoic fluid (Woods *et al.*, 1971), and in plasma (Kalliecharan and Hall, 1974; Marie, 1981; Tanabe *et al.*, 1986) indicate that adrenocortical function commences about midway through the embryonic period and increases at first gradually and then abruptly, approaching hatch. In addition, alterations in domestic fowl adrenocortical function during the transition from embryonic to postembryonic life have been studied at the cellular level (Carsia *et al.*, 1987). Both basal and maximal ACTH-induced corticosteroid production by isolated adrenocortical cells increases abruptly from day 18 of embryonic life to hatch and then declines gradually thereafter. Other results suggest an abrupt increase in the intracellular pool of available steroidogenic enzymes during the perihatch period. Furthermore, cellular sensitivity to ACTH increases nearly three times from day 18 of embryonic life to 1 day posthatching and then declines. These changes in cellular sensitivity to ACTH appear to be due to alterations in ACTH-cell interaction, since increases in cellular sensitivity to substances which bypass the plasma membrane and increase corticosterone production, 8-bromo-cyclic AMP and 25-hydrxcholesterol, occur at later ages (1 day and 1 week posthatching, respectively) and then remain constant thereafter. Thus, alterations in adrenocortical cell function *per se*, in part, are responsible for alterations in *in vivo* corticosteroid parameters during domestic fowl development and posthatching maturation, and are consistent with experiments with adrenocortical tissue (Gonzalez *et al.*, 1983) and adrenocortical homogenates (Nakamura *et al.*, 1978). Presumably, these changes in adrenocortical function are induced by ACTH (Woods *et al.*, 1971). It is interesting to note that, as seen in developing anuran adrenocortical tissue (Hsu *et al.*, 1980), the appearance and peak (days 3.5 and 8 *in ovo*) of HSD activity in the domestic fowl antedates the secretion of corticosteroids (Domm and Erickson, 1972).

Although availabe information is restricted to eutherian mammals, the evidence suggests that changes in adrenocortical function that accompany the transition from embryonic life (either *in ovo* or premetamorphic stages) to postembryonic or postmetamorphic life of piscine, amphibian and avian apecies are conserved in mammals during development. In several species, as for example, the rat (Dupouy *et al.*, 1975; Nathanielsz, 1978), rabbit (Nathanielsz, 1978; Malee and Marotta, 1982), guinea pig (Dalle and Delost, 1976), sheep (Elsner *et al.*, 1980; Rose *et al.*, 1982; Kitts *et al.*, 1984; Saez *et al.*, 1984), monkey and man (Murphy, 1975), there is an increase in adrenocortical activity near term. The subsequent rise in fetal plasma corticosteroids is a requirement for the maturation of many fetal organ systems (Liggins, 1976) and is instrumental in the initiation of parturition in ruminants (Thorburn and Challis, 1979). A few studies *in vivo* indicate alterations in adrenocortical function during the transition from fetal to neonatal life. There is evidence for an increase in adrenal responsiveness (Wintour *et al.*, 1975) and sensitivity (Rose *et al.*, 1982) to ACTH near term. The results of *in vivo* work are corroborated by *in vitro* work (see review by Saez *et al.*, 1984; Durand *et al.*, 1984). There is evidence for increases in ACTH receptor concentration, and the activity of ACTH-sensitive adenylyl cyclase, stimulatory guanine nucleotide-binding coupling protein, and key steroidogenic enzymes.

Little is known about the development of adrenocortical stress response in vertebrates. Work with avian species (pigeon, domestic fowl) suggests that after hatch there is a stress non-responsive period (Ramade and Baylé, 1980; Freeman and Flack, 1981; Freeman, 1982; Freeman and Manning, 1984) which is similar to that which occurs in the rat after parturition (Schapiro *et al.*, 1962; Milković and Milković,

1963), albeit in avian species it is of shorter duration (about 48 hr.) (Ramade and Baylé, 1980; Freeman, 1982) than in the rat (about 7 days). Work with isolated fowl adrenocortical cells suggests that the stress non-responsive period is not due to diminished adrenocortical responsiveness to ACTH, since cellular corticosteroid production and sensitivity to ACTH are maximal at 1 day posthatch (Carsia *et al.*, 1987). Indeed, *in vivo* work with the domestic fowl (Wise and Frye, 1973; Freeman and Manning, 1984) and the rat (Tang and Phillips, 1977; Guillet and Michaelson, 1978) suggests that the stress non-responsive period is due to a transient failure in the hypothalamo-hypophyseal axis, possible involving inhibition of CRF release (Walker *et al.*, 1986).

Information on the development and maturation of a circadian rhythm of adrenocortical function among the vertebrates is scant. In the domestic fowl, a diurnal rhythm is not apparent until 11 weeks posthatching and does not change to an adult diurnal rhythmic pattern until 17 weeks posthatching (Webb and Mashaly, 1985). In mammals, a diurnal rhythm is not apparent until several weeks [rat (Allen and Kendall, 1967; Hiroshige and Sato, 1970; Honma *et al.*, 1984)] or months [man (Vermes *et al.*, 1980)] after parturition. However, in rats, the diurnal rhythm expressed postpartum may be entrained to the maternal diurnal rhythm during early gestation (Honma *et al.*, 1984). It is generally thought that diurnal pituitary ACTH secretion establishes adrenocortical diurnal rhythm. However, work with mammals suggests that pituitary ACTH may play only a permissive role, whereas other unknown factors, possibly including the adrenocortical innervation, are the actual zeitgeber of adrenocortical diurnal rhythmicity (Meier, 1976; Taylor *et al.*, 1976; Holaday *et al.*, 1977; Ottenweller *et al.*, 1978; Ottenweller and Meier, 1982).

As far as maturation is concerned, little information exists on the alterations in the adrenocortical function of vertebrates below Aves. On the one hand, work with amphibians (Jaffe, 1981; Krug *et al.*, 1983; Setoguti *et al.*, 1985), avian species (Kalliecharan and Hall, 1974; Holmes and Kelly, 1976; Nakamura *et al.*, 1978; Freeman, 1983; Carsia *et al.*, 1985; Carsia and Weber, 1986; Tanabe *et al.*, 1986), and mammals (Dvořák, 1972; Dalle and Delost, 1974; Greiner *et al.*, 1976; Hirose, 1977; Vermes, 1980) suggests a decrease in adrenocortical function to a level that remains fairly constant until senescent changes occur. This is evident at the cellular level. Sensitivity to ACTH of adrenocortical cells from maturing domestic fowl (2 days to 26 weeks posthatching) decreases by one fortieth; this decrease appears to be due to an alteration in a corticosteroidogenic step prior to cAMP formation (Carsia *et al.*, 1985). However, no further decline in cellular sensitivity is apparent up to 2 years posthatching (Carsia and Weber, unpublished observations). On the other hand, maturation does not appear to alter the plasma corticosteroid diurnal rhythm in reptiles [turtle (Mahapatra *et al.*, 1987)], birds (Assenmacher and Jallageas, 1980) and mammals [rat (Allen and Kendall, 1967; Ramaley, 1973)].

B. Senescence

1. Structural Features

Information on the structural alterations in senescent adrenocortical tissue is limited to mammals. Absolute adrenal weight appears to increase (Haensly and Getty, 1968; Tang and Phillips, 1978), although relative adrenal weight remains fairly constant (Haensly and Getty, 1965; Tang and Phillips, 1978). Histological studies show an increase in nodular hyperplasia [man (Kreiner, 1975); rat (Attia, 1985)], irregular zonation [man (Kreiner, 1975)], fatty degeneration [rat (Leathem, 1974)], and infiltration with macrophages [rat (von Seebach *et al.*, 1975)] and fibroblasts [rat (Tang and Phillips, 1978)].

A hallmark of adrenocortical senescence is an increase in lipofuscin bodies within adrenocortical cells, not only in the zona reticularis where they are commonly present before senescence, but also in the zona fasciculata [mouse (Samorajski and Ordy, 1967); Mongolian gerbil (Nickerson, 1979), Golden hamster (Nickerson, 1979) rat (Reichel, 1968; Szabó *et al.*, 1970; von Seebach *et al.*, 1975; Malamed *et al.*, 1981); man (Kreiner, 1975)]. Lipofuscin bodies contain lipid peroxidation products and are thought to be formed in response to free radical generation by degenerating senescent mitochondria (Miquel *et al.*, 1978). Indeed, untrastructural studies reveal changes in mitochondria with senescence. In the rat, from 6 to 24 months of age, there is an increase in mitochondrial volume (161%) and number (68%) per cell which precedes the increase in lipofuscin body volume (60%) and number (150%) per cell (Malamed *et al.*, 1981). In addition, giant mitochondria and mitochondria with lipofuscin and paracrystalline inclusions have been observed (Murakoshi *et al.*, 1985). The occurrence of paracrystalline inclusions (cholesterol), together with an increase in lipid droplet volume (100%) and number (232%) per cell (Malamed *et al.*, 1981) suggests a decrease in function (Szabó *et al.*, 1982). The functional significance of other organelle changes is unclear. These include an increase in agranular (smooth) endoplasmic reticulum [rat: 192% increase from 6 to 24 months of age (Malamed et al., 1981); mouse: increase in lamellar forms (Setoguti et al., 1979)] and an increase in lysosomal volume per cell [rat: 531% increase from 6 to 24 months of age (Malamed et al., 1981)].

2. Functional Features

Here again, as with structural features, studies of functional features of senescent adrenocortical tissue have been limited to mammals. Numerous studies in vivo indicate that basal glucocorticoid secretion is unchanged with advanced age [rat (Hess and Riegle, 1970; Tang and Phillips, 1978; Sonntag et al., 1987); domestic ruminants (Riegle and Nellor, 1967; Riegle et al., 1968)], whereas basal aldosterone [man (Hegstad et al., 1983)] and dehydroepiandrosterone [man (Parker et al., 1983)] secretion decline. In addition, glucocorticoid circadian rhythmicity is unaltered with senescence [rat (Sonntag et al., 1987); man (Touitou et al., 1982)].

The effect of senescence on in vivo adrenocortical response to ACTH or stress is equivocal. Excluding humans, glucocorticoid response to ACTH or stress, in general, declines with advancing age [rat (Hess and Riegle, 1970); domestic ruminants (Riegle and Nellor, 1967; Riegle et al., 1968); dog (Breznock and McQueen, 1970); monkey (Bowman and Wolf, 1969). However, other reports suggest no change [rat (Tang and Phillips, 1978)] or an increase in response [rat (Sonntag et al., 1987)]. In contrast, aldosterone response to ACTH, angiotensin II and K^+ appears to increase in rats (Frolkis et al., 1985). Interestingly, in man, cortisol response to ACTH is unchanged, whereas dehydroepiandrosterone response is depressed (Yamaji and Ibayashi, 1969; Vermeulen et al., 1982; Ohashi et al., 1986) thus suggesting a selective inhibition of zona reticularis function. However, the amount of zona reticularis tissue is unaltered with advancing age (Parker et al., 1983).

In contrast to equivocal results in vivo, results in vitro are quite consistent. Advancing age decreases basal and ACTH-induced corticosterone production by isolated adrenocortical cells [rat (Pritchett et al., 1979; Malamed and Carsia, 1983; Popplewell et al., 1986; Popplewell and Azhar, 1987)], and basal and ACTH-, angiotensin II- and K^+-induced aldosterone production by isolated zona glomerulosa cells [cow (Potter and Goodfriend, 1987)]. However, the relation of senescence to the inhibition of adrenocortical function is not clear. There are reports suggesting a decrease in ACTH receptor concentration and/or function

(Pritchett *et al.*, 1979; Malamed and Carsia, 1983). However, a recent study suggests that ACTH receptor concentration and function is unaltered (Popplewell *et al.*, 1986). The most recent evidence suggests an impairment of postreceptor events, after the formation of cAMP and prior to cholesterol side-chain cleavage, most likely the steps regulating the mobilization of cholesterol for steroidogenesis (Popplewell and Azhar, 1987; Popplewell *et al.*, 1987; Potter and Goodfriend, 1987). Prior to these senescent-related alterations of rat adrenocortical function (which occur at and after 12 months of age), there is a great decrease (about 50%) in the resting membrane potential (becoming more positive at 6 months of age) (Lymangrover *et al.*, 1978).

It is puzzling that reports conflict on adrenocortical function during senescence *in vivo*, whereas *in vitro*, the evidence consistently supports a decrease in function. Possibly *in vivo*, extrapituitary factors, such as adrenomedullary catecholamines (Ottenweller *et al.*, 1978; Engeland *et al.*, 1980), which increase with age (Banerji *et al.*, 1984) and are secreted in greater amounts in response to stress with age (Chiueh *et al.*, 1980; McCarty, 1984, 1986), and the adrenocortical innervation (Ottenweller and Meier, 1982; Edwards *et al.*, 1986) increase adrenocortical sensitivity to ACTH and thus compensate for the intrinsic cellular decrement of function during senescence (Sonntag *et al.*, 1987).

IV. DISCUSSION AND CONCLUSIONS

Little is known about the comparative aspects of the ontogeny, maturation (embryonic and postembryonic development) and senescence of adrenomedullary chromaffin and adrenocortical cell function and structure of vertebrates. Interpretations of the results of work with poikilotherms are hindered because of physiologic adaptations to the environment and relatively complicated life histories. For example, the effects of the seasons causing abatement and recrudescence of adrenocortical cell function, and the physiologic events associated with hibernation, estivation, smoltification, spawning and metamorphosis, that are inextricably linked with adrenocortical cell function, are superimposed on the concomitant changes in cellular function accompanying ontogeny, maturation and senescence. In addition, in some species, life span after sexual maturity is relatively short, and studies of interrenal/adrenal gland function may reveal events associated with the processes of death rather than those of maturation and senescence. In contrast to poikilotherms, homeotherms are more amenable to studies of interrenal/adrenal gland function because in general, they have fewer glandular adaptations to the environment and have a protracted life span after sexual maturity. Thus, among the vertebrates, birds and mammals have been most studied.

Alterations in adrenomedullary chromaffin cell function are highly conserved among the vertebrates. In general, mature pheochromocytes are formed during the embryonic period, albeit in mammals, complete maturation of chromaffin cells may not occur until the postnatal period. In species that undergo metamorphosis, pheochromocyte formation is delayed until early metamorphosis. However, matamorphosis may be regarded as the end of a protracted posthatching embryonic period. With maturation and aging, cellular function increases. In mammals, the advantage of having adrenomedullary chromaffin tissue centralized within the adrenal gland is unclear. Although there is a tendency for greater E secretion in mammals compared to lower vertebrates its preponderance varies among species. Perhaps centralization is more important for inhibiting the transformation of adrenomedullary chromaffin cell secretory function to neuronal function. Such an inhibition might be intrinsic in cells or might be controlled by unknown extrinsic factors in lower vertebrates that have been lost with evolution to mammals. This notion has not been investigated.

Age related alterations in adrenocortical cell function also have been conserved among the vertebrates. There is a consistent increase in cellular function at the end of the embryonic/fetal period, which peaks near hatch or parturition, and then declines. Here again, as with adrenomedullary chromaffin cell function, in species undergoing metamorphosis, this increase in adrenocortical function is delayed until the metamorphic period, peaks at climax, and then declines. Whether the cellular mechanisms operating in avian and mammalian species during the transition from embryonic/fetal life to postembryonic/neonatal life also operate in lower vertebrates awaits investigation. However, an interesting finding is that the expression of HSD activity precedes the secretion of corticosteroids.

Information on the alterations in adrenocortical function with senescence is limited to mammals. Although results _in vivo_ are equivocal, the results _in vitro_ are consistent; they indicate a decline in function. In addition, hypotheses for the significance of adrenocortical-adrenomedullary chromaffin tissue association can be drawn from these studies. It may be that the increase in catecholamine secretion with age increases adrenocortical sensitivity to ACTH _in vivo_ and thus compensates for the intrinsic decline in adrenocortical function thereby maintaining normal adrenocortical diurnal rhythms. In fact it can be hypothesized that adrenocortical-adrenomedullary chromaffin tissue association is needed for normal function of both tissues. A corollary to this hypothesis is that in lower vertebrates, where the two types of tissue are relatively separate, factors other than physical contiguity serve to define normal functional status. According to this notion, such factors are poorly conserved with evolution to mammals. A similar hypothesis can be raised concerning adrenocortical zonation in mammals. Among other intriguing puzzles that have received little attention is the relationship between the tubular arrangement of mitochondrial membranes characteristic of steroidogenic cells and the function of these cells.

REFERENCES

Accordi, F., and Grassi Milano, E. (1977). Gen. Comp. Endocrinol. 33, 187.

Accordi, F., Mastrolia, L., Grassi Milano, E., and Manelli, H. (1975). Riv. Biol. 68, 155.

Albano, J.D.M., Jack, P.M., Joseph, T., Gould, R.P., Nathanielsz, P.W., and Brown, B.L. (1976). J. Endocrinol. 71, 333.

Allen, C., and Kendall, J.W. (1967). Endocrinology 80, 926.

Assenmacher, I., and Jallageas, M. (1980). In "Avian Endocrinology" (A. Epple and M.H. Stetson, eds.), p.391. Academic Press, New York.

Attia, M.A. (1985). Arch. Toxicol. 57,77.

Banerji, T.K., Parkening, T.A., and Collins, T.J. (1984). J. Gerontol. 39, 264.

Bartholet, J.Y. (1980). J. Endocrinol. 87,1.

Bidlingmaier, F., Dörr, H.G., Eisenmenger, W., Kuhnle, U., and Knorr, D. (1986). J. Clin. Endocrinol. Metab. 62,331.

Black, I.B. (1982). Science 215,1198.

Bourne, G.H. (1949). "The Mammalian Adrenal Gland." Oxford University Press, London, New York.

Bowman, R.E., and Wolf, R.C. (1969). Proc. Soc. Exp. Biol. Med. 130,61.

Breznock, E.M., and McQueen, R.D. (1970). Amer. J. Vet. Res. 31,1269.

Cameron, G., Jones, T., Anderson, A., and Griffith, K. (1969). J. Endocrinol. 45,215.

Carmichael, S.W. (1987). In "Stimulation-Secretion Coupling in Chromaffin Cells" (K. Rosenheck and P.I. Lelkes, eds.), p.2. CRC Press, Inc., Boca Raton, Florida.
Carmichael, S.W., Spagnoli, D.B, Frederickson, R.G., Krause, W.J., and Culberson, J.L. (1987). Amer. J. Anat. 179,211.
Carsia, R.V., Morrin, M.E., Rosen, H.D., and Weber, H. (1987). Proc. Soc. Exp. Biol. Med. 184,436.
Carsia, R.V., Scanes, C.G., and Malamed, S. (1985). Proc. Soc. Exp. Biol. Med. 179,279.
Carsia, R.V., and Weber, H. (1986). Proc. Soc. Exp. Biol. Med. 183,99.
Castañe, P., Saliban, A., Zylbersztein, C., and Herkovits, J. (1987). Comp. Biochem. Physiol. 86A,667.
Chester Jones, I. (1957). "The Adrenal Cortex". Cambridge University Press, London.
Chester Jones, I. (1976). J. Endocrinol. 71,3P.
Chester Jones, I., and Phillips, J.G. (1986). In "Vertebrate Endocrinology: Fundamentals and Biomedical Implications" (P.K.T. Pang and M.P. Schreibman, eds.), Vol. 1, p.319. Academic Press, Orlando.
Chiueh, C.C., Nespor, S.M., and Rapoport, S.I. (1980). Neurobiol. Aging 1,157.
Chuvilina, O.Y., and Kirillov, O.I. (1984). Z. Mikrosk. Anat. Forsch. 98,213.
Ciaranello, R.D., Wooten, G.F., and Axelrod, J. (1976). Brain Res. 113,349.
Couch, R.M., Muller, J., and Winter, J.S.D. (1986). J. Clin. Endocrinol. Metab. 63,613.
Coupland, R.E. (1965). "The Natural History of the Chromaffin Cell". Longmans, Green and Co., London.
Coupland, R.E., and Weakley, B.S. (1968). J. Anat. 102,425.
Crivello, J.F., Hornsby, P.J., and Gill, G.N. (1983). Endocrinology 113,235.
Cutler, Jr., G.B., and Loriaux, D.L. (1980). Federation Proc. 39,2384.
Dalle, M., and Delost, P. (1974). J. Endocrinol. 63,483.
Dalle, M., and Delost, P. (1976). J. Endocrinol. 70,204.
Dallman, M.F., Engeland, W.C., and Holzwarth, M.A. (1977). Ann. N.Y. Acad. Sci. 297,373.
Dickerman, Z., Grant, D.R., Faiman, C., and Winter, J.S.D. (1984). J. Clin. Endocrinol. Metab. 59,1031.
Domm, L.V., and Erickson, G.C., (1972). Proc. Soc. Exp. Biol. Med. 140,1215.
Dupouy, J.P., Coffigny, H., and Magre, S. (1975). J. Endocrinol. 65,347.
Dupouy, J.P., and Dubois, M.P. (1975). Cell Tissue Res. 161,373.

Durand, Ph., Cathiard, A.M., and Saez, J.M. (1985). Mol. Cell. Endocrinol. 39,145.
Dvořák, M. (1972). J. Endocrinol. 54,473.
Edwards, A.V., Jones, C.T., and Bloom, S.R. (1986). J. Endocrinol. 110,81.
El-Maghraby, M., and Lever, J.D. (1980). J. Anat. 131,103.
Elsner, C.W., Magyar, D.M., Fridshal, D., Eliot, J., Klein, A., Glatz, T., Nathanielsz, P.W., and Buster, J.E., (1980). Endocrinology 107,801.
Enders, A.C., and Schlafke, S. (1967). Amer. J. Anat. 120,185.
Engeland, W.C., and Dallman, M.F. (1975). Neuroendocrinology 19,352.
Engeland, W.C., Siedenburg, F., Wilkinson, C.W., Shinsako, J., and Dallman, M.F. (1980). Endocrinology 106,1410.
Faucheux, B.A., Bourliere, F., Baulon, A., and Dupuis, C. (1981). Gerontology 27,313.
Fawcett, D.W. (1981). "The Cell". p.448. W.B. Saunders Co., Philadelphia.
Freeman, B.M. (1982). Comp. Biochem. Physiol. 72A,251.
Freeman, B.M. (1983). In "Physiology and Biochemistry of the Domestic Fowl" (B.M. Freeman, ed.), Vol. 4, p.191. Academic Press, New York.
Freeman, B.M., and Flack, I.H. (1981). Comp. Biochem. Physiol. 70A,257.
Freeman, B.M., and Manning, A.C.C. (1984). Comp. Biochem. Physiol. 78A,267.
Frolkis, V.V., Verkhratsky, N.S., and Magdich, L.V. (1985). Gerontology 31,84.
Gonzalez, C.B., Cozza, E.N., De Benders, M.E.O., Lantos, C.P., and Aragones, A. (1983). Gen. Comp. Endocrinol. 51,384.
Gorbman, A. (1986). In "Vertebrate Endocrinology: Fundamentals and Biomedical Implications" (P.K.T. Pang and M.P. Schreibman, eds.), Vol. 1, p.465. Academic Press, Orlando.
Greiner, J.W., Kramer, R.E., and Colby, H.D. (1976). J. Endocrinol. 70,127.
Grynszpan-Winograd, O. (1974). J. Neurocytol. 3,341.
Guillet, R., and Michaelson, S.M. (1978). Neuroendocrinology 27,119.
Haensly, W.E., and Getty, R. (1965). J. Gerontol. 20,544.
Hall, B.K., and Hughes, H.P. (1970). Z. Zellforsch. Mikrosk. Anat. 108,1.
Hanke, W., and Pehlemann, F.W. (1969). Gen. Comp. Endocrinol. 13,509.
Harrison, P.G., and Hoey, M.J. (1960). "The Adrenal Circulation". Blackwell Scientific Publication, Oxford.

Hegstad, R., Brown, R.D., Jiang, N.-S., Kao, P., Weinshilboum, R.M., Strong, C., and Wisgerhof, M. (1983). Amer. J. Med. 74,442.
Hervonen, A. (1971). Acta Physiol. Scand. [Suppl.]368, 1.
Hess, G.D., and Riegle, G.D. (1970). J. Gerontol. 25,354.
Hillman, J.R., Seliger, W.G., and Epling, G.P. (1975). Gen. Comp. Endocrinol. 25,14.
Hirose, T. (1977). Acta Endocrinol. 84,349.
Hiroshige, T., and Sato, T. (1970). Endocrinology 86,1184.
Holaday, J.W., Martinez, H.M., and Natelson, B.H. (1977). Science 198,13.
Holmes, W.N., and Cronshaw, J. (1984). J. Exp. Zool. 232,627.
Holmes, W.N., and Dickson, A.D. (1971). J. Anat. 108,159.
Holmes, W.N., and Kelly, M.E. (1976). Pflugers Arch. 365,145.
Honma, S., Honma, K.-I., Shirakawa, T., and Hiroshige, T. (1984). Endocrinology 114,1791.
Hopwood, D., (1971). Prog. Histochem. Cytochem. 3,1.
Hornsby, P.J. (1982). Endocrinology 111,1092.
Hornsby, P.J., and Crivello, J.F. (1983). Mol. Cell. Endocrinol. 30,123.
Hotta, M., and Baird, A., (1987). Endocrinology 121,150.
Hsu, C.-Y., Yu, N.-W., and Chen, S.-J. (1980). Gen. Comp. Endocrinol. 42,167.
Hullinger, R.L. (1978). Anat. Histol. Embryol. 7,1.
Ito, K., Sato, A., Sato, Y., and Suzuki, H. (1986). Neurosci. Lett. 69,263.
Jaffe, R.C. (1981). Gen. Comp. Endocrinol. 44,314.
Kalliecharan, C. (1981). J. Submicrosc. Cytol. 13,627.
Kalliecharan, R., and Hall, B.K. (1974). Gen. Comp. Endocrinol. 24,364.
Kalliecharan, R., and Hall, B.K. (1976). Gen Comp. Endocrinol. 30,404.
Kikuta, A., and Murakami, T., (1982). Amer. J. Anat. 164,19.
Kleitman, N., and Holzwarth, M.A. (1985a). Cell Tissue Res. 241,139.
Kleitman, N., and Holzwarth, M.A. (1985b). Amer. J. Physiol. 248,E261.
Kreiner, E. (1975). Verh. Dtsch. Ges. Pathol. 59,419.
Krug, E.C., Honn, K.V., Battista, J., and Nicoll, C.S. (1983). Gen. Comp. Endocrinol. 52,232.
Lagercrantz, H., and Slotkin, T.A. (1986). Scientific American 254,100.
Leathem, J.H. (1974). In "Epidemiology of Aging" (A.M. Ostfeld and D.C. Gibson, eds; C.P. Donnelly, tech. ed.), p.182. U.S. Department of Health Education and Welfare, Publication No. (NIH) 75-711, Bethesda.

LeDouarin, N.M., and Teillet, M.-A. (1971). C.R. Acad. Sci. 272,481.
LeDouarin, N., and Teillet, M. (1974). Dev. Biol. 41,162.
Leist, K.H., and Hanke, W. (1969). Gen. Comp. Endocrinol. 13,517.
Levine, J., Wolfe, L.G., Schiebinger, R.J., Loriaux, D.L., and Cutler, Jr., G.B. (1982). Endocrinology 111,1797.
Liggins, G.C. (1976). Amer. J. Obstet. Gynecol. 126,931.
Lymangrover, J., Saffran, M., and Matthews, E.K. (1978). Mech. Ageing Dev. 8,377.
Magalhães, M.M., Breda, J.R., Magahães, M.C., and Reis, J. (1981). J. Ultrastructural Res. 76,215.
Mahapatra, M.S., Mahata, S.K., and Maiti, B.R. (1987). Gen. Comp. Endocrinol. 67,279.
Mahata, S.K., and Ghosh, A. (1986). Boll. Zool. 53,63.
Majchrzak, M., and Malendowicz, L.K. (1983). Cell Tissue Res. 232,457.
Malamed, S. (1975). In "Adrenal Cortex" (G. Sayers ed.), Handbook of Physiology, Section 7: Endocrinology, Vol. 6, p.25. American Physiological Society, Bethesda.
Malamed, S., and Carsia, R.V. (1983). J. Gerontol, 38,130.
Malamed, S., Carsia, R.V., and Weiner, B. (1981). Anat. Record 199, 158A.
Malamed, S., Gibney, J.A., Yamasaki, D.S., and Carsia, R.V. (1984). Anat. Record 208, 113 A.
Malee, M.P., and Marotta, S.F. (1982). Proc. Soc. Exp. Biol. Med. 169, 355.
Marie, C. (1981). J. Endocrinol. 90, 193.
Martin, K.O., and Black, V.H. (1982). Endocrinology 110, 1749.
Martin, K.O., and Black, V.H. (1983). Endocrinology 112, 573.
Mason, J.I., Carr, B.R., and Rainey, W.E. (1986). Endocrine Res. 12, 447.
Mastrolia, L., and Manelli, H. (1969). Acta Embryol. Exp. 1969, 257.
McAllister, J.M., and Hornsby, P.J. (1987). Endocrinology 121, 1908.
McCarty, R. (1984). Neurobiol. Aging 5, 285.
McCarty, R. (1986). Gerontology 32, 172.
Meier, A.H. (1976). Endocrinology 98, 1475.
Migally, N. (1979). Anat. Record 194, 105.
Milković, K. and Milković, S. (1963). Endocrinology 73, 535.
Miquel, J., Lundgren, P.R., and Johnson, Jr., J.E. (1978). J. Gerontol. 33, 5.
Murakoshi, M., Osamura, Y., and Watanabe, K. (1985). Tokai J. Exp. Clin. Med. 10, 531.

Murakoshi, M., Osamura, Y., and Watanabe, K. (1987). Cell Structure and Function 12, 181.
Murphy, B.E.P., Patrick, J., and Denton, R.L. (1975). J. Clin. Endocrinol. Metab. 40, 164.
Nagele, R.G., Hunter, E., Busch, K., and Lee, H. (1987). J. Exp. Zool. 244, 425.
Nakamura, T., Tanabe, Y., and Hirano, H. (1978). Gen. Comp. Endocrinol. 35, 302.
Nathanielsz, P.W. (1978). Sem. Perinatol. 2, 223.
Nickerson, P.A. (1977). J. Anat. 124, 383.
Nickerson, P.A. (1979). Amer. J. Pathol. 95, 347.
Nikicicz, H., Kasprzak, A., Malendowicz, L.K. (1984). Cell Tissue Res. 235, 459.
Nishikawa, T., and Strott, C.A. (1984). Endocrinology 114, 486.
Norr, S. (1973). Dev. Biol. 34, 16.
Novy, M.J. (1977). In "The Fetus and Birth". CIBA Foundation Symposium 47, p. 259. Elsevier Excerpta Medica North Holland, Amsterdam.
Ohashi, M., Kato, K.-I., Nawata, H., and Hiroshi, I. (1986). Gerontology 32, 43.
Ostlund, E. (1954). Acta Physiol. Scand. [Suppl.]31, 1.
Ottenweller, J.E., and Meier, A.H. (1982). Endocrinology 111, 1334.
Ottenweller, J.E., Meier, A.H., Ferrell, B.R., Horseman, N.D., and Proctor, A. (1978). Endocrinology 103, 1875.
Parker, L.N., Lifrak, E.T., Ramadan, M.B., and Lai, M.K. (1983). Arch. Androl. 10, 17.
Pearce, R.B., Cronshaw, J., and Holmes, W.N. (1978). Cell Tissue Res. 192, 363.
Pelto-Huikko, M., Salminen, T., Partanen, M., Toivanen, M., and Hervonen, A. (1985). Anat. Record 211, 458.
Phillipe, M. (1983). Amer. J. Obstet. Gynecol. 146, 840.
Piezzi, R.S. (1965). Acta Physiol. Lat. Amer. 15, 96.
Pillai, A.K., Salhanick, A.I., and Terner, C. (1974). Gen. Comp. Endocrinol. 24, 152.
Pohorecky, L.A., and R.J. Wurtman (1971). Pharmacol. Rev. 23, 1.
Popplewell, P.Y., and Azhar, S. (1987). Endocrinology 121, 64.
Popplewell, P.Y., Butte, J., and Azhar, S. (1987). Endocrinology 120, 2521.
Popplewell, P.Y., Tsubokawa, M., Ramachandran, J., and Azhar, S. (1986). Endocrinology 119, 2206.
Potter, C.L., and Goodfriend, T.L. (1987). Gerontology 33, 77.
Pritchett, J.F., Sartin, J.L., Marple, D.N., Harper, W.L., and Till, M.L. (1979). Hormone Res. 10, 96.

Ramade, F., and Baylé, J.-D. (1980). J. Physiol. (Paris) 76, 283.
Ramaley, J.A. (1973). Steroids 21, 433.
Reichel, W. (1968). J. Gerontol. 23, 145.
Riegle, G.D., and Nellor, J.E. (1967). J. Gerontol. 22, 83.
Riegle, G.D., Przekop, F., and Nellor, J.E. (1968). J. Gerontol. 23, 187.
Rose, J.C., Meis, P.J., Urban, R.B., and Greiss, Jr., F.C. (1982). Endocrinology 111, 80.
Saez, J.M., Durand, Ph., and Cathiard, A.M. (1984). Mol. Cel. Endocrinol. 38, 93.
Samorajski, T., and Ordy, J.M. (1967). J. Gerontol. 22, 253.
Schapiro, S., Geller, E., and Eiduson, S. (1962). Proc. Soc. Exp. Biol. Med. 109, 937.
Seiler, K., and Seiler, R. (1973). Gengebaurs Morph. Jahrb. Leipzig 119, 796.
Seiler, K., Seiler, R., and Claus, R. (1981). Endokrinologie 78, 297.
Seiler, K., Seiler, R., Claus, R., and Sterba, G. (1983). Gen. Comp. Endocrinol. 51, 353.
Seiler, K., Seiler, R., and Hoheisel, G. (1973). Gengebaurs Morph. Jahrb. Leipzig 119, 823.
Setoguti, T., Satou, Y., and Yoshiki, G. (1979). Arch. Histol. Japan 42, 95.
Setoguti, T., Shin, M., Inoue, Y., Matsumura, H., and Chen, H.-S. (1985). Arch. Histol. Japan 48, 199.
Shaposhnikov, V.M. (1985). Mech. Ageing Dev. 30, 123.
Shaposhnikov, V.M., and Bezrukov, V.V. (1985). J. Submicrosc. Cytol. 17, 75.
Sivaram, S. (1969). Acta Histochem. (Jena) 32, 253.
Smail, P.J., Faiman, C., Hobson, W.C., Fuller, G.B., and Winter, J.S.D. (1982). Endocrinology 111, 844.
Sonntag, W.E., Goliszek, A.G., Brodish, A., and Eldridge, J.C. (1987). Endocrinology 120, 2308.
Spagnoli, D.B., Frederickson, R.G., Robinson, R.L., and Carmichael, S.W. (1987). Amer. J. Anat. 179, 220.
Stanton, H.C., and Woo, S.K. (1978). Amer. J. Physiol. 234, E137.
Stark, E., Gyévai, A., Bukulya, B., Szabó, D., Szalay, K.S., and Mihály, K. (1975). Gen. Comp. Endocrinol. 25, 472.
Stern, S., Biggers, J.D., and Anderson, E. (1971). J. Exp. Zool. 176, 179.
Suzuki, M.R., and Kikuyama, S. (1983). Gen. Comp. Endocrinol. 52, 272.
Szabó, D., Dzsinich, C., Ökrös, I., and Stark, E. (1970). Exp. Gerontol. 5, 335.

Szabó, D., Somogyi, J., and Mitro, A. (1982). Endokrinologie 79, 76.
Tanabe, Y., Saito, N., and Nakamura, T. (1986). Gen. Comp. Endocrinol. 63, 456.
Tang, F., and Phillips, J.G. (1977). J. Endocrinol. 75, 183.
Tang, F., and Phillips, J.G. (1978). J. Gerontol. 33, 377.
Taylor, A.N., Lorenz, R.J., Turner, B.B., Ronnekleiv, O.K., Casady, R.L., and Branch, B.J. (1976). Psychoneuroendocrinology 1, 291.
Teillet, M.-A., and LeDouarin, N.M. (1974). Arch. Anat. Microsc. Morphol. Exp. 63, 51.
Thorburn, G.D., and Challis, J.R.G. (1979). Physiol. Rev. 59, 863.
Tischler, A.S., DeLellis, R.A., Perlman, R.L., Allen, J.M., Costopoulos, D., Lee, Y.C., Nunnenmacher, G., Wolfe, H.J., and Bloom, S.R. (1985). Lab. Invest. 53, 486.
Touitou, Y., Sulon, J., Bogdan, A., Touitou, C., Reinberg, A., Beck, H., Sodoyez, J.-C., Demey-Ponsart, E., and Van Cauwenberge, H. (1982). J. Endocrinol. 93, 201.
Ungar, F., and Stabler, T.A. (1980). J. Steroid Biochem. 13, 23.
Unzicker, K. (1969). Z. Zellforsch. Mikrosk. Anat. 95, 608.
Vermes, I., Dohanics, J., Tóth, G., and Pongrácz, J. (1980). Hormone Res. 12, 237.
Vermeulen, A., Deslypere, J.P., Shelfhout, W., Verdonck, L., and Rubens, R. (1982). J. Clin. Endocrinol. Metab. 54, 187.
Vilee, D.B. (1975). "Human Endocrinology: A Developmental Approach". W. B. Saunders Co., Philadelphia.
Vinson, G.P., Whitehouse, B.J., Goddard, C., and Sibley, C.P. (1979). J. Endocrinol. 81, 5P.
von Seebach, H.B., Lützen, L., Kreiner, E., Ueberberg, H., Pappritz, G., and Dhom, G. (1975). Verh. Dtsh. Ges. Pathol. 59, 414.
Walker, C.-D., Perrin, M., Vale, W., and Rivier, C. (1986). Endocrinology 118, 1445.
Webb, M.L., and Mashaly, M.M. (1985). Poultry Sci. 64, 744.
Weisbart, M. (1975). Gen. Comp. Endocrinol. 26, 368.
Weisbart, M., and Youson, J.H. (1975). Gen. Comp. Endocrinol. 27, 517.
Weiss, M. (1984). Comp. Biochem. Physiol. 79B, 173.
Weiss, M., and Ford, V.L. (1977). Comp. Biochem. Physiol. 57B, 15.
Weiss, M., and Ford, V.L. (1982). Gen. Comp. Endocrinol. 46, 168.
Wexler, B.C. (1981). Paroi. Arterielle 7, 121.

Wintour, E.M., Brown, E.H., Denton, D.A., Hardy, K.J., McDougall, J.G., Oddie, C.J., and Whipp., G.T. (1975). Acta Endocrinol. (Copenh) 79, 301.

Wise, P.M., and Frye, B.E. (1973). J. Exp. Zool. 185, 277.

Woods, J.E., DeVries, G.W., and Thommes, R.C. (1971). Gen. Comp. Endocrinol. 17, 407.

Wurtman, R.J., and Axelrod, J. (1966). J. Biol. Chem. 241, 2301.

Yamaji, T., and Ibayashi, H. (1969). J. Clin. Endocrinol. Metab. 29, 273.

Yeasting, R.A. (1986). *In* "The Adrenal Gland" (P.J. Mulrow, ed.), p. 45. Elsevier, New York.

Zajicek, G., Ariel, I., and Arber, N. (1986). J. Endocrinol. 111, 477.

20

DEVELOPMENT, MATURATION AND SENESCENCE OF SYMPATHETIC INNERVATION OF SECONDARY IMMUNE ORGANS

David L. Felten, Suzanne Y. Felten, Kelley S. Madden,
Kurt D. Ackerman and Denise L. Bellinger

Department of Neurobiology & Anatomy
University of Rochester School of Medicine
Rochester, New York 14642

I. INTRODUCTION: A. Evidence for Neural Interactions with the Immune System.

Epidemiological evidence supports the existence of mutual communication between the nervous system and the immune system. This evidence derives largely from investigations of the effects of psychological factors on the outcome of disease, including peptic ulcers, essential hypertension, ulcerative colitis, hyperthyroidism, regional enteritis, rheumatoid arthritis, and bronchial asthma (Ader, 1981; Miller, 1983; Weiner, 1977), as well as infectious diseases and neoplasia (Fox, 1981; Plaut _et al._, 1981; Riley, 1981). These studies have demonstrated that: (1) many of the classical "psychosomatic" diseases listed above involve autoimmune phenomena in their etiology (Geschwind _et al._, 1982; Solomon _et al._, 1974; Solomon, 1981); (2) acute flare-ups of autoimmune diseases commonly are precipitated by emotional upheaval (Solomon _et al._, 1974; Solomon, 1981); (3) a variety of stressors can alter immunocompetence in response to neoplasms (Fox, 1981; Riley, 1981; Solomon _et al._, 1974); and (4) psychosocial factors such as bereavement, depression, and marital separation are associated with diminished measures of immune responses (Bartrop _et al._, 1977; Stein, 1985).

A wide range of studies in laboratory animals point towards interactions between the nervous and immune systems. Ader and colleagues (Ader _et al._, 1975; Ader _et al._, 1982; Ader _et al._, 1979; Ader _et al._,

1985; Bovbjerg et al., 1982; Cohen et al., 1979) and others (Rogers et al., 1976; Wayner et al., 1978; Gorczynski et al., 1982; Kusnecov et al., 1983) have demonstrated classical behavioral conditioning of immune responses, and of the progression of immune mediated diseases. Studies of psychological stressors have demonstrated that these factors can influence morbidity and mortality from tumors and pathogens (Plaut et al., 1981; Riley, 1981; Sklar et al., 1979). Neuroanatomical studies employing discrete lesions of hypothalamic regions, limbic forebrain structures, brain stem autonomic and reticular regions, and the cerebral cortex have shown transient or chronic alterations in immune reactivity (Szentivany et al., 1958; Cross et al., 1980; Roszman et al., 1985; Ader et al., 1987); these same regions appear to respond to immunization with alterations in their electrical activity (Besedovsky et al., 1977) and monoamine metabolism (Besedovsky et al., 1987; Carlson et al., 1987). This combined evidence suggests an integrated circuitry of cerebral cortex, limbic forebrain, hypothalamus and brain stem autonomic nuclei that can modulate immune reactivity in the periphery.

Two major efferent routes exist by which the brain can regulate the viscera in the periphery: (1) neuroendocrine outflow from the hypothalamo-pituitary axis; and (2) autonomic outflow from the preganglionic neurons of the brain stem and spinal cord. A huge body of literature, reviewed in (Berczi, 1986; Berczi, 1987) points towards the neuroendocrine route as one important regulatory control system interacting with the immune system. A strong emphasis has been placed on glucocorticoid regulation of immune proliferation and responsiveness (Besedovsky et al., 1986) as an important neuroendocrine loop interconnecting the nervous and immune systems. Our laboratories have investigated the autonomic interconnections with organs of the immune system, and have found extensive sympathetic postganglionic noradrenergic innervation of both smooth muscle compartments and the parenchyma in these organs, and have demonstrated a prominent influence of this neural connection on many measures of immune responsiveness. This chapter explores some of these neural-immune interactions in development, maturation, and senescence; we propose unique interactions and communication routes at these different periods of life.

B. Evidence for Sympathetic Noradrenergic Innervation of Organs of the Immune System

Findings from our laboratories (Williams et al., 1981; Williams et al., 1981; Felten et al., 1987; Felten et al., 1987; Felten et al., 1984; Felten et al., 1981; Felten et al., 1987; Felten et al., 1987; Ackerman et al., 1987; Bellinger et al., 1987; Livnat et al., and others (Giron et al., 1980; Bulloch et al., 1984; Calvo, 1968; Calvo et al., 1969; Walcott et al., 1985; Sergeeva, 1974; Reilly et al., 1979) have revealed direct autonomic innervation of the vasculature and the parenchyma in both primary and secondary lymphoid organs. In primary organs, the noradrenergic (NA) postganglionic sympathetic nerve fibers travel with the vasculature and then branch from plexuses along the vessels into the parenchyma, ending among cellular elements of these immune organs. In bone marrow, these fibers distribute among cells within the substance of the marrow. In the thymus, the NA fibers distribute mainly into the cortical region, and end among thymocytes. In secondary immune organs, including the spleen, lymph nodes, and gut-associated lymphoid tissue (GALT), the NA innervation is regional and specific. The nerve fibers generally distribute to zones of T lymphocytes and macrophages, and do not enter follicular or nodular regions of B lymphocytes directly. In the spleen, NA fibers follow the central artery and its branches into the white pulp, and course into the periarteriolar lymphatic sheath, a zone composed predominantly of T lymphocytes in adults. These fibers distribute into the parenchyma of the PALS, along the marginal sinus and into the marginal zone, and course along parafollicular sites [see (Felten et al., 1985; Felten et al., 1987) for detailed descriptions]. In lymph nodes, the NA fibers enter the hilar region with the vasculature, distribute through the medullary cords, and branch abundantly among T lymphocytes in the cortical and paracortical regions surrounding the non-innervated follicles. In GALT, the NA fibers enter with the vasculature along the muscular layers, turn inward between the follicles, traverse the T-dependent zones of lymphocytes, and arborize among the cells of the lamina propria.

In the spleen, the NA nerve fibers have been investigated with a variety of anatomical, neurochemical, and immunological techniques to ascertain whether the transmitter, norepinephrine, fulfills the criteria for neurotransmission with lymphocytes or other immune cells in the spleen as specific targets. These criteria include: (1) localization of NA nerve fibers in specific compartments of the spleen adjacent to the target cells; (2) release and availability of NE in concentrations that are adequate to permit interactions with receptors on target cells; (3) presence of adrenoceptors on lymphocytes and other immune cells within the innervated splenic compartment; and (4) pharmacologically predictable functional responses obtainable from pharmacological manipulation or denervation. These criteria have been met in the spleen [see (Felten et al., 1987) for a detailed discussion], suggesting that NE functions as a neurotransmitter with lymphocytes as the target cell. The mode of this communication, whether by direct synaptic interaction or by paracrine secretion will be discussed below.

This chapter discusses only the role of NE as a sympathetic postganglionic neurotransmitter during development, maturation, and senescence. By no means do we consider this neurotransmitter to be the only neurally active molecule to be available for interaction with lymphocytes. We have found neuropeptide-Y containing nerve terminals directly abutting lymphocytes at the electron microscopic level, and have some preliminary evidence that many neuropeptides can be found in nerve fibers in secondary immune organs. A further discussion of these peptides is beyond the scope of this discussion, and can be found elsewhere (Payan et al., 1985; Rogers et al., 1976).

II. DEVELOPMENT OF SYMPATHETIC INNERVATION OF THE SPLEEN: A POSSIBLE ROLE IN COMPARTMENTATION AND DEVELOPMENT OF IMMUNE COMPETENCE

Noradrenergic innervation of primary and second lymphoid organs has been observed in a variety of mammals (Ayers et al., 1972; Fillenz, 1970; Gillespie et al., 1965; Reilly, 1985), but originally was considered to be confined in function to the regulation of blood flow and contraction of smooth muscle. Findings from our laboratories, noted above, indicate a prominent role for the noradrenergic innervation of secondary lymphoid organs in the regulation of immune responses and proliferative activity of lymphocytes. In adults, both the vascular compartment and the parenchyma are innervated abundantly, making separation of these different functions difficult to approach experimentally. We have investigated early postnatal development of the innervation of the spleen (Ackerman et al., 1987; Ackerman et al., 1987; Ackerman et al., 1988), and have found evidence for exclusive parenchymal compartmentation of the NA fibers during the first postnatal week of life in rats.

Although the NA nerve fibers appear to reach the spleen from the superior mesenteric/coeliac ganglion in nerves that travel with the vasculature, at postnatal day 1, these NA fibers distribute only into the parenchyma. These nerve fibers encircle the small, developing PALS (Fig. 1), and are distant from the vascular at this time. At day 1, the central arteriolar system does not have smooth muscle cells along the vasculature, and there is no evidence of NA nerve fibers along this developing central arteriolar system. The NA fibers distribute along a zone that is predictive of the location of the marginal sinus at a later age, a zone that separates the PALS on the inside from the marginal zone on the outside. We have examined this outer zone of the PALS with electron microscopic immunocytochemical observations that suggested direct contracts in the parenchyma.

Several possibilities come to mind for the role these nerve fibers may play at this early postnatal development stage. First, these NA nerves may act as a "guide fiber" system in the development of the PALS. These are present early, and compartmentation of the white pulp appears to develop around these nerve fibers; the lymphocytes migrate into the PALS, between the NA fibers and the central artery, while the macrophages (ED3+) line up along the NA fibers at a site predictive of where the marginal sinus will form later. Second, the NA nerve fibers may be interacting transiently with a population of B lymphocytes in the PALS; these early contacts do not persist into adulthood. In view of our finding of enhanced B lymphocyte proliferation in adulthood as a consequence of denervation, and the existence of beta-adrenoceptors on B lymphocytes at 1 day of age and into adulthood, the neurotransmitter NA may interact with receptors on the developing B lymphocytes and regulate early developmental activities, perhaps proliferation or the development of immunocompetence. Third, the NA fibers may interact with T lymphocytes and influence their development. In preliminary studies in our laboratories (Ackerman et al., 1988), 10 day old rats, subjected to chemical sympathectomy at birth with 6-hydroxydopamine (6-OHDA) showed decreased spontaneous lymphocyte proliferation, and a decreased inhibitory response to the T cell mitogen concanavalin A (Con A). In addition, neonatal 6-OHDA treatment diminished natural killer (NK) cell activity in 10 day old spleens. Seven week old rats, sympathectomized at birth, demonstrated increased NK cell activity, with control levels of spontaneous and Con A-induced proliferation. These results suggest that NE in the developing spleen may be interacting transiently with a population of neonatal suppressor cells as an additional target for noradrenergic sympathetic neurotransmission.

By postnatal day 7, TH+noradrenergic fibers persist at the outer zone of the PALS, and many fibers traverse the PALS itself, and end adjacent to the central arteriolar system (Fig. 2). Smooth muscle cells begin to appear at this time. It is possible that these vascular nerve fibers are beginning to regulate blood flow through the white pulp. However, many of the TH+profiles, both within the PALS and in the adventitia at the innermost zone of the PALS, end adjacent to lymphocytes. During the first 2 weeks of life, FD3+(antigen-presenting) macrophages line up prominently along the marginal sinus, adjacent to the TH+fibers that have persisted from birth (Fig. 3). By the end of week 3, the B lymphocytes have clustered together in the outer regions of the PALS, and have formed follicles. The NA fibers, for the main part, avoid direct innervation of these follicles, and traverse the outside parafollicular edges. By 28 days of age, when the follicles are well-formed, the distribution of NA nerve fibers has taken on the adult pattern (Fig. 4), and the trabecular/capsular system of smooth muscle is innervated. Thus, during development, specific transient relationships exist between the NA nerves and specific populations of immune cells. These nerves are suggested to influence the functional events of lymphocyte migration and compartmentation within the developing white pulp, growth and proliferation, and the development of immunocompetence.

III. MATURATION OF SYMPATHETIC INNERVATION OF SPLEEN AND LYMPH NODES

A. Neurotransmission and the Mode of Communication Between Nerves and Lymphocytes

In the adult spleen, NE fulfills the criteria for neurotransmission with lymphocytes as the target (Felten et al., 1987). Anatomically, the TH+noradrenergic nerve fibers are found in specific compartments of the spleen (Fig. 5), including the PALS, the marginal sinus and marginal zone, and the parafollicular zones (Felten et al., 1987; Ackerman et al., 1987; Felten et al., 1987; Felten et al., 1985; Livnat et al., 1985). It is clear that the NE is associated with nerves, because superior mesenteric/coeliac ganglionectomy depletes the spleen of 95% of its NE content, eliminates fluorescent nerve profiles, and eliminates TH immunocytochemical staining. Electron microscopic immunocytochemistry (Felten et al., 1987) has revealed direct contacts between TH+nerve terminals and lymphocytes in the PALS in rats. These contacts are 6 nm neuroeffector junctions, often indented in the lymphocyte membrane; they are present along the inner PALS at the adventitial junction, deeper within the parenchyma of the PALS, and along the marginal sinus. These direct contacts provide the anatomical basis for interaction of released neurotransmitter with the target cells with which they form contacts.

The concentration of NE available in the spleen has been measured directly using in vivo dialysis (Felten et al., 1986). The extracellular fluid concentrations of NE were in the micromolar range, suggesting very high availability of the neurotransmitter for potential interactions with cells of the immune possessing receptors to permit ligand-receptor interactions.

A variety of studies have demonstrated the presence of adrenoceptors, particularly of the beta-2 subclass, on lymphocytes (both T and B), monocytes and macrophages, and granulocytes (Bishopric et al., 1980; Coffey et al., 1985; Hadden et al., 1970; Landmann et al., 1981; Landmann et al., 1985). We have found beta-adrenoceptors on both T and B lymphocytes, with greater numbers on B lymphocytes than on T lymphocytes, perhaps reflecting the greater distance between NA nerve fibers and B lymphocytes within the follicles.

The fulfillment of these first three criteria for neurotransmission suggests several possibilities for the mode of communication by which NE might influence functional activities of immunocytes. The most obvious mode is direct neurotransmission, as described in detail elsewhere (Felten et al., 1987); since NE is present in nerves found in specific compartments, is released into the extracellular fluid, and can interact with high-affinity beta-adrenoceptors on lymphocytes and other cells of the immune system, this possibility seems likely, and is supported by functional evidence, reviewed below. What is not obvious at this time is whether the process of NA neurotransmission with lymphocytes as the target cells takes place via paracrine secretion into the extracellular fluid, with subsequent diffusion, via direct contacts at neuroeffector junctions, or via both modes. It is possible that the direct contacts between TH+nerve terminals and lymphocytes provides sites of particularly high concentration of neurotransmitter, superimposed on the background paracrine availability of NE that is able to diffuse through the system.

An additional possibility also must be considered. The sites of direct nerve-lymphocyte contacts may provide a mechanism for cytokines released from the lymphocyte to act upon the nerve terminal and modulate its release of neurotransmitter. The findings of Besedovsky and colleagues (Besedovsky et al, 1979), confirmed by our laboratory, indicate that splenic NE levels are diminished selectively during the peak of an immune response, despite the remarkable stability of splenic NE content in the face of altered immune compartmentation (Carlson et al., 1987). Thus, it appears that cytokine-derived interactions are capable of regulating NE secretion in the spleen during the peak of an immune response, suggesting lymphocyte-nerve interactions. The neuroeffector junction may be the site of bi-directional communication. It is likely that hormones, neurotransmitters, and cytokines all are present in the local microenvironment of the splenic white pulp, and are capable of interacting with any cell type, immune or neural, that possesses receptors to permit an intracellular response.

A further possibility must be considered, that NE released from nerve terminals, in addition to interacting with adrenoceptors on lymphocytes, may interact with accessory cells, regulating their activities and secretions, thereby providing secondary influences on lymphocytes. Regulation of histamine secretion from mast cells, or serotonin secretion from enterochromaffin cells of the gut, are examples of this potential secondary effect.

B. Regulation of Mature Immune Responses

The final criterion for substantiating noradrenergic neurotransmission with lymphocytes as the target cells is a demonstration of predictable functional roles for the neurotransmitter, using pharmacological and denervation strategies. We have carried out a large number of functional studies utilizing chemical sympathectomy with 6-OHDA (Felten *et al.*, 1987; Felten *et al.*, 1985; Felton *et al.*, 1984; Livnat *et al.*, 1985; Livnat *et al.*, 1987; Madden *et al.*, 1986). These studies suggest a very complex role for NA innervation of the spleen and lymph nodes, and suggests that many cells may be targets of neurotransmission for the NA nerves.

We have explored the effects of denervation on proliferative responses of B and T lymphocytes. In lymph nodes, denervation is followed by enhanced B cell proliferation, both spontaneous and mitogen-induced. This finding is consistent with NE interactions with beta-adrenoceptors, initiating a second messenger response involving cAMP (Singh *et al.*, 1979; Singh *et al.*, 1976), producing a diminished proliferative response, denervation would be expected to disinhibit these responses. However, T lymphocyte proliferative responses were diminished in these same denervated structures, indicating that not all lymphocyte responses to denervation were the same. Furthermore, there were regional difference among different lymph nodes, and the splenic proliferative responses were either not present, or much less robust than their counterparts in lymph nodes. Furthermore, preliminary studies in development suggest that denervation may shift the sensitivity of the mitogen-response curve rather than only the magnitude, emphasizing the necessity of carrying out complete dose-response studies rather than the usual "maximal" single dose studies of mitogens.

We have also investigated primary and secondary immune responses in mice to antigen challenge with sheep red blood cells or KLH antigen. In denervated popliteal lymph nodes challenged by foot pad immunization, the primary immune response was diminished by 97%, despite enhanced B cell proliferation. In denervated spleen challenged systemically, the primary immune response was diminished by 80%. These findings suggest that NA nerves are necessary in these secondary immune organs for immunocompetence. Secondary immune responses also were diminished in spleen and lymph nodes when denervation was carried out prior to boosting. These findings are consistent with the detailed studies of immune responses with catecholamine antagonists and agonists by Sanders and Munson (Sanders *et al.*, 1984; Sanders *et al.*, 1984; Sanders *et al.*,

1985; Sanders et al., 1985), suggesting that beta-adrenoceptor stimulation enhances primary antibody responses. These findings reinforce the importance of the NA innervation of secondary immune organs for competent immune responses, and suggest that use of proliferation or mitogen responses, often equated with "immune responsiveness" and suggested to represent some predictive notion of response in a functioning immune system, can be highly misleading. If primary immune responses can be severely diminished in the face of enhanced B lymphocyte proliferation, this suggests that earlier events, perhaps at the level of antigen processing and presentation, or T lymphocyte responsiveness, might be affected by denervation, thereby diminishing the overall response regardless of what happens to B lymphocyte proliferation. In order to make accurate predictions about what might happen to functional immune responses in vivo, it is necessary to measure them in vivo, and not to depend upon a potentially misleading and simplistic measurement.

We have looked further at other immune responses in mice denervated with 6-OHDA. Delayed-type sensitivity responses to a contact sensitizing agent were diminished by 50% in denervated mice. Cytotoxic T lymphocyte responses in denervated draining lymph nodes primed with TNCB and stimulated with hapten-modified syngeneic spleen cells, were diminished by 50%, accompanied by approximately 50% decrease in IL2 production. This suppression of the efferent phase of the response may be due to a deficit in the lymphokine-producing capability of T lymphocytes following denervation. Natural killer cell activity also was investigated in denervated mice. Both spontaneous and poly I:C-induced NK cell activity were enhanced following denervation, indicating that not all functional responses show the same directionality or magnitude of response following denervation.

A further possibility for NE interactions with the functioning immune system was explored in lymphocyte migration studies. LN cells from denervated C3H mice showed markedly decreased location 1 hour after infusion in LNs of control syngeneic recipients (migration to spleen and liver was unchanged), while LN cells from controls showed markedly enhanced location 1 hour after infusion into LNs of sympathectomized mice. These findings are consistent with past evidence that catecholamines can regulate lymphocyte migration (Ernstrom et al., 1973; Ernstrom et al., 1975), and with findings from Ottaway (Ottaway, 1984; Ottaway, 1985) that expression of VIP receptors on T lymphocytes can influence their migratory patterns to GALT.

The picture that emerges in the mature immune system is a complex role for NE in modulation of immune responses. It is likely that NE is available to interact with many cell types, and can influence proliferation, differentiation, migration, cytokine secretion, and other intracellular events selectively in each cell type, perhaps selectively in each organ.

C. Plasticity of Mature Innervation in Response to Changing Immune Compartmentation

The NA innervation of the splenic white pulp shows specific compartmentation in adult rodents. We challenged the spleen with lymphocyte depleting drugs, cyclophosphamide or hydrocortisone (Carlson *et al.*, 1987), reducing the size of the organ, the size of the white pulp, and the number of lymphocytes present in the white pulp. Despite this reduction in the white pulp, the NA nerves retained their compartmentation, retracted with the diminishing geometry of the white pulp, showed an increased density of innervation, but remained unchanged in total splenic NE content. This suggests that the innervation remains in its appropriate compartment, and perhaps is available in the diminished compartment in higher concentration, due to increased density of terminals; we do not yet know whether actual turnover remains constant under these conditions. As a physiological model of this drug-induced phenomenon, we have made preliminary observations of immediate postpartum female rat spleens, and have found a similar diminution of the white pulp, with increased density of NA innervation due to the changing geometry. Thus, it is possible that a lymphocyte is exposed to varying concentrations of neurotransmitter, depending on the current conditions of the immune organ and the proximity to NA nerve terminals.

IV. SENESCENCE OF SYMPATHETIC INNERVATION OF THE SPLEEN

In the spleen of aged Fischer 344 rats, the NA innervation shows a decline, measured both by chemical analysis and by histochemical evaluation (Felten *et al.*, 1987; Bellinger *et al.*, 1987). At 12 months of age, the TH+innervation and the compartmentation of B and T lymphocytes and macrophages has the general appearance of patterns in young adults (Fig. 6), but starting at approximately 18 months of age, and becoming more pronounced by 27 months of age, the TH+innervation is diminished (Figs. 7,8), and the compartmentation of macrophages and lymphocytes is disrupted. The PALS in 27 month old rats shows diminished numbers of T lymphocytes (OX-19+), and the marginal sinus shows markedly reduced numbers of ED3+macrophages.

Studies in man and laboratory animals have demonstrated a decline in normal immune functions with aging. Age-related immune dysfunction generally is associated with an increased frequency of autoimmune disorders, cancer, and viral and fungal infections. In particular, normal aging processes seem to affect preferentially the T lymphocyte mediated immunity. As a concise summary of a vast body of literature, the decline in immune function with age is reflected by: (1) decreased T helper cell and cytotoxic T cell activity; (2) increased suppressor T lymphocyte activity; (3) decreased T lymphocyte proliferative responses to mitogens and antigens; (4) decreased production of a variety of lymphokines; (5) decreased responsiveness of T lymphocytes to thymic hormones; and (6) decreased resistance to tumor cell challenge [reviewed in (Makinodan, 1976; Weksler *et al.*, 1984; Weindruch *et al.*, 1982)].

We do not know if there is a causal relationship between diminished NA innervation of secondary lymphoid organs and diminished T lymphocyte functions. The decline in NA innervation is selective for spleen and lymph nodes, and is not seen in thymus or heart. Furthermore, the extent of decline in NA innervation of secondary immune organs may depend upon the history of antigenic exposure of the particular rodents; one group of rodents raised in a standard vivarium showed a greater decline in innervation of spleen that did another group of rodents of the same strain raised in a barrier facility. It is tempting to speculate that the denervation studies in young adults (showing markedly diminished primary immune responses) have a natural aging counterpart, in which diminished innervation of spleen and LNs contributes to a decline in T lymphocyte functions in those organs. This is a testable hypothesis that we now are investigating. At present, however, we do not know whether the diminished innervation of spleen and lymph nodes seen with age is: (1) a cause of diminished immune responsiveness; (2) a consequence of diminished immune responsiveness and altered immune compartmentation; (3) an epiphenomenon that has nothing to do with altered immune responsiveness; or (4) an unpredictable phenomenon that does not permit predictive experimental manipulation. However, in view of the striking influences that NE is able to exert on some functional capabilities of lymphocytes in young adults, it now seems appropriate to pursue the use of neuro-active agents in an attempt to enhance some specific immune functions in vivo.

Acknowledgements

This work was supported by N00014-84-K-0488 from the Office of Naval Research; R01 NS25223, T32 GM07356, and F32 NS07980 from NIH: R01 MH42073, K01 MH00572, and F31 MH09356 from NIMH; and a John D. and Catherine T. MacArthur Foundation Prize Fellowship to DLF.

References

Ackerman, K.D., Felten, S.Y., Bellinger, D.L. and Felten, D.L (1987). Noradrenergic sympathetic innervation of the spleen: III. Development of innervation in the rat spleen. J. Neurosci. Res. 18, 49-54.

Ackerman, K.D., Felten, S.Y., Bellinger, D.L., Livnat, S. and Felten, D.L. (1987). Noradrenergic sympathetic innervation of spleen and lymph nodes in relation to specific cellular compartments. Prog. Immunol. 6, 588-600.

Ackerman, K.D., Felten, S.Y., Felten, D.L. and Livnat, S. (1988). Presence of beta-adrenergic receptors on developing spleen cells and alterations in immunologic reactivity following neonatal sympathetic denervation. Soc. Neurosci. Abstr. 14, (Abstract).

Ader, R. (1981). Psychoneuroimmunology. Academic Press, New York.

Ader, R. and Cohen, N. (1975). Behaviorally conditioned immunosuppression. Psychosom. Med. 37, 333-340.

Ader, R. and Cohen, N. (1982). Behaviorally conditioned immunosuppression and murine systemic lupus erythematosus. Science 215, 1534-1536.

Ader, R. and Cohen, N. (1985). CNS-Immune system interactions: Conditioning phenomena. Behav. Brain Sci. 8, 379-395.

Ader, R., Cohen, N. and Felten, D.L. (1987). Brain, behavior, and immunity. Brain Behav. Immun. 1, 1-6.

Ader, R., Cohen, N. and Grota, L.J. (1979). Adrenal involvement in conditioned immunosuppression. Int. J. Pharm. 1, 141-145.

Ayers, A.B., Davies, B.N. and Withrington, P.G. (1972). Responses of the isolated perfused human spleen to sympathetic nerve stimulation, catecholamines and polypeptides. Br. J. Pharmac. 44, 17-30.

Bartrop, R.W., Luckhurst, E., Laarus, L., Kiloh, L.G. and Penny, R. (1977). Depressed lymphocyte function after bereavement. Lancet i, 834-836.

Bellinger, D.L., Felten, S.Y., Collier, T.J. and Felten, D.L. (1987). Noradrenergic sympathetic innervation of the spleen: IV. Morphometric analysis in adult and aged F344 rats. J. Neurosci. Res. 18, 55-63.

Berczi, I. (1986). Pituitary Function and Immunity. CRC Press, Boca Raton.

Berczi, I. (1987). Hormones and Immunity. MTR Press.

Besedovsky, H.O., del Rey, A.E., Sorkin, E., Burri, R., Honegger, C.G., Schlumpf, M. and Lichtensteiger, W. (1987). T lymphocytes affect the development of sympathetic innervation of mouse spleen. Brain Behav. Immun. 1, 185-193.

Besedovsky, H.O., del Rey, A.E., Sorkin, E., Da Prada, M. and Keller, H.H. (1979). Immunoregulation mediated by the sympathetic nervous system. Cell. Immunol. 48, 346-355.

Besedovsky, H.O., del Rey, A.E., Sorkin, E. and Dinarello, C.A. (1986). Immunoregulatory feedback between interleukin-1 and glucocorticoid hormones. Science 233, 652-654.

Besedovsky, H.O., Sorkin, E., Felix, D. and Haas, H. (1977). Hypothalamic changes during the immune response. Eur. J. Immunol. 7, 325-328.

Bishopric, N.H., Cohen, H.J. and Lefkowitz, R.J. (1980). Beta adrenergic receptors in lymphocyte subpopulations. J. Allergy Clin. Immunol. 65, 29-33.

Bovbjerg, D.H., Ader, R. and Cohen, N. (1982). Behaviorally conditioned suppression of a graft-versus-host response. Proc. Natl. Acad. Sci. (USA) 79, 583-585.

Bulloch, K. and Pomeranta, W. (1984). Autonomic nervous system innervation of thymic related lymphoid tissue in wild-type and nude mice. J. Comp. Neurol. 228, 57-68.

Calvo, W. (1968). The innervation of the bone marrow in laboratory animals. Am. J. Anat. 123, 315-328.

Calvo, W. and Forteza-Vila, J. (1969). On the development of bone marrow innervation in newborn rats as studied with silver impregnation and electron microscopy. Am. J. Anat. 126, 355-359.

Carlson, S.L., Felten, D.L., Livnat, S. and Felten, S.Y. (1987). Alterations of monoamines in specific central autonomic nuclei following immunization in mice. Brain Behav. Immun. 1, 52-63.

Carlson, S.L., Felten, D.L., Livnat, S. and Felten, S.Y. (1987). Noradrenergic sympathetic innervation of the spleen: V. Acute drug-induced depletion of lymphocytes in the target fields of innervation results in redistribution of noradrenergic fibers but maintenance of compartmentation. J. Neurosci. Res. 18, 64-69.

Coffey, R.G. and Hadden, J.W. (1985). Neurotransmitters, hormones, and cyclic nucleotides in lymphocyte regulation. Fed. Proc. 44, 112-117.

Cohen, N., Ader, R., Green, N. and Bovbjerg, D.H. (1979). Conditioned suppression of a thymus-independent antibody response. Psychosom. Med. 41, 487-491.

Cross, R.S., Markesberry, W.R., Brooks, W.H. and Roszman, T.L. (1980). Hypothalamic-immune interactions. I. The acute effect of anterior hypothalamic lesions on the immune response. Brain Res. 196, 79.

Ernstrom, U. and Sandberg, G. (1973). Effects of alpha- and beta-receptor stimulation on the release of lymphocytes and granulocytes from the spleen. Scand. J. Haematol. 11, 275-286.

Ernstrom, U. and Soder, O. (1975). Influence of adrenaline on the dissemination of antibody-producing cells from the spleen. Clin. Exp. Immunol. 21, 131-140.

Felten, D.L., Ackerman, K.D., Wiegand, S.J. and Felten, S.Y. (1987). Noradrenergic sympathetic innervation of the spleen: I. Nerve fibers associated with lymphocytes and macrophages in specific compartments of the splenic white pulp. J. Neurosci. Res. 18, 28-36.

Felten, D.L., Felten, S.Y., Bellinger, D.L., Carlson, S.L., Ackerman, K.D., Madden, K.S., Olschowka, J.A. and Livnat, S. (1987). Noradrenergic sympathetic neural interactions with the immune system: structure and function. Imm. Rev. 100, 225-260.

Felten, D.L., Felten, S.Y., Carlson, S.L., Olschowka, J.A. and Livnat, S. (1985). Noradrenergic and peptidergic innervation of lymphoid tissue. J. Immunol. 135, 755s-765s.

Felten, D.L., Livnat, S., Felten, S.Y., Carlson, S.L., Bellinger, D.L. and Yeh, P. (1984). Sympathetic innervation of lymph nodes in mice. Brain Res. Bull. 13, 693-699.

Felten, D.L., Overhage, J.M., Felten, S.Y. and Schmedtje, J.F. (1981). Noradrenergic sympathetic innervation of lymphoid tissue in the rabbit appendix: further evidence for a link between the nervous and immune systems. Brain Res. Bull. 7, 595-612.

Felten, S.Y., Bellinger, D.L., Collier, T.J., Coleman, P.D. and Felten, D.L. (1987). Decreased sympathetic innervation of spleen in aged Fischer 344 rats. Neurobiol. Aging 8, 159-165.

Felten, S.Y., Housel, J. and Felten, D.L. (1986). Use of in vivo dialysis for evaluation of splenic norepinephrine and serotonin. Soc. Neurosci. Abstr. 12, 1065.

Felten, S.Y. and Olschowka, J.A. (1987). Noradrenergic sympathetic innervation of the spleen: II. Tyrosine hydroxylase (TH)-positive nerve terminals form synaptic-like contacts on lymphocytes in the splenic white pulp. J. Neurosci. Res. 18, 37-48.

Fillenz, M. (1970). The innervation of the cat spleen. Proc. Roy. Soc. Lond. 174, 459-468.

Fox, B.H. (1981). Psychosocial factors and the immune system in human cancer. In: Psychoneuroimmunology Ader, R. ed. Academic Press, New York.

Geschwind, N. and Behan, P. (1982). Left-handedness- Association with immune disease, migraine, and developmental learning disorder. Proc. Natl. Acad. Sci. (USA) 79, 5097-5100.

Gillespie, J.S. and Kirpekar, S.M. (1965). The localization of endogenous and infused noradrenaline in the spleen. J. Physiol. 179, 46P.

Giron, L.T., Crutcher, K.A. and Davis, J.N. (1980). Lymph nodes - a possible site for sympathetic neuronal regulation of immune responses. Ann. Neurol. 8, 520-522.

Gorczynski, R.M., Macrea, S. and Kennedy, M. (1982). Conditioned immune response associated with allogenic skin grafts in mice. J. Immunol. 129, 704-709.

Hadden, J.W., Hadden, E.M. and Middleton, E. Jr. (1970). Lymphocyte blast transformation-I. Demonstration of adrenergic receptors in human peripheral lymphocytes. Cell. Immunol. 1, 583-595.

Kusnecov, A.W., Wivyer, M., King, M.G., Husband, A.J., Cripps, A.W. and Clancy, R.L. (1983). Behaviorally conditioned suppression of the immune response by antilymphocyte serum. J. Immunol. 130, 2117-2120.

Landmann, R., Bittiger, H. and Buhler, F.R. (1981). High affinity beta-2-adrenergic receptors in mononuclear leucocytes: Similar density in young and old subjects. Life Sci. 29, 1761-1771.

Landmann, R., Burgisser, E., West, M. and Buhler, F.R. (1985). Beta adrenergic receptors are different in subpopulations of human circulating lymphocytes. J. Recept. Res. 4, 37-50.

Livnat, S., Felten, S.Y., Carlson, S.L., Bellinger, D.L. and Felten, D.L. (1985). Involvement of peripheral and central catecholamine systems in neural-immune interactions. J. Neuroimmunol. 10, 5-30.

Livnat, S., Madden, K.S., Felten, D.L. and Felten, S.Y. (1987). Regulation of the immune system by sympathetic neural mechanisms. Prog. Neuro-psychopharmacol. & Biol. Psychiat. 11, 145-152.

Madden, K.S., Felten, D.L., Felten, S.Y. and Livnat, S. (1986). Sympathetic nervous system modulates natural and T cell immunity. Proc. 6th Int'l Cong. Immunol. 476.

Makinodan, T. (1976). Immunology of aging. J. Am. Geriatr. Soc. 24, 249-252.

Miller, N.E. (1983). Behavioral medicine - Symbiosis between laboratory and clinic. Ann. Rev. Pschol. 34, 1-31.

Ottaway, C.A. (1984). In vitro alteration of receptors for vasoactive intestinal polypeptide changes the in vivo localization of mouse T cells. J. Exp. Med. 160, 1054-1069.

Ottaway, C.A. (1985). Evidence for local neuromodulation of T cell migration in vivo. Adv. Exp. Med. 186, 637-645.

Payan, D.G. and Goetzl, E.J. (1985). Modulation of lymphocyte function by sensory neuropeptides. J. Immunol. 135, 783s-786s.

Plaut, S.M. and Friedman, S.B. (1981). Psychosocial factors in infections disease. In: Psychoneuroimmunology Ader, R. ed. pp. 3-30. Academic Press, New York.

Reilly, F.D. (1985). Innervation and vascular pharmacodynamics of the mammalian spleen. Experientia 41, 187-192.

Reilly, F.D., McCuskey, P.A., Miller, M.L., McCuskey, R.S. and Meineke, H.A. (1979). Innervation of the periarteriolar lymphatic sheath of the spleen. Tissue and Cell 11, 121-126.

Riley, V. (1981). Psychoneuroendocrine influences on immunocompetence and neoplasis. Science 212, 1100-1109.

Rogers, M.P., Reich, P., Strom, T.B. and Carpenter, C.B. (1976). Behaviorally conditioned immunosuppression: replication of a recent study. Psychosom. Med. 38, 447-452.

Roszman, T.L., Cross, R.J., Brooks, R.J., Brooks, W.H. and Markesberry, W.R. (1985). Neuroimmunonomodulation - Effects of neural lesions on cellular immunity. In: Neural Modulation of Immunity Guillemin, R., Cohn, M. and Melnechuk, T. eds. pp. 95-107. Raven Press, New York.

Sanders, V.M. and Munson, A.E. (1984). Kinetics of the enhancing effect produced by norepinephrine and terbutaline on the murine primary antibody response in vitro. J. Pharm. Exp. Ter. 231, 527-531.

Sanders, V.M. and Munson, A.E. (1984). Beta-adrenoceptor mediation of the enhancing effect of norepinephrine on the murine primary antibody response in vitro. J. Pharm. Exp. Ther. 230, 183-192.

Sanders, V.M. and Munson, A.E. (1985). Role of alpha adrenoceptor activation in modulating the murine primary antibody response in vitro. J. Pharm. Exp. Ther. 232, 395-400.

Sanders, V.M. and Munson, A.E. (1985). Norepinephrine and the antibody response. Pharmacol. Rev. 37, 229-248.

Sergeeva, V.E. (1974). Histotopography of catecholamines in the mammalian thymus. Bull. Exp. Bio. Med. 77, 456-458.

Singh, U., Millson, S., Smith, P.A. and Owen, J.J.T. (1979). Identification of beta-adrenergic adrenoceptors during thymocyte ontogeny in mice. Eur. J. Immunol. 9, 31-35.

Singh, U. and Owen, J.J.T. (1976). Studies on the maturation of thymus stem cells - The effects of catecholamines, histamine, and peptide hormones on the expression of T alloantigens. Eur. J. Immunol. 6, 59-62.

Sklar, L.S. and Anisman, H. (1979). Stress and coping factors influence tumor growth. Science 205, 513-515.

Solomon, G.F. (1981). Emotional and personality factors in the onset and course of autoimmune disease, particularly rheumatoid arthritis. In: Psychoneuroimmunology Ader, R. ed. Academic Press, New York.

Solomon, G.F., Amkraut, A.A. and Kasper, P. (1974). Immunity, emotions and stress, with special reference to the mechanisms of stress effects on the immune system. Ann. Clin. Res. 6, 313-322.

Stein, M. (1985). Bereavement, depression, stress, and immunity. In: Neural Modulation of Immunity Guillemin, R., Cohn, M. and Melnechuk, T. eds. pp. 29-41. Raven Press, New York.

Szentivany, A. and Filipp, G. (1958). Anaphalaxis and the nervous system - Part II. Ann. Allergy 16, 143-151.

Walcott, B. and McLean, J.R. (1985). Catecholamine-containing neurons and lymphoid cells in a lacrimal gland of the pigeon. Brain Res. 328, 129-137.

Wayner, E.A., Flannery, G.R. and Singer, G. (1978). The effect of taste aversion conditioning on the primary antibody response to sheep red blood cells and Brucella abortus in the albino rat. Physiol. Behav. 21, 995-1000.

Weindruch, R.H. and Walford, R.L. (1982). In: The Reticuloendothelial System. A Comprehensive Treatise. v. 3. Phylogeny and Ontogeny Cohen, N. and Sigel, M.M. eds. pp. 713-748. Plenum Press, New York.

Weiner, H. (1977). Psychobiology and Human Disease. Elsevier, New York.

Weksler, M.E. and Siskind, G.W. (1984). The cellular basis of immune senescence. Monogr. Dev. Biol. 17, 110-121.

Williams, J.M. and Felten, D.L. (1981). Sympathetic innervation of murine thymus and spleen: A comparative histofluorescence study. Anat. Rec. 199, 531-542.

Williams, J.M., Peterson,R.G. Shea, P.A., Schmedtje, J.F., Bauer, D.C. and Felten, D.L. (1981). Sympathetic innervation of murine thymus and spleen: Evidence for a functional link between the nervous and immune systems. Brain Res. Bull. 6, 83-94.

Figure Legend

1. Tyrosine hydroxylase (TH)-positive nerve fibers in the spleen of a 1 day old Fischer 344 rat. Nerve fibers (arrowheads) are located mainly at the outer edge of the developing perarteriolar lymphatic sheath, PALS (p), surrounding the central artery (a) of the white pulp. Double label immunocytochemistry for TH nerve fibers (black) and OX-19, a pan-T lymphocyte marker (brown). The PALS also contains B lymphocytes at this early stage, not stained in this micrograph. X 185.
2. Tyrosine hydroxylase (TH)-positive nerve fibers in the spleen of a 7 day old Fischer 344 rat. These fibers, stained black with nickel enhancement, are located along the outer edge of the PALS (p) (arrowheads), and at this stage have moved inward along the central arteriolar system (a). TH+nerve fibers are scattered throughout the PALS. Doubles label immunocytochemicstry for TH+nerve fibers and OX-19+T lymphocytes (brown). X185.
3. Tyrosine hydroxylase (TH)-positive nerve fibers in the spleen of a 14 day old Fischer 344 rat. These fibers, stained black with nickel enhancement, are located along the developing marginal sinus (s) (large arrowheads), within the PALS (p), along the central arteriolar system (a), and within the marginal zone (z) (small

arrowheads). Triple label immunocytochemistry for TH+nerve fibers (black), ED3+macrophage along the marginal sinus (brown), and OX-19+T lymphocytes in the PALS (brown). X 185.

4. Tyrosine hydroxylase (TH)-positive nerve fibers in the spleen of a 28 day old Fischer 344 rat. These fibers, stained with nickel enhancement, are particularly prominent along the inner edge of a follicle (f) (arrowheads), in addition, to their usual location within the PALS (p), along the central arteriolar system (a) and along the marginal sinus. z = marginal zone. Double label immunocytochemistry for TH+nerve fibers (black) and IgM+cells (mainly B lymphocytes) (brown). X185.
5. Tyrosine hydroxylase (TH)-positive nerve fibers in the spleen of a 3 month old Fischer 344 rat. These fibers, stained with nickel enhancement, are located around the central arteriolar system (a), within the PALS (p), along the outer edge of the PALS at the marginal sinus (arrowheads), and along the large veins (v) associated with the white pulp. This represents the mature adult pattern of innervation. Double label immunocytochemistry for TH+nerve fibers (black) and OX-19+T lymphocytes (brown). X 185.
6. Tyrosine hydroxylase (TH)-positive nerve fibers in the spleen of a 12 month old Fischer 344 rat. These fibers, stained with nickel enhancement, are located around the central arteriolar system (a), within the PALS (p), along the marginal sinus at the outer edge of the PALS, and in parafollicular locations (arrowheads). f=follicle, z=marginal zone. These 12 month old animals retain the adult pattern of innervation. Double label immunocytochemistry for TH+nerve fibers (black) and OX-19+T lymphocytes (brown). X 92.
7. Tyrosine hydroxylase (TH)-positive nerve fibers in the spleen of a 27 month old Fisher 344 rat. These fibers, stained with nickel enhancement, are sparse in number, but are present in the usual adult compartments. They are present along the central arteriolar system (a), within the PALS (p), and along the marginal sinus. However, at this age, fewer T lymphocytes are present in the PALS compared with younger ages. Double immunocytochemistry for TH+ nerve fibers (black) and OX-19+T lymphocytes (brown). X 185.
8. Tyrosine hydroxylase (TH)-positive nerve fibers in the spleen of a 27 month old Fischer 344 rat. This is a lower magnification photomicrograph of the PALS (p) at 27 months of age demonstrating diminished innervation in a zone possessing numerous T lymphocytes. a = central arteriolar system. Double label immunocytochemistry for TH+nerve fibers (black) and OX-19+T lymphocytes. X 92.

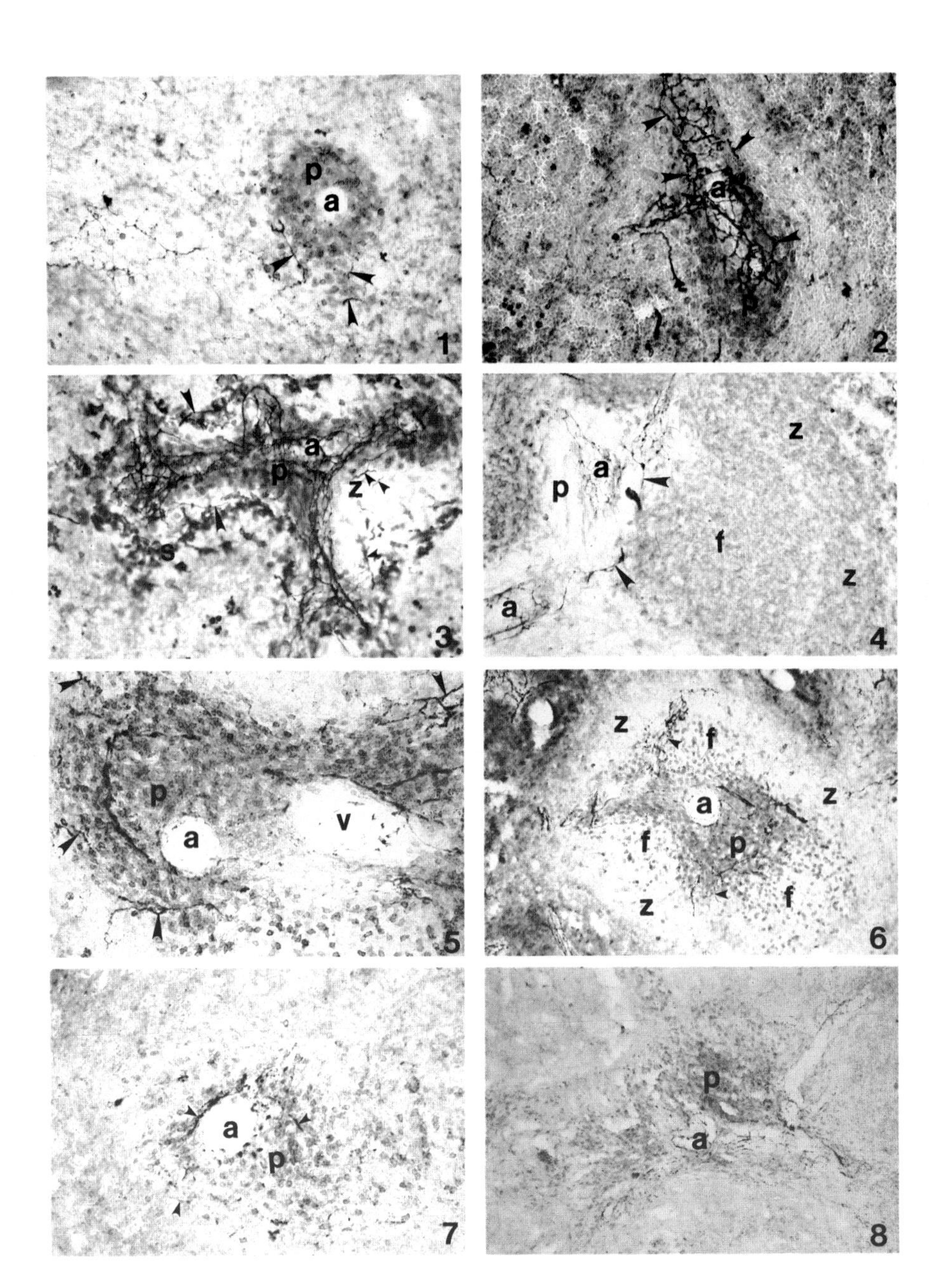
p
a
1
a
2
a
p
z
s
3
z
p
a
f
z
a
4
p
a
v
5
z
f
a
z
f
p
z
f
6
a
p
7
p
a
8

21

IS SENESCENCE OBLIGATORY in EUKARYOTIC CELLS ?

Caleb E. Finch

Andrus Gerontology Center
and the Dept. of Biological Sciences
University of Southern California
Los Angeles CA 90089-0191

I. Introduction

This essay examines the question: is senescence obligatory in eukaryotic cells? Although many biologists believe that cell senescence is general, inevitable, and underlies organismic senescence, I propose to the contrary that all somatic cell lineages may not be inevitably destined for senescence. Moreover, some manifestations of cell and organismic senescence represent the outcomes of specific patterns of gene regulation established during differentiation and development and/or epiphenomena from various aspects of function or exposure to risk during adult life. The strategy in this brief inquiry is compare aspects of senescence in species selected from diverse phyla and to give examples of the manipulability of senescence in others.

The term senescence is often applied to age-related changes that need have no mechanism in common. At a population level, senescence is often detected by exponential increases in the mortality rate as a function of age (Comfort, 1979) which are attributable to physiological (intrinsic) rather than to ecological factors. At the organ level, senescence is accompanied by dysfunctions and diseases which increase an exponential rate, roughly in advance of the increase of mortality (Simms and Berg, 1957). The contributions of dysfunctions in different systems to mortality risk are usually difficult to resolve, because many factors may make small contributions to mortality risk which are additive or multiplicative. At the cell level, even more difficulties arise in identifying the causes of organismic senescence, because finite lifespans are a normal feature in the life cycles of many mammalian cells (e.g. erythrocytes, exfoliating epithelia) and need not be related to organismic at all senescence. Moreover, consider the postmitotic status of human diploid fibroblasts after their finite number of divisions in vitro, which is a widely studied model in gerontology (Hayflick, 1977). I suggest that clonal senescence has no direct bearing on cell senescence, because we should then describe postmitotic neurons of the newborn human as senescent, which seems inappropriate. Even in vitro, the postmitotic status of fibroblasts does not imply loss of many other cell functions, as judged by their ability to support viral infections (Holland et al., 1973; Tomkins et al., 1974) and to survive for many months in a stationary state (Bell et al., 1978). The postmitotic status of terminally postmitotic fibroblasts may be consequent to the presence of a few species of mRNA and hence the activities of a small set of genes (Lumpkin et al., 1986). There is no evidence for a general failure of genomic functions in terminally postmitotic diploid fibroblasts.

As other examples of the maintained function of postmitotic cells at advanced ages, consider the hypothalamic capacity to secrete vasopressin at advanced ages in humans; this system becomes even more responsive to increases of blood osmolality (Helderman et al., 1978). Similarly, the capacity to secrete corticosteroids and/or ACTH during stress is undiminished at ages beyond the average lifespan in humans (Blichert-Toft, 1975) and in rodents (Finch et al., 1969;

Sapolsky et al., 1986). In regard to effects of age on brain messenger RNA populations, we showed that total rat brain polysomal poly(A)RNA does not change in amount or nucleotide sequence complexity at least through the average lifespan (Colman et al., 1980); this study does not preclude changes in select neurons or glia (see below), but sets an upper limit of ca. 10% on the average extent of change throughout the brain. Although these examples indicate that specific functions of some postmitotic cells can be maintained for prolonged times, other functions and cells do show age-related degeneration that are properly called senescent. These examples suggest that the postmitotic status does not inevitably cause cell senescence. As described next, some diploid cell lines of invertebrates have unlimited capacities for proliferation.

II. Asexual reproduction and the capacity for cell proliferation

Limiting numbers of divisions that are characteristics of fibroblasts and many other mammalian cells *in vitro* as well as of differentiating neurons, erythroblasts, and many other cells *in vivo*. However, many species of plants and animals reproduce asexually and agametically by variations of somatic cell cloning. Major phyla with asexually reproducing species include: annelids, bryozoans, coelenterates, ctenophores, echinoderms (starfish), mesozoa, nimertines, platyhelminthes, pogonophores, porifera, sipunculids, and tunicates (Bell, 1982), which together include a huge number of species that can reproduce asexually. It is very interesting to find asexual reproduction in the tunicates (Brien et al., 1948; Freeman, 1964), which are primitive chordates with a two-chambered heart, an intestine, and immune system. Sabbadin (1979) comments on the absence of an 'Hayflick limit' to clonal proliferation in *Botryllus schlosseri* during asexual reproduction for hundreds of generations. These facts are well known to comparative geneticists and comparative reproductive physiologists (Bell, 1982), and give a different perspective on the finite proliferative capacity of diploid mammalian fibroblasts *in vitro*. In view of the capacities for asexual reproduction from somatic cells in starfish and tunicates, I propose that *in vitro* clonal senescence, as shown in fibroblasts from many vertebrates, represents a specific pattern of gene expression in fibroblasts that emerged in select deuterostome phylogenetic lineages. Alternatively, asexual reproduction could have arisen in deuterostomes from mutations that altered the patterns of gene regulation in regard to clonal senescence of certain somatic cells.

Nonetheless, organismic senescence may occur in species that reproduce asexually. In Botryllus the parental cells simultaneously degenerate in colonies after the asexual buds mature through a mechanism involving autoreactive blood cells that may be triggered by histocompatibility antigens (Harp et al., in press). Other asexually reproducing species such as porifera can be maintained for many decades with no signs of senescence (Comfort, 1979). Thus, there does not seem to be an obligatory link between organismic senescence and the presence or absence of asexual reproduction.

Many other organisms with very similar organization at the tissue, cellular, and subcellular level cannot reproduce asexually. Major phyletic groups without asexual reproduction include: arthropods, brachiopods, caetognaths, echiuroids, gastrotrichs, gordiaceans, molluscs, nematodes, rotifers, tardigraves, and all vertebrates. The absence of asexual reproduction in these species can not be attributed to obvious differences in cell characteristics, but may be sought in many specific differences that arise during differentiation. For example, nematodes as a rule lack proliferating cells excepting those in the gonads; some species (Ascaris) even loose genomic totipotency in most somatic cells by chromosomal diminution (Davidson, 1986). The rigid and irreplaceable exoskeleton and postmitotic status of most somatic cells of insects and rotifers would also limit the potential for asexual reproduction.

The limited capacity of many other cell types for proliferation may be viewed as an aspect of differentiation, rather than a fundamental limitation of higher cells. As noted above, activity of small number of genes appears to determine the proliferative arrest of diploid human fibroblast cultures. In mammals, cells have widely varying capacities for proliferation that emerge during development. Most neurons are permanently postmitotic in mammals, while neurons in the olfactory bulb continue to be formed in 6 month old rats (Bayer, 1983). In some birds, neurons proliferate seasonally and become incorporated into functioning circuits (Paton and Notebohm, 1984). At the other extreme, some cells have vast proliferative capacity, e.g. the stem cells for hemopoiesis. As shown by serial transplantation (Harrison, 1985), bone marrow does not become depleted of hemopoietic cells at ages far beyond the average mouse lifespan. The possibility that some somatic stem cells in mammals have truly unlimited proliferation would be consistent with the unlimited asexual reproduction by somatic cells in their distant relative, the protochordate tunicates.

III. Typologies in senescence

An overview of the natural history of senescence in animals and plants shows great diversity in the organ systems and functions that cause senescent death. It is useful to classify patterns of senescence according to a spectrum ranging from rapid to gradual to negligible senescence. Species manifesting rapid or gradual senescence generally show exponential age-related increases of mortality rate and maximum lifespans. However, the lifespan alone does not always indicate the type of senescence, as some organisms show rapid involution after very long lives. For example, many species of bamboo remain in a juvenile vegetative phase that lasts for decades, even beyond a century, before they flower and rapidly senesce (Janzen, 1976). As another example showing the difficulties of interpreting the maximum reported lifespan, many species of birds have very high overall (age-independent) mortality rates (up to 70% per year) that dwarf the age-related mortality rate (Botkin and Miller, 1974). Apparently short lifespans may thus give no indication about the rate of senescence. More about birds later.

Rapid senescence is typified by species like the spawning Pacific salmon, the octopus, marsupial mice, or aphagous insects which manifest nearly synchronous senescence with specific characteristics of senescence as a population (Table 1). Each of these species has different pathophysiological events (reviewed in Finch, 1987). Reproduction with neuroendocrine involvement often triggers these changes (illustrated below). The definable causes in most of these species implicate a senescent changes in a limited number of cells. By inference a limited number of genes are the generators of senescence. There is little doubt that most of the other cells in these organisms could survive far longer than the usual lifespan. In some cases this can be proven by experimental manipulations which prevent the onset of senescence and allow survival to a supranatural age, as illustrated by castration of Pacific salmon before spawning, which prevents the adrenal cortex from becoming hyperactive and causing senescence in association with Cushingoid elevations of blood corticosteroids (Robertson, 1961). Rapid senescence in the mouse-sized marsupial _Antechinus_ is also associated with elevated corticosteroids; interventions permit survival beyond a year to a more usual lifespan for mammals of this size (Lee et al., 1982). A different example is given by some species of _Octopus_, which commonly cease to eat after mating and appear to die from starvation (Wodinsky, 1977). In at least one species (_O. hummelincki_) this involution was prevented by removal of the optic ganglion before spawning (Wodinsky, 1977). Other examples of lifespan manipulation are available for rapidly senescing species (Finch, 1976, 1987). In many such instances, senescence can be considered nonobligate. Moreover, geographic and climatic variations can permit survival of _Antechinus_ and other annual species to exceed their usual lifespans. Thus, the position in the spectrum of senescence is subject to environmental and experimental modifications. In other species, senescence is obligate, e.g. in many insects which lack mouth parts (see below) or in the inability to replace wings and other body parts that wear out through mechanical usuage.

Most mammals show _gradual senescence_, with more widely varying characteristics between individuals in a population, even in highly inbred laboratory rodents (Nelson et al., 1982). The exponential mortality constants vary over a 30-fold range and humans have the smallest values (Sacher, 1977). No manipulations of lifespan or of the patterns of senescence have been achieved that are as dramatic as for the rapidly senescing species. However, the exponential mortality rate can be slowed (Sacher, 1977) and many pathophysiological changes of senescence can be postponed in rodents by dietary restriction (Yu et al., 1982). Specific manipulations of neuroendocrine age changes in rodents are described in Section V.

Negligible senescence is shown by sponges, tortoises, and possibly some long-lived fish and birds which do not have a clearly defined pathophysiological senescence and definite maximum lifespan. Many woody plants also have extremely long-lifespans that would suggest negligible senescence. Unfortunately, there is little reliable data on population age structures of these long-lived organisms for studying how mortality rate changes at advanced ages, with the exception of data from banded wild birds. The long lifespans and very small age-related mortality coefficients of some marine birds (Botkin and Miller, 1974; Witten and Finch, in prep.) contrast with the short lifespans and manifestations of senescence in domestic fowl (Comfort, 1979).

TABLE I. Typology in Sudden Death and Rapid Senescence

A. <u>Obligate Rapid Senescence</u>	<u>Examples</u>
1. Aphagy due to congenital defects of mouth or gut organs.	Mayflies, moths
2. Mechanical sensecence due to wearing out of wings or other replaceable organs.	winged insects
3. Birth-related maternal death.	orthonectids, a few nematodes and flatworms
4. Virus- or episome-induced	some fungal strains
5. Hormonally triggered by sexual maturation.	many plants, Pacific salmon
6. Aphagy triggered by mating.	some spiders
B. <u>Non-obligate rapid senescence</u>	
1. Stress-induced during breeding	marsupial mice species that can be either iteroparous or semelparous
2. Seasonally-induced	many plants can be annuals or perennials

IV. Epigenetic determinants of the patterns of senescence

As described above, the presence or absence of asexual reproduction in some species of the same phlogenetic lineage (e.g. tunicates vs. vertebrates) appears to be an outcome of the particulars of gene expression that differ between species. The genetic determinants for the expression of asexual reproduction are not known at present, but would require genomic totipotency and unlimited clonal expansion in some cell lineages. While premature to speculate on the details of these epigenetic mechanisms, they may not be much different from the mechanisms of transdifferentiation manifested during regeneration in lower vertebrates (Davidson, 1986).

A different aspect of this issue concerns the extent of variations in lifespan that can arise from the same genotype. Social insects provide important examples of alternate programming of the same genome to yield adults with radically different lifespans and characteristics of senescence, short-lived worker bees with exponential increases of mortality and characteristics of rapid senescence vs. much longer-lived queens, with very gradual or negligible senescence. Both arise from the same clutches of diploid eggs. If bee larvae are not fed a special diet, they become workers with lifespans of months and exponential increases of mortality. If fed bee milk, they become queens with lifespans of least 5 years (Comfort, 1979). Termites have similar caste differentiation in lifespan; in a few cases, termite queens continued to lay eggs for at least a decade (Haskins, 1960). Thus the same genome can be epigenetically programmed to create adults with lifespans that differ at least by 10-fold.

Other examples that imply epigenetic programming of the lifespan are the thousands of butterfly and moth species which lack parts of their mouths or digestive tracts as adults, and male rotifers which universally lack a digestive tract as adults (references in Finch, 1987). Their inability to eat as adults clearly limits the lifespan potential and inevitably leads to obligate rapid senescence from starvation if not from some other cause. Few such species have adult lifespan of more than one months, although development may take much longer.

Mammals do not offer corresponding examples of major differences in lifespan or patterns of senescence within a species or genotype that arise from epigenetic programming.

However, an example pertinent to a major physiological change, female reproductive senescence, is the developmental determination of the numbers of ovarian oocytes. The lifelong stock in rodents and humans is fixed by birth and declines progressively thereafter. According to the species, the onset of menopause or its equivalent is associated with the near or complete exhaustion of these cells (Vom Saal and Finch, 1988). In general contrast to mammals, bony fish continue to produce new oocytes after maturity (Tokartz, 1978). Moreover, the mass of eggs produced scale with the size of the fish and shows no sign of decreasing at advanced sizes e.g. in the Atlantic cod (Mays, 1967). In the case of mammals, the genes limiting the proliferation of oogonia are evidently set to terminate proliferation sooner than in fish, where the potential may never be lost in some species. There is an interesting parallel with neuronal proliferation here. The loss of cell regenerative capacity in the brain, ovary, and other organs of mammals can be considered to result from events during differentiation for which alternate patterns exist in the lower vertebrates with continuing oogenesis. Because damage to these and other postmitotic cells can occur during aging, the causes of senescence in postmitotic cells may be obligate only insofar as they are consequent to aspects of differentiation that limit regeneration.

V. Mechanisms of senescence in nondividing cells that may not be obligate

The brain is a crucial locus of senescence in mammals because of the irreplaceable nature of its neurons and because of its position in the physiological hierarchy. Neuroendocrine mechanisms in mammalian senescence can be demonstrated in several brain loci, but the consequences are not as dire or fatal as observed in the species of salmon, marsupial mice, and octopus that show rapid senescence. The recurrence of neuroendocrine mechanisms in rapid senescence (Table 1) adds to interest in identifying which aspects of mammalian senescence could be linked to neuroendocrine functions. Evidence is growing to suggest that at least some age changes in neuronal functions results from influences extrinsic to those cells. Two brain systems show impact of endogenous steroids that account for subsets of senescent changes: estrogens acting on the hypothalamus and pituitary, and adrenal steroids acting on the hippocampus. Age-related changes in the brain that may also be influenced by diet and by degeneration of afferent and efferent pathways give other examples of nonobligate and modifiable cell senescence.

A. Ovarian steroids and female reproductive senescence

Laboratory rodents at middle age show marked declines of fertility because of changes in at least two loci, the ovary and the hypothalamus (Finch et al., 1984; Finch, 1987). Although their ovaries are approaching depletion of oocytes, the number of ova shed at ovulation remains remarkably constant. At the same time, estrous (fertility) cycles lengthen and the size of the estrogen-dependent surge of luteinizing hormone at proestrus decreases markedly. Within a space of a few months, estrous cycles will cease entirely. Replacement of old ovaries by young grafts to beneath the kidney capsule does not restore lengthened cycles or reinitiate cycles once they are lost. The inability of young grafts to reverse these changes is consistent with the importance of hypothalamic dysfunctions in the reproductive senescence of female rodents. These and other markers of reproductive senescence from my lab and others are summarized in Table II (for details, see Finch et al., 1984; Felicio et al., 1983; Gordon et al., 1988; Mobbs et al., 1985; Nelson et al., 1982; Telford et al., 1986).

So far no specific neurons have been identified as primary loci of change. For example, the number and cytological appearance of LHRH-containing neurons is unchanged well beyond the age when cycles have stopped (Hoffman and Finch, 1986). This negative finding illustrates an important issue in the physiology of senescence: that major functional changes can occur without drastic loss of cells. Nonetheless, the glial hyperactivity in the arcuate nucleus of the hypothalamus at middle age in female rats and mice (Schipper et al., 1981) suggests of neuronal damage; efforts continue to find neuronal damage or loss.

Although the cellular mechanisms of reproductive senescence in the brain are poorly documented, the physiological mechanisms are becoming well understood. One of the major factors in hypothalamic-pituitary changes is the impact of ovarian steroids. This possibility first emerged from Aschheim's pioneering study in which he grafted young ovaries into old rats that were ovariectomized when they were young (Aschheim, 1965). In contrast to old intact controls in which the young grafts did not reactivate cycles, the long-term ovariectomized rats showed some cycles. We have extensively studied this phenomenon, which implies that the presence of the ovary induces damage in the hypothalamus and pituitary. Nearly all aspects of neuroendocrine reproductive senescence are delayed or reduced by removing the ovaries of young laboratory mice and rats (Table II). For these parameters,

TABLE II. Manipulation of Reproductive Aging: Ovarian- and E_2-induced Neuroendocrine Aging Syndromes of Rodents

Markers of Reproductive Aging	Delayed, by chronic OVX in old	Accelerated by chronic-E_2 in young
Ovarian cycles lengthened	yes	?
Ovarian cycles lost	yes	yes
Smaller E_2-induced LH surge	yes	yes
Pulsatile release of LH	yes	yes
Smaller postOVX LH elevations	yes	yes
Glial hyperactivity in arcuate n.	yes	yes
Reduced pituitary stalk blood DA	yes	yes
Increased pituitary DA	yes	yes
Increased pituitary glucose-6 phosphate dehydrogenase	yes	?
Lactotrope adenomas	yes	yes
Increased number of lactotropes	yes	yes
Prolactinemia	yes	yes

For references see Finch _et al_. (1980,1984), Gordon _et al_. (1988), Mobbs _et al_. (1985) and Telford _et al_. (1986).

manifestations of senescence depend on the cumulative or continuing presence of ovarian steroids. Because the damage leading to acyclicity appear to occur in the final stages of ovarian oocyte depletion, we conclude that the limited stock of ovarian oocytes in the major pacemaker of the ovary-dependent neuroendocrine senescence.

Estradiol appears to be the ovarian steroid responsible for the ovary-dependent hypothalamic-pituitary syndrome of reproductive senescence. When administered to either intact or ovariectomized young mice, the parameters that were delayed by chronic ovariectomy are then prematurely induced (Table II). Each parameter may have different requirements for the dose and duration of steroid exposure, and show reversibility to varying degrees. For example, the glial hyperactivity in the arcuate nucleus is the least reversible by ovariectomy in middle aged mice (Schipper et al., 1981) while impairments in the regulation of luteinizing hormone are moderately reversible (Mobbs et al., 1985), and the changes of anterior pituitary dopamine (Telford et al., 1986) and glucose-6-phosphate dehydrogenase (Gordon et al., 1988) are freely reversible. Approximately 12 weeks of sustained daily exposure to high physiological levels of estradiol in the drinking water causes ovariectomized mice to permanently loss the ability to sustain estrous cycles when given young ovarian grafts (Kohama et al., 1986), which is equivalent to about 3000 pg-days/ml plasma of estradiol exposure.

Taken together, these phenomena imply an unexpected facet of senescent dyfunctions in neurons, that an endogenous hormone may cause damage during the course of normal physiological functions. These phenomena require modification of the view that gonadal steroids have permanent or organizing effects on the mammalian brain only during development. Moreover, they support the search for other neuronal changes that may be linked to endogenous molecules.
The generalization of these mechanisms to humans is very uncertain in view of the more passive role of the hypothalamus in the ovulatory cycles as compared with laboratory rodents. However, studies on Turner's syndrome indicate that estrogens also have some irreversible effects on the human hypothalamus. Some adult Turners' with complete gonadal dysgenesis have been treated with estradiol, and after withdrawal of steroid then developed hot flashes for the first time (Yen, 1977).

Yet another hormonal manipulation of female reproductive senescence is the effects of hypophysectomy in retarding the rate of oocyte loss in mice (Jones and Krohn, 1961). Hormones have crucial roles in age-related dysfunctions at the ovary, whose hormones influence neuroendocrine senescence, at neuroendocrine loci, whose control of gondotrophins influences ovarian senescence.

B. Corticosteroid-hippocampal interactions.

Adrenal corticosteroids are also implicated in senescent changes in the rat hippocampus. The age-related loss of large neurons with receptors for corticosteroids and the hyperactive glia in the hippocampus among other changes are retarded by chronic adrenalectomy and prematurely induced in young rats by sustained exposure to corticosteroids (Sapolsky et al., 1986; Landfield et al., 1978). These findings suggest that under some circumstances sustained stressors have adverse effects on brain neurons. Again, the generalization to human neuronal senescence is not established. The molecular and cellular mechanisms are not known. We are studying stress and corticosteroid-sensitive messenger RNA in the hippocampus that may be useful in identifying transitions from the reversible to irreversible damage in neurons during chronic exposure to steroids (Nichols et al., 1986). The occurrence of corticosteroid- or stress-associated senescent death in salmon and marsupial mice (Table I) suggests that these mechanisms represent a general vulnerability in vertebrates whose consequence vary widely between genotypes.

C. Nutritional influences

Influences of diet on the patterns of senescence and longevity are being extend to the cell level in many laboratories, and are briefly mentioned here to identify them as important factors in the environment of cells that may contribute to nonobligate senescence. Diet restriction in rodents is the best established example of reducing the exponential mortality rate through environmental interventions (Sacher, 1977; Yu, 1985). At the cell level, moderate diet restriction of female laboratory mice can reduce the rate of ovarian oocyte loss (Nelson et al., 1985). The mechanism is unclear, but may involve reductions in stimulation by gonadotrophins; oocyte loss is also reduced by hypophysectomy (see above). A different example concerns the age-related decrease of dopamine (D-2) receptors in the striatum, which we showed to begin before midlife in rodents and humans (Morgan et al., 1987). The decrease of D-2 receptors could result from loss of neurons and/or neuronal projections (axons or dendrites). Again, diet restriction slows the extent of D-2 loss (Levin et al., 1981). Besides the possibility of dietary influences on hormones that could directly or indirectly affect virtually all neurons, Cerami (1987) showed cumulative damage by blood glucose on slowly replaced proteins through nonezymatically glycosylation. The many neurological complications of diabetes suggests that endogenous glucose could be an important general factor in neuronal deterioration durng senescence that is not obligate.

D. Deafferentation syndromes

A new candidate for endogenous influences that alter neuron function during age-related neurological diseases is the degeneration of afferent pathways. An extensive cytological literature documents atrophy of cell bodies and nucleoli in some types of neurons in age-related neurological diseases such as Alzheimer's and Parkinsonism (Mann, 1985). Neuronal atrophy also occurs to some extent in the neurologically normal elderly, e.g. in the substantia nigra (Mann and Yates, 1979). Experimental lesions of projection systems can also induce similar changes (Arendash, 1987; Pearson et al., 1983). We recently found that long-term consequences of lesions in the substantia nigra of rats include the shrinkage of nucleoli and a selective loss of the messenger RNA for tyrosine hydroxylase, to a much greater extent than for ß-tubulin (Pasinetti et al., 1987). We hypothesize that the atrophy of the remaining nigral neurons reflects a deafferentation syndrome, in which the reduced nigrostriatal input causes a transynaptic effect, possibly through decreased trophic factors, at the striatonigral terminals. The potential involvement of growth factors in cell senescence in the basal ganglia and elswhere is indicated by the reversal of striatal neuron atrophy in senescent rats by local infusions of nerve growth factor (Fischer et al., 1987). Changes of messenger RNA in the atrophy of select neurons during senescence would be difficult to detect in studies of whole brain (Colman et al.,1980).

VI Summary

From these briefly presented examples, I argued that eukaryotic somatic cells do not obligatorily show clonal senescence, as illustrated by the many species of deuterostomes and protostomes that reproduce from somatic cells. In these species, the successive vegetatively produced generations are equivalent to somatic cell clones and demonstrate that clonal senescence need not be an obligate characteristic of cell differentiation. Among organisms that show well-defined senescence with exponential increases of mortality, senescent dysfunctions in many organs and cells may also be nonobligate, and can be extensively manipulated through hormones (Table 2), diet, and other factors extrinsic to cells. Many rapidly senescing species (Table 1) are open to major extensions of lifespan and modification of senescence. Thus, many features of senescence at the organismic, organ, and cell levels are not intrinsic to aging in the sense that they are not obligate consequences of time passing. Although environmental influences on mammalian lifespan and mortality rates are more modest (<2-fold in rodents), than the 30-fold differences in lifespan and mortality rates between species, the very low mortality rates of some long-lived birds, which is even lower than for humans (Botkin and Miller, 1974; Witten and Finch, in prep.) suggests a major potential for increasing the lifespan potential of other warm-blooded vertebrates.

REFERENCES

Arendash, G.W., Millard, W.J., Dunn, A.J., and Meyer, E.M. (1987): Long-term neuropathological and neurochemical effects of nucleus basalis lesions in the rat. Science, 238:952-956.

Bell, E., Marek, L.F., Levinstone, D.S., Merrill, C., Sher, S., Young, I.T., and Eden, M. (1978). Loss of division potential in vitro: Aging or differentiation. Science, 202:1158-1163.

Bell, G. (1982). "The Masterpiece of Nature: The Evolution and Genetics of Sexuality." University of California Press, Berkeley.

Blichert-Toft, M. (1975): Secretion of corticotrophin and somatotrophin by the senescent adenohypophysis in man. Acta Endocrinol., 78:15-154.

Botkin, K. and Miller, R. (1974): Mortality rates and survival of birds. Am. Nat., 108:181-192.

Brien, P. (1948): Embranchement des tuniciers. Morphologie et reproduction. In "Traite de Zoologie," edited by P.P. Grasse, Vol. II, pp. 553-894. Masson, Paris.

Cerami, A., Vlassera, H. and Brownlee, M. (1987): Glucose and aging. Sci. Am., 256:90-96.

Colman, P.D., Kaplan, B.B., Osterburg, H.H., and Finch, C.E. (1980): Brain poly(A)RNA during aging: Stability of yield and sequence complexity in two rat strains. J. Neurochem., 34:335-345.

Comfort, A. (1979): The Biology of Senescence, 3d Ed. Churchill Livingstone, Edinburgh.

Felicio L.S., Nelson, J.F., and Finch, C.E. (1983): Restoration of ovulatory cycles by young ovarian grafts in aging mice: Potentiation by long-term ovariectomy decreases with age. Proc. Nat. Acad. Sci., 80:6076-6080.

Finch, C.E. (1976): The regulation of physiological changes during mammalian aging. Q. Rev. Biol., 51:49-83.

Finch, C.E. (1987): Neural and endocrine determinants of

senescence: investigaton of causality and reversibility of laboratory and clinical interventions. In "Modern Biological Theories of Aging," edited by E. Knobil, Raven Press, New York, pp. 261-306.

Finch, C.E., Felicio, L.S., Mobbs, C.V., and Nelson, J.F. (1984): Ovarian and steroidal influences on neuroendocrine aging processes in female rodents. Endocr. Rev., 5:467-497.

Finch, C.E., Foster, J.R., and Mirsky, A.E. (1969): Aging and the regulation of cell activities during exposure to cold. J. Gen. Physiol., 54:690-712.

Freeman, G. (1964): The role of blood cells in the process of asexual reproduction in the tunicate Perophora viridis. J. Exp. Zool., 156:157-184.

Gordon M.N., Mobbs, C.V., and Finch, C.E., (1988): Pituitary of hypothalamic glucose-6-phospate dehydrogenase: effects of estradiol and age in C57BL/6J mice. Endocrinology. 122, in press.

Haskins, C.P. (1960): Note on the natural longevity of fertile females of Aphaenogaster picea. New York Ent. Soc., 68:66-67.

Harp, J.A., Tsuchida, C.B., Weisman, I.L. and Scofield, V.L. (1988): Autoreactive Blood cells and programmed cell death in growth and development of protochordates. J. Exp. Zoology., in press.

Harrison, D.M. (1985): Cell and tissue transplantation: A means of studying the aging process. In "Handbook of the Biology Aging," edited by C.E. Finch and E.L. Schneider, pp. 322-356. Van Nostrand, New York.

Hayflick, L. (1977): The cellular basis for biological aging. In "Handbook of the Biology of Aging," edited by C.E. Finch and E.L. Schneider, pp. 159-188. Van Nostrand, New York.

Helderman, J.H., Vestal, R.E., Rowe, J.W., Tobin, J.D., Andres, R., and Robertson, G.L. (1978): The response of arginine vasopressin to intravenous ethanol and hypertonic saline in man: The impact of aging. J. Immunol., 129:2673-2677.

Hoffman, G.E. and Finch, C.E. (1986): LHRH neurons in the

femal C57BL/6J mice: No loss up to middle-age. Neurobiol. Aging, 7:45-48.

Holland, J.J., Kohne, D. and Doyle, M.V. (1973): Analysis of virus replication in ageing human fibroblast cultures. Nature, 245:316-318.

Janzen, D.H. (1976): Why bamboos wait so long to flower. Annu. Rev. Ecol. Systematics, 7:347-391.

Jones, E.C., and Krohn, P.L. (1961): The effect of hypophysectomy on age change in the ovaries of mice. J. Endocrinol., 20:497-508.

Kohama, S., May, P., and Finch, C.E. (1986): Oral administration of estradiol induces age-like reproductive acyclicity in C57BL/6J mice. Soc. Neurosci. Abstr., 12:1466.

Landfield, P.W., Lynch, G., and Waymire, J.C. (1978): Hippocampal aging and adrenocorticoids: Quantitative correlations. Science, 202:1098-1102.

Lee, A.K., et al. (1982): Life history strategies of Dasyurid marsupials. In "Carnivorous Marsupials," edited by M. Archer, pp. 1-11. Royal Zoological Society of New South Walse, Mosman, Austria.

Levin, P., Janda, J.K., Joseph, J.A. Ingram, D.K., and Roth, G.S. (1981): Dietary restriction retards the age-associated loss of rat striatal dopamenergic receptors. Science, 214:561-562.

Lumpkin, C.K., McClung, J.K., Pereira-Smith, O.M., and Smith, J. R. (1987): Existence of high abundance antiproliferative mRNA's in senescent human diploid fibroblasts. Science, 232:393-395.

Mann, D.M.A. (1985): The neuropathology of Alzheimer's disease: a review with pathogenetic aetiological, and therapeutic considerations. Mech. Aging Devel., 31:213-255.

Mann, D.M.A. and Yates, P.O. (1979): The effects of ageing on the pigmented nerve cells of the human locus caeruleus and substantia nigra. Acta Neuropathol., 47:93-97.

Mays, A.W. (1967): Fecundity of Atlantic cod. J. Fish Res. Bd. Can. 24:1531-1551.

Maurizio, A. (1959): Factors influencing the lifespan of bees. CIBA Found. Colloq. Aging, 5:231-234.

Mobbs, C.V., Flurkey, K., Gee, D.M., Yamamoto, K., Sinha, Y.N. and Finch, C.E. (1984): Estradiol-induced anovulatory syndrome in female C57BL/6J mice: Age-like neuroendocrine, but not ovarian impairments. Biol. Reprod., 30:556-563.

Morgan, D.G., May P.C. and Finch, C.E. (1987): Dopamine and serotonin systems in human and rodent brain: effects of age and degenerative disease. J. Am. Ger. Soc., 35:334-345.

Nelson, J.F., Felicio, L.S., Randall, P.K., Simms, C., and Finch C.E. (1982): A longitudinal study of estrous cyclicity in aging C57BL/6J mice. 1. Cycle frequency, length, and vaginal cytology. Biol. Reprod., 27:327-339.

Nelson, J.F., Gosden, R.G, and Felicio, L.S. (1985): Effect of dietary restriction on estrous cyclicity and follicular reserves in aging C57BL/6J mice. Biol. Reprod., 32:515-522.

Nichols, N.R., Masters, J.N., May, P.C., Millar, S.L., and Finch, C.E. (1976): Altered RNA abundence in rat hippocampus in respnse to corticosterone. Endocrine Society 68th Ann. Meeting (Abstracts) p. 265.

Pasinetti, G.M., Lerner, S.P., Johnson, S.A., Morgan, D.G., Telford N.A., Myers, M.M., and Finch, C.E. (1987): Chronic lesions differentially decrease messenger RNA in dopaminergic neurons of rat substantia nigra. Soc. Neurosci. Abstr., 13:378.

Paton, J.A., and Nottebohm, F.N. (1984): Neurons generated in the adult brain are recruited in functional circuits. Science, 225:1046-1048.

Pearson, R.C.A., Gatter, K.C., and Powell, T.P.S. (1983): Retrograde cell degeneration in the basal nucleus in monkey and man. Brain Research, 261:321-326.

Robertson, O.H. (1961): Prolongation of the lifespan of Kokanee salmon (O. nerka kennerlyi) by castration before beginning development. Proc. Natl. Acad. Sci., 47:609-621.

Sabbadin, A. (1979): Colonial structure and genetic patterns in ascidians. In "Biology and Systematics of Colonial

Maurizio, A. (1959): Factors influencing the lifespan of bees. CIBA Found. Colloq. Aging, 5:231-234.

Mobbs, C.V., Cheyney, D., Sinha, Y.N., and Finch C.E. (1985): Age-correlated and ovary-dependent changes in relationships between plasma estradiol and luteinizing hormone, prolactin, and growth hormone in female C57BL/6J mice. Endocrinology, 116:813-820.

Morgan, D.G., May P.C. and Finch, C.E. (1987): Dopamine and serotonin systems in human and rodent brain: effects of age and degenerative disease. J. Am. Ger. Soc., 35:334-345.

Nelson, J.F., Felicio, L.S., Randall, P.K., Simms, C., and Finch C.E. (1982): A longitudinal study of estrous cyclicity in aging C57BL/6J mice. 1. Cycle frequency, length, and vaginal cytology. Biol. Reprod., 27:327-339.

Nelson, J.F., Gosden, R.G, and Felicio, L.S. (1985): Effect of dietary restriction on estrous cyclicity and follicular reserves in aging C57BL/6J mice. Biol. Reprod., 32:515-522.

Nichols, N.R., Masters, J.N., May, P.C., Millar, S.L., and Finch, C.E. (1976): Altered RNA abundence in rat hippocampus in respnse to corticosterone. Endocrine Society 68th Ann. Meeting (Abstracts) p. 265.

Pasinetti, G.M., Lerner, S.P., Johnson, S.A., Morgan, D.G., Telford N.A., Myers, M.M., and Finch, C.E. (1987): Chronic lesions differentially decrease messenger RNA in dopaminergic neurons of rat substantia nigra. Soc. Neurosci. Abstr., 13:378.

Paton, J.A., and Nottebohm, F.N. (1984): Neurons generated in the adult brain are recruited in functional circuits. Science, 225:1046-1048.

Pearson, R.C.A., Gatter, K.C., and Powell, T.P.S. (1983): Retrograde cell degeneration in the basal nucleus in monkey and man. Brain Research, 261:321-326.

Robertson, O.H. (1961): Prolongation of the lifespan of Kokanee salmon (O. nerka kennerlyi) by castration before beginning development. Proc. Natl. Acad. Sci., 47:609-621.

Sabbadin, A. (1979): Colonial structure and genetic patterns in ascidians. In "Biology and Systematics of Colonial Organisms," edited by G. Larwood and B.R. Rosen, pp. 433-444. Academic Press, New York.

Sacher, G.A. (1977): Life table modification and life prolongation. In "Handbook of the Biology of Aging," edited by C.E. Finch and L. Hayflick, pp. 582-638. Van Nostrand, New York.

Sapolsky, R.M., Krey, L.C., and McEwen, B.S. (1986): The neuroendocrinology of stress and aging: The glucocorticoid cascade hypothesis. Endocr. Rev., 7:284-301.

Schipper, H., Brarner, J.R., Nelson, J.F., Felicio, L.S., and Finch, C.E. (1981): The role of gonadal steroids in the nestologic aging of the arcuate nucleus. Biol. Repro., 25:413-419.

Simms, H.S. and Berg B.N. (1957): Longevity and the onset of lesions in male rats. J. Gerontol., 12:244-252.

Telford, N., Mobbs, C.V., and Finch, C.F. (1986): The increase of anterior pituitary dopamine in aging C57BL/6J female mice is caused by ovarian steroids, not intrinsic pituitary aging. Neuroendocrinology, 43:135-142.

Tokarz, R.R. (1978): Oogonial proliferation, oogenesis, and folliculogenesis in nonmammalian vertebrates. In "The Vertebrate Ovary: Comparative Biology and Evolution," edited by R.E. Jones, Plenum Press, New York.

Tomkins, G.A., Stanbridge, E.J., and Hayflick, L. (1974): Viral probes of aging in the human diploid cell strain, W.I. 38. Proc. Soc. Exp. Biol. Med., 146:385-390.

vom Saal, F.S. and Finch, C.E. (1988): Reproductive senescence: phenomena and mechanisms in mammals and selected vertebrates. In, "Physiology of Reproduction," edited by E. Knobil, Raven Press, New York, in press.

Wellinger, R., and Guigoz, Y. (1986): The effect of age on the induction of tyrosine aminotransferase and tryptophan oxygenase genes by physiological stress. Mech. Ageing Dev., 34:203-217.

Wodinsky, J. (1977): Hormonal inhibition of feeding in death of Octupus: Control by optic gland secretion. Science,

198:948-951.

Yen, S.S.C. (1977): The biology of menopause. J. Reprod. Med., 18:287.

Yu, B.P. (1985): Recent advances in dietary restriction and aging. Rev. Biol. Res. Aging, 2:435-444.

Index

A

E

R

T

V